Estimating and Measurement for Simple Building Works in Hong Kong

This book is an introductory text on building measurement and estimating for simple buildings in Hong Kong, based on the Hong Kong Standard Method of Measurement of Building Works 4th Edition Revised 2018 (HKSMM4 Rev 2018). It provides a toolkit for students and surveying technicians who are new to the subject.

This second edition updates the contents in line with the HKSMM4 Rev 2018 and incorporates the latest industry developments such as BIM. The main text is divided into five parts following the development of a typical project.

- Part 1, Building the project team, introduces the team setup for a typical project.
- Part 2, Deciding the procurement strategy, explains the various procurement decisions to be made by an employer before any cost estimating and measurement work takes place.
- Part 3, Preparing for tender, covers the tendering methods, tender documentation and approximate estimating techniques used by Quantity Surveyors.
- Part 4, Measuring quantities, introduces measurement principles and HKSMM4 Rev 2018, followed by a detailed review of the measurement methods for each major trade, with worked examples.
- Part 5, Estimating unit rates, explores the basic techniques for unit rate preparation.

The book contains worked examples from real Hong Kong building projects, self-assessment questions, reminders and points of note. It is essential reading for Hong Kong construction and surveying students, international Quantity Surveyors working in the local area and those wanting international examples of Quantity Surveyors practice.

Caroline T. W. Chan is a lecturer in the City University of Hong Kong, teaching sub-degree students in the Architectural, Engineering and Construction (AEC) programmes. Caroline is a qualified Quantity Surveyor and, before joining academia, she practised in large Quantity Surveyor consultancy firms and with large contractors in Hong Kong.

Estimating and Measurement for Simple Building Works in Hong Kong

Second Edition

Caroline T. W. Chan

Routledge
Taylor & Francis Group
LONDON AND NEW YORK

Second edition published 2021
by Routledge
2 Park Square, Milton Park, Abingdon, Oxon, OX14 4RN

and by Routledge
605 Third Avenue, New York, NY 10017

Routledge is an imprint of the Taylor & Francis Group, an informa business

First edition published by Pearson Education Asia Limited 2014

British Library Cataloguing-in-Publication Data
A catalogue record for this book is available from the British Library

Library of Congress Cataloging-in-Publication Data
Names: Chan, Caroline T. W. (Caroline Tak-wa), author.
Title: Estimating and measurement for simple building works in Hong Kong/Caroline T.W. Chan.
Description: Second edition. | Abingdon, Oxon ; New York, NY : Routledge, 2021. | Includes bibliographical references and index.
Identifiers: LCCN 2020025940 (print) | LCCN 2020025941 (ebook) | ISBN 9780367862367 (hardback) | ISBN 9780367862329 (paperback) | ISBN 9781003017837 (ebook)
Subjects: LCSH: Building–Estimates–China–Hong Kong.
Classification: LCC TH113.H85 C53 2021 (print) | LCC TH113.H85 (ebook) | DDC 692/.5095125–dc23
LC record available at https://lccn.loc.gov/2020025940
LC ebook record available at https://lccn.loc.gov/2020025941

ISBN 13: 978-0-367-86236-7 (hbk)
ISBN 13: 978-0-367-86232-9 (pbk)

Typeset in Baskerville
by KnowledgeWorks Global Ltd.

Visit the eResources: www.routledge.com/9780367862329

This book was previously published by Pearson Education Asia Limited

Contents

President's foreword, Hong Kong Institute of Surveyors

As the President of the Hong Kong Institute of Surveyors (HKIS) this year, it is my honour to say a few words about this book.

At HKIS we understand the need for surveying students in tertiary institutions in Hong Kong to learn about the industry via local topics, cases, and examples. Thus, in 2010, our Board of Education launched its textbook sponsorship scheme to encourage our experienced members and university faculty members from relevant disciplines to write textbooks, or reference books, leveraging their expertise.

Dr Caroline Tak-wa Chan, at City University of Hong Kong, rose to the challenge seven years ago, sharing her insights on estimating and building measurement from a local perspective. While assessments of project costs and project quantities are fundamental skills for Quantity Surveyors, such a local textbook answered the need to address the differences in measurement rules and industry practices in Hong Kong from books printed for overseas readers.

As a lecturer herself, Dr Chan is aware of the necessity to keep textbooks for architecture, engineering, and construction (AEC) students and young practitioners up to date in order to encompass evolving trends and practices. This year, she again responded to the call by undertaking a second edition, which was timely given that another edition was due after the issuing and authorisation of the Hong Kong Standard Method of Measurement of Building Works 4th Edition Revised 2018 (HKSMM4 Rev 2018).

With building information modelling (BIM) becoming increasingly significant to surveyors, we are pleased that this new edition includes applications and necessary precautions related to BIM. To ensure relevance, Dr Chan also addresses feedback from readers and students. Most importantly, she reiterates the significance of understanding the underlying principles despite the availability of new technologies. Procurement, tendering, contract administration, financial feasibility, and computer applications in the local context, just to name a few, all contribute to the proactive and indispensable services performed by Quantity Surveyors.

Always keen to nurture young professionals and students, we are delighted to support the production of this new edition. We look forward to more contributions from Dr Chan and other seasoned surveyors to nurture our young talents.

Sr Winnie SHIU
President 2019–2020
The Hong Kong Institute of Surveyors

Acknowledgements

I would like to thank the Hong Kong Institute of Surveyors for its generous sponsorship to support the preparation of the manuscripts.

I would like to thank the many people who have helped me with this book. Special thanks must go to Ed Needle for coordinating, reviewing and giving advice to the proposal and manuscripts. I would also like to thank the editorial team of Routledge, particularly Patrick Hetherington, for the editing and prompt feedback. Many thanks go to the reviewers for their constructive comments. I would also like to extend my gratitude to my students who often inspire me with creative questions and comments. My students have all shaped my thinking and writing.

Acronyms

ASD	Architectural Services Department (HKSAR)
BIM	Building Information Modelling
BOT	Build, Operate and Transfer
BQ	Bills of Quantities
BS	British Standard
CAD	Computer-aided Drafting
CFA	Construction Floor Area
D&B	Design and Build
DBFO	Design, Build, Finance and Operate
DBO	Design, Build and Operate
DLP	Defects Liability Period
dpc	Damp proof course
GFA	Gross Floor Area
GMP	Guaranteed Maximum Price
HKIS	Hong Kong Institute of Surveyors
HKSAR	Hong Kong Special Administrative Region
HKSMM	Hong Kong Standard Method of Measurement
JV	Joint Venture
MC	Management Contracting
MPF	Mandatory Provident Fund
nr.	number
NSC	Nominated Subcontractors
P.C.	Prime Cost
PFI	Private Finance Initiative
PPP	Public-Private Partnership
QS	Quantity Surveyor
R.C.	Reinforced concrete
RICS	Royal Institute of Chartered Surveyors
SMM	Standard Method of Measurement of Building Works
T.A.L.	Tension Anchorage Length
T.L.L.	Tension Lap Length
UKSMM	U.K. Standard Method of Measurement of Building Works

Part 1

Building the project team

1 Introduction

Quantity surveying

Quantity surveyors (QSs) emerged as a response to the need for cost planning and for measurement and valuation of work in progress. Hudson (1994) gives the following definition of Quantity Surveyor:

> whose business consists in taking out in detail the measurements and quantities from plans prepared by an architect for the purpose of enabling builders to calculate the amounts for which they would execute the plans.

The above description portrays a simple but clear picture of the most representative work performed by a QS – to carry out estimating and measurement of construction works. With the growing complexity of construction technologies and designs, QSs are required to provide advice and services related to a wide range of aspects including contract administration, financial advice, dispute resolution and contractual advice. Estimating and measurement are the primary skills to deliver these contract and financial services satisfactorily to the project owners.

Estimating and measurement

When an organisation has an idea to construct a project, many questions related to project cost have to be answered: Is the project financially feasible? How much does the project cost? Are there any alternative proposals which cost less? How much is the life-cycle cost of the project? All these questions are answered by a QS through the estimating and measurement processes.

Estimating calculates the likely cost of a project (or items of work) to be incurred. Agreement of the estimated cost with the actual cost, or the accuracy of the estimated cost, depends on the availability and understanding of project details and the use of appropriate estimating method.

Measurement calculates the quantities of work required for various trades from drawings, sketches and specifications prepared by designers, principally architects and engineers. Accurate measurement of the work enables realistic cost estimates and cost control to be carried out.

From the project owner's perspective, a QS consultant is employed to act as his/her project adviser. The QS reports to the project owner and advises the project estimated cost from inception to completion as shown in **Figure 1.1**. In addition,

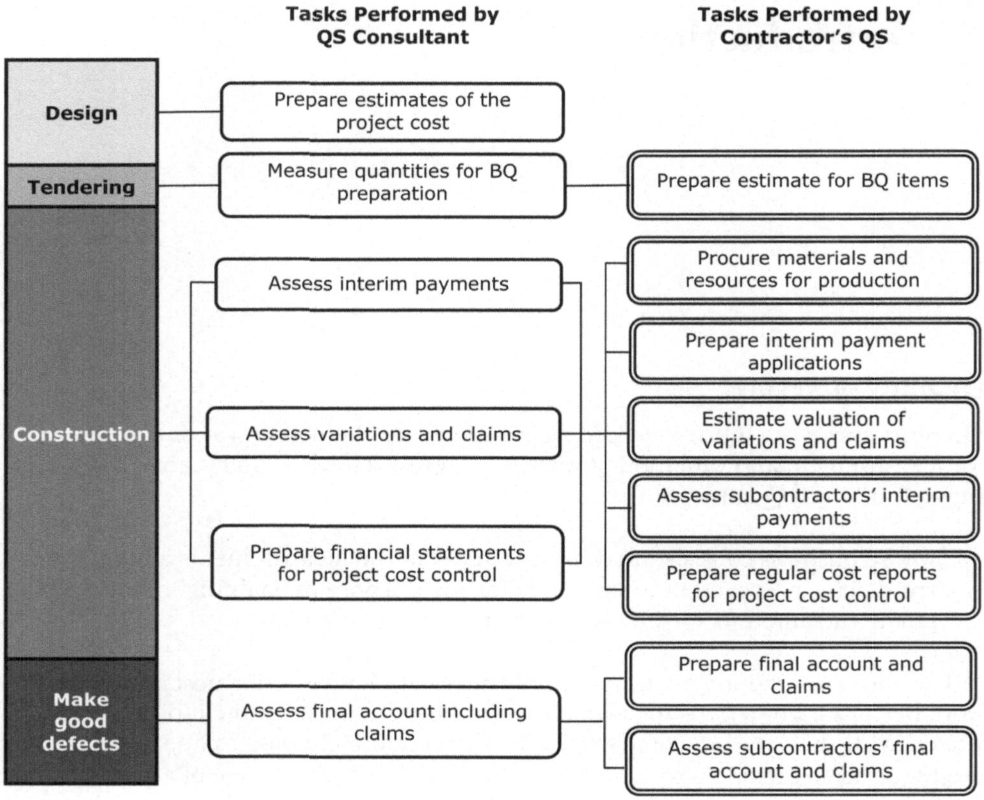

Figure 1.1 Estimating and measurement tasks performed by QS consultant and contractor's QS.

the QS manages the tendering process, contract award and commercial interfaces. S/he measures the work in accordance with the standard method of measurement, prepares the bills of quantities (BQ) for tender and evaluates the tenders received. During the construction period, the QS continues to review and control the project cost, prepares necessary measurements and determines variations and monthly payments to the contractors. When the project completes, the QS assesses the measurements and prices for final account and claims.

On the contractor side, contractor's QS uses his/her expertise to bid a contract and secure reasonable profit. During the tender stage, a contractor's QS, usually employed in the position of an estimator, prepares an estimate (or price) for each BQ item. The estimator studies the project conditions and design to estimate the cost implication. A base estimate which represents all the likely direct costs will then be worked out. The contractor's management will decide on the allocation of profit and overheads in order to finalise the tender bid. If the contract is awarded to the contractor, the contractor's QS(s) will be recruited as core member(s) of the project team to handle the procurement and cost control activities. Quantities of work will be measured for materials ordering. Measurement of work quantities and cost estimation will be performed at regular intervals to assess subcontractors' payments and to prepare interim payment applications.

The quantity surveying operations performed by both parties (the project owner and the contractor) are all related to building measurement and cost estimation. These two skills contribute to the core function of a QS, to plan and manage the project cost. Although information technology in recent years has turned a lot of labour-intensive tasks into automated processes, measurement and estimating remains a vital knowledge for project cost management.

2 Principal stakeholders of a project

Introduction

Construction projects are intricate and fragmented in nature. The project owner, who initiates the project and contracts it out, interacts with many companies and stakeholders in the whole project life cycle. During the design and construction period, the project owner employs a diverse group of professionals or companies to collaborate at different stages to complete the project in a timely and cost-effective manner. These companies form an integrated team (as shown in **Figure 2.1**), with each of them playing specific roles. It is only through the concerted effort and contribution of all team members that the project completes.

In a traditional designer-led project, the project owner (or employer) engages a lead designer (or design leader) who is usually an architectural consultant or engineering consultant. Other design consultants like building services engineer, structural engineer or landscape designer will also be appointed by the project owner to form a design team under the direction of the lead designer. The lead designer will coordinate the work of all these consultants and monitor the contractor's work to ensure compliance with design and schedule.

As shown in **Figure 2.1**, the dotted lines linking the architect, structural engineer, building services engineer and landscape designer represent the close working relationship between these parties to complete the overall design. In some projects, the project owner may engage a multi-disciplinary design consultant to prepare the full range of design services including architectural, structural and building services design. The project owner also employs a quantity surveying consultant and a main contractor at appropriate time for the execution of project work.

Points to note

Every project is unique and may have different team models based on the project nature, scope and delivery method. For instance, in a project delivered by a design-and-build approach, no separate architect will be employed by the developer. Details of the project team set-up under different project delivery methods will be further discussed in **Chapter 3**. For illustration purpose, the major roles of each project party are explained with reference to a traditional designer-led project.

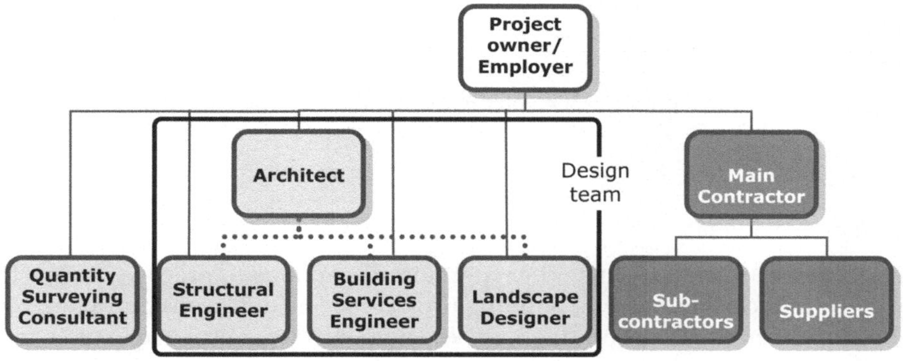

Figure 2.1 Typical project team model of a designer-led project.

Principal team members

The project owner

A project owner (or an employer), is an organisation that initiates the project. Examples are the government, Hospital Authority, Swire Properties, MTRC, and so forth. Commission for Architecture and the Built Environment, U.K. (2013) provides a detailed guide on the roles of a project owner throughout the project life cycle. To summarise, a project owner is the overall decision-maker and coordinator of a project. A project owner has to recruit a team of professionals or consultants, give clear directions and set operational criteria for the project. In addition, a project owner will also decide on budgets, project completion date and quality thresholds. To ensure construction team members play their respective roles, the project owner has to grant possession of site to the contractors on the agreed date and minimise disruption to the contractors' work. This includes but not limited to supplying necessary information promptly to the various parties and making timely payments for design and construction.

The architect

An architect is an organisation or individual whose primary role is to design a facility. An architect is usually the lead designer in the project team of a building project. Starting from project initiation, an architect must collaborate with other project team members to develop feasible and satisfactory designs and specifications that meet the project owner's requirements, local codes and regulations.

In most projects, the project owner delegates the contract administration duties such as issuance of certificates to the architect. As a result, the architect has to oversee the construction to ensure meeting of the specifications that are laid down in the construction documents. The architect will prepare written assessments/approvals of contractor's work at regular intervals during construction as well as after completion of the project. Other duties, like review of the contractors' shop drawings and acquisition of permits, are likely to be performed by the architect.

Question 2.1

In a tunnel project, which professional will be the lead designer of the design team?

Other design consultants

A project owner often appoints other specialist consultants to prepare designs for a project. Design consultants of a building project may include a structural engineer, a geotechnical engineer, a building services engineer, an acoustic engineer, a traffic engineer, a landscape architect and an interior designer.

The design consultants will cooperate with each other to produce the design information for production. With the growing maturity of building information modelling (BIM) in the industry, design consultants may be required to prepare design modelling outputs using BIM software. The tasks of design consultants also include monitoring, organising, commissioning and assessing the contractor's work to ensure satisfactory completion.

The quantity surveying consultant

A quantity surveying consultant is usually appointed at an early stage of design to provide financial and contractual services for a project. The QS consultant provides expert advice on the selection of procurement route, contract type and tendering procedures. In addition, the consultant has to coordinate the tender process, including the preparation of tender and contract documentation. Another primary duty of the QS consultant is to estimate the likely project cost. This estimate continues to change when the design is evolving and refining. It is the responsibility of the QS consultant to update the cost estimates and inform the project owner and lead designer regularly. Should the cost requirement reach the budgetary limit, the QS consultant has to alert the architect or other design consultants to consider alternative solutions so as to lower the cost requirement. This is how the QS consultant assists the design team to develop the most economical design through life-cycle cost analysis.

During the construction stage, the QS consultant is responsible for the valuation of contractors' work and assessment of variations and financial claims. Contractual advice is also provided by him/her in cases where disputes arise between the contractor and the project owner. Upon project completion, the QS consultant will prepare and settle the final account.

The main contractor

Here, a main contractor is a company that has signed a direct contract with the project owner for the construction of a project. In many cases, the project owner appoints only a single contractor to carry out all the construction work so as to save coordination effort. However, some project owners may choose to employ several independent contractors to carry out different portions of the project work simultaneously. This approach allows the project owner to control the performance of contractors directly and therefore, the quality of work. However, the burden of coordinating these independent contractors will rest on the project owner.

Once a main contractor is engaged in a direct contract with a project owner, it is responsible for the performance of all the work in accordance with the contract documents. The contractor must plan and coordinate all necessary resources for production and establish an effective control system to meet the time, cost, quality, safety and environmental requirements.

The subcontractors and suppliers

Labour, materials and equipment are the three main types of resources required for construction. General contractors usually sublet most of the tasks to subcontractors because construction projects involve too many made-to-order or engineered-to-order components. If a general contractor has to employ direct labour and equipment to carry out all these customised operations, they must maintain a wide range of specialties at all times (O'Brien, Formoso, Ruben, & London, 2010). Therefore, a general contractor usually employs multiple subcontractors and suppliers to provide the necessary products and services.

A subcontractor is usually a firm specialised in a specific trade/aspect of work, for instance, painting, plumbing or plastering. A subcontractor must carry out his work diligently and collaboratively with the rest of the project team. Depending on the agreement with the main contractor, the subcontractor may provide both labour and materials or just labour to complete the work. The firms that provides materials only are the suppliers. Suppliers have to deliver the specified materials/products according to the agreed schedule.

Question 2.2

i What is a domestic subcontractor?
ii What is a nominated subcontractor? Is it the same as a named subcontractor?

Other external project stakeholders

> A stakeholder is an individual, group, or organization who may affect, be affected by, or perceive itself to be affected by a decision, activity, or outcome of a project.
> (Project Management Institute, 2013).

All the team members of a project, including the team member's own company, are the essential stakeholders that one must deal with.

Reminder

Employees and shareholders of an individual team member's company are important stakeholders as well. Their interest is critically affected by the success of a project (Friedman & Miles, 2006).

Since the population is high and the land is scarce in Hong Kong, many building projects such as the Hong Kong Palace Museum, the Xiqu Centre and the redevelopment of the Ho Tung Garden attracted intense debate and controversy during the consultation period. Costs-and-benefits of the project, construction nuisance, sustainable development, construction quality and many other aspects can trigger huge reactions such as strikes, demonstrations from the public and judicial reviews.

Besides the project team members highlighted in the previous sections, common external stakeholders of a construction project include pressure groups, shareholders, financial institutes, government departments, official bodies, district councils, trade unions, future users/occupants, mass media, neighbours and the public. To ensure smooth execution and satisfactory completion of the project, the project owner and his project team should pay full attention to manage the various stakeholders during the whole project life cycle. Good communication between the project team and these stakeholders is indispensable.

Question 2.3

Besides the project owner and the project team, list the possible external project stakeholders of a proposed hospital project in the New Territories from the main contractor's perspective.

Suggested answers

Question 2.1

A civil engineering consultant should be appointed as the lead designer of a tunnel project.

Question 2.2

i A domestic subcontractor is a subcontractor who is selected by the main contractor to carry out part of the main contract work. While the project owner plays no part in the selection process of the subcontractor, the main contractor is fully responsible for any default of the domestic subcontractor.

ii A nominated subcontractor (NSC) is a subcontractor selected by the project owner to carry out specialist work (such as building services, fitting-out work etc.) provided for by the prime cost sum (details of the prime cost sums can be found in **Chapter 7**). When the architect issues an instruction to the main contractor to nominate the subcontractor for the expenditure of a prime cost sum, the main contractor is obliged to accept the nomination, subject to a limited right of objection. The main contractor shall then enter into a subcontract with this contractor, who shall become his nominated subcontractor. It should be noted that the NSC is imposed on the main contractor after the award of the main contract.

A named subcontractor is a subcontractor whose name is included in the tender document of the main contract for the completion of a specific portion of work. Occasionally, the project owner may shortlist a few subcontractors and put their names in the tender document for a specific task/trade. The main contractor needs to choose one from the shortlist and enter into a subcontract with it. This subcontract will be the same as any other domestic subcontract signed between the main contractor and his domestic subcontractor.

One of the main differences between the two types of subcontracts is the extent to which the main contractor may be held liable for the subcontractor's default such as defective work. In case of an NSC's default, the main contractor may not be held liable depending on the nature of default. However, in case of a named subcontractor's default, the main contractor will be held full responsibility regardless of whether the work is subcontracted or not.

Question 2.3

Possible external project stakeholders of a proposed hospital project in the New Territories from the main contractor's perspective:

- Approving bodies/government departments such as the Hospital Authority, Fire Services Department, Buildings Department etc.
- Legislative council and Legco members
- Political parties
- District council and district councillors
- Political pressure groups
- Other hospitals and community health centres
- Utility companies such as CLP Power
- Residents of the neighbourhood
- Owners and users of adjacent buildings
- Patients' organisations
- Charities and volunteers
- Trade unions of labour, medical and health care practitioners etc.
- Media
- General public

Part 2

Deciding the procurement strategy

3 Project delivery methods

Introduction

We have reviewed the various project parties in **Chapter 2**, but how does the project owner recruit the contractor(s) and consultants to carry out the project work? This chapter will illustrate the various project delivery methods, i.e. the various routes to procure these parties to complete a project.

In the *Construct for Excellence* report (Construction Industry Review Committee, 2001), the Construction Industry Review Committee recommended that project owners of local construction projects should adopt alternative procurement methods (which means methods other than the conventional designer-led approach) to achieve better value for money. Since then, the Environment, Transport and Works Bureau of the Hong Kong Special Administrative Region (HKSAR) has seriously investigated the various project delivery methods and made attempts to increase the adoption of alternative procurement methods in government projects. The initiative has spread to the private sector.

Project delivery options

Project delivery options are differentiated by describing the responsibilities allocated to the general contractor. In general, we can broadly classify them into five approaches:

- Designer-led
- Management Contracting
- Design and Build
- Design, Build and Operate
- Design, Build, Finance and Operate

Each delivery method has its own advantages and disadvantages. Some methods are better suited for certain project conditions or constraints.

Designer-led

The designer-led method is also known as the design-bid-build method in other countries such as Australia. In many countries, the designer-led method is the most popular procurement method (Chakra and Ashi, 2019) and is often regarded as the 'traditional' method. This project delivery method was first introduced to

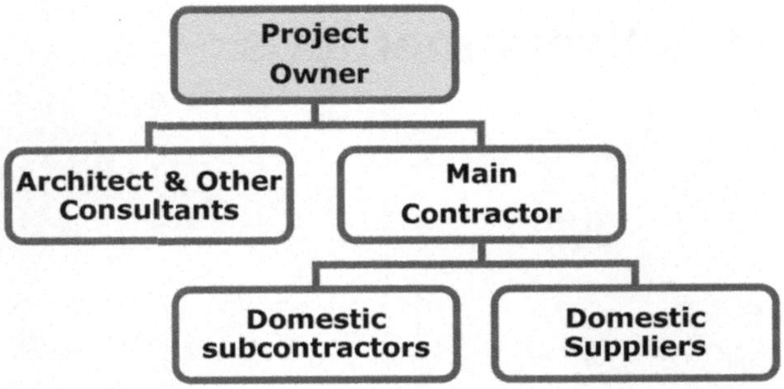

Figure 3.1 Contractual relationships in the designer-led delivery method.

Hong Kong a hundred years ago and is still the most popular approach in the construction industry (Environment, Transport and Works Bureau, HKSAR, 2004a).

The designer-led method is characterised by the separation of design and construction responsibilities. As shown in **Figure 3.1**, the lead designer, usually an architect or an engineer, prepares the design which is then given to the contractor to construct. A QS consultant is appointed to administer the project cost and contracts. Design and construction of works are carried out separately in identifiable stages.

Advantages and disadvantages

Advantages:

- Employer has good control over design.
- Reliable price information is available before construction starts.
- Bills of quantities, which provide a good basis for valuation of variations, are usually available.
- A system of checks and balances is provided because the architect and the contractor are in a position to find errors made by the other.

Disadvantages:

- The overall project period is generally longer because the design and construction stages do not overlap.
- Design input from the contractor is not possible. This may limit the effectiveness and constructability of the design.
- Tension is often created between the designer and the contractor. The system is criticised as promoting poor relationship between the two parties.
- In case of dispute, the architect and the contractor cannot sue each other except if the other has been negligent.

Local standard form of contracts

Since designer-led is the most popular method in the local construction industry, standard forms of contract have been developed by the government and professional institute for different types of main contract works:

The government of the HKSAR standard contracts for public sector:

- General Conditions of Contract for Civil Engineering Works 1999 Edition,
- General Conditions of Contract for Building Works, 1999 Edition, and
- General Conditions of Contract for E & M Engineering Works, 1999 Edition.

The Hong Kong Institute of Surveyors standard contracts for private sector:

- Standard Form of Building Contracts, Agreement & Schedule of Conditions of Building Contracts, without quantities, 2006 Edition, and
- Standard Form of Building Contracts, Agreement & Schedule of Conditions of Building Contracts, with quantities, 2005 Edition.

Management contracting

To avoid some of the problems associated with designer-led method, such as lack of design input from contractor, the management contracting approach has emerged as an alternative method. Here, a management contractor is engaged by the project owner at the design stage to provide management services for a fee. The management contractor acts as the project owner's agent to assist the development of design, cost estimates and schedule. With technical and management expertise, the management contractor can review the designs for constructability, administer and negotiate tender bids, and coordinate works contractors on site. Architects and other design consultants are employed to provide the design service. The underlying philosophy of this approach is to allow the management contractor to become part of the project owner's team so that the management functions can be carried out in partnership with other design team members.

Normally, the management contractor does not involve in any construction activities. The construction work is done by works contractors who have direct contracts signed with the employer, as shown in **Figure 3.2**. The management contractor may provide project overheads such as mechanical plant and other site facilities for the works contractors.

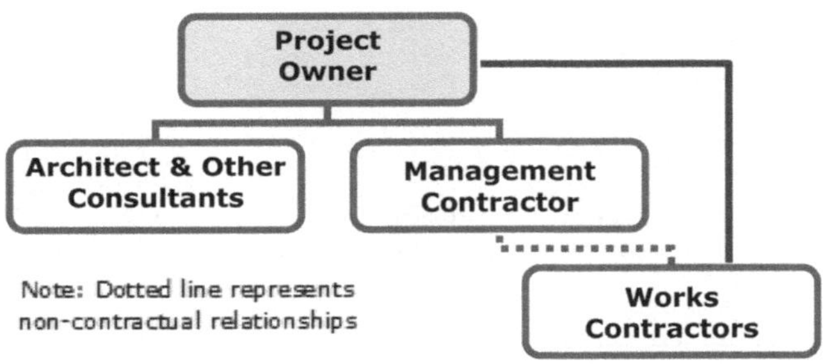

Figure 3.2 Contractual relationships in the traditional management contracting method.

Points to note

In practice, many local project owners are resistant to sign direct contracts with so many works contractors. In that case, the project owner will require the management contractor to sign works contracts with the works contractors, and at the same time ask each works contractor to provide a warranty to the project owner.

Some local examples of management contracting projects

Although management contracting is comparatively less popular in Hong Kong, there are some well-known projects which used this delivery method (as shown in **Table 3.1**). These projects shared some common difficulties including high complexity and tight construction schedule.

Advantages and disadvantages

Advantages:

- The project period can usually be shortened, as different work packages can be awarded and commenced at different times before the full design is completed.
- There is greater flexibility for the project owner to defer decisions on design aspects and to modify his requirements as long as the work packages concerned have not yet commenced.
- The integration of the contractor's knowledge into planning and design processes improves construction efficiency.

Table 3.1 Examples of Hong Kong Projects Delivered by the Management Contracting Approach.

Projects	*Project Owner*	*Management Contractor*	*Project Description*
M+ museum[1]	West Kowloon Cultural District Authority	Gammon Construction Ltd.	Oversee the completion of the M+ project after the termination of the main contractor.
Tai Kwun (Central Police Station Conservation and Revitalisation Project)[2]	Hong Kong Jockey Club	Gammon Construction Ltd.	Manage the revitalisation of 16 old buildings including the police station, quarters and prisons for adaptive re-use of these heritage buildings into art gallery and auditorium.
Zero Carbon Building, Hong Kong[2]	Construction Industry Council	Gammon Construction Ltd.	Manage the construction of a two-storey office and exhibition building including basement.

Sources: [1] West Kowloon Cultural District Authority, HKSAR, 2018

[2] Gammon Construction Limited, 2019

Disadvantages:

- The professional fee charged by the management contractor is an additional cost to the project owner.
- The project owner can only confirm the total project cost when the contract is signed with the last works contractor.
- The prices of later work packages may be higher to ensure no delay to the work that has been started on site.
- The project owner has to assume all risks for project schedule and cost.
- The project owner has to enter into contracts with multiple works contractors instead of a single main contractor. This increases the administration and communication workload of the project owner.

Design and build

Although the construction management approach successfully integrates a contractor's expertise into the design stage, the burden of engaging multiple works contractors makes project owners reluctant to use that approach.

To streamline project coordination, the design and build (D&B) approach presents an attractive alternative for project owners. The contractual relationships between the project owner and other parties are substantially reduced. As seen in **Figure 3.3**, the project owner only engages a single party, the D&B contractor, to design and construct the facility.

Normally, the project owner provides the project brief which includes the outline design requirement, level of workmanship/material, time frame and other key requirements of the project. The tenderers will submit a design proposal and the price required to complete the project. The successful contractor will then complete the detailed design and procure his subcontractors after the contract is awarded. In this way, all the design and construction risks are transferred to the D&B contractor.

> **Reminder**
>
> The project brief, also known as the client's brief, refers to the list of requirements prepared by a project owner for the development of initial designs.

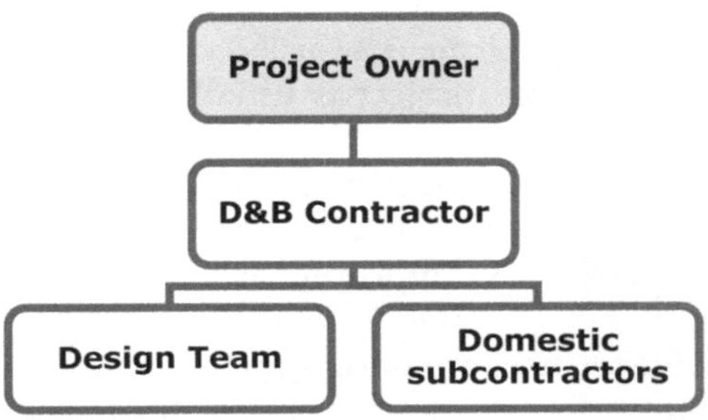

Figure 3.3 Contractual relationships in the design and build method.

Points to note

According to the Environment, Transport and Works Bureau (2004a), the D&B method can be further divided into three approaches based on the relationship between the designer and the contractor:

1 Client's designer novated – the D&B contractor takes over from the client (project owner) a previous contract for the design work, completes the design and builds the work accordingly.
2 Independent designer – an independent design team is employed by the project owner to develop the preliminary design and the contractor prepares the detailed design followed by construction of the work. There is no contractual relationship between the project owner's designer and the D&B contractor.
3 Contractor's designer – the D&B contractor provides complete design and constructs the work.

In other countries, D&B is often sub-divided into package deal and turnkey approaches. Details can be referred to in Chan and Sin (2009), p. 38–39.

Some local examples of design and build projects

Many project owners perceive design and build method as a fast-track and lesser risk option. In the United States, the design and build method has become more popular than the designer-led method (Duggan and Patel, 2014). The use of design and build in the local market has also increased substantially over the past decade. As shown in **Table 3.2**, some of the large-scale building projects completed recently are procured by the design and build method. Even mega-infrastructural projects such as the Hong Kong/Zhuhai/Macao Bridge (including the main bridge, the associated link roads (the Hong Kong Link Road and the Tuen Mun-Chek Lap Kok Link) and the Hong Kong Boundary Crossing Facilities) are procured under the design and build approach (Highways Department, HKSAR, 2019).

Advantages and disadvantages

Advantages:

- Single point of responsibility for design and construction.
- Agreed project cost when contract is awarded.
- Greater cost certainty as the project is usually less susceptible to delays and variations.
- Reduces disputes between the designer and the builder when compared with the traditional method.
- Buildability of the project is improved as the design is developed by the contractor.
- Project period is shorter because design and construction stages overlap. The method also enhances teamwork and improves efficiency in the review of design, budget and schedule.

Table 3.2 Examples of Local Building Projects Delivered by the Design and Build Approach.

Projects	*Project Owner*	*Design and Build Contractor*	*Project Description*
Kai Tak Cruise Terminal Building[1]	Architectural Services Department	Dragages Hong Kong Ltd.	Design and build a three-storey terminal building and ancillary facilities.
New Civil Aviation Department Headquarters[1]	Architectural Services Department	Dragages Hong Kong Ltd.	Design and build a new air traffic control centre building, an office and training building, and a facilities building.
North Lantau Hospital[2]	Architectural Services Department	Leighton Asia and Able Engineering Co Ltd. (JV)	Design and build an eight-storey hospital block including connecting road.
Tamar Development Project[3]	The government of the HKSAR	Gammon Construction Ltd. and Hip Hing Construction Ltd. (JV)	Design and build two central government complex office blocks, one LegCo complex building block and open space.

Sources: [1] Dragages Hong Kong Ltd., 2013
[2] Leighton Holdings, 2013
[3] China Trend Building Press Ltd., 2008

Disadvantages:

- Employer's requirements have to be clearly defined in the tender stage to ensure cost and time certainty.
- The system of checks and balances may be lost if there is no independent designer to oversee the D&B contractor's work.
- Tendering cost is much higher to the bidders and such cost is often transferred to the bids.
- The bids are difficult to compare since each design will be different. In addition, the project schedule and prices will also be different between bidders.
- The contractor may develop a design which revolves around his favourite or the most profitable method of construction, and as a result, quality of the building may be sacrificed.

Local standard form of contracts

Compared with management contracting, the design and build method is more popular in Hong Kong. To boost the use of this approach, the *General Conditions of Contract for Design and Build Contract*, 1999 Edition has been published by the HKSAR government for use in the government projects.

Design, build and operate

The design and build approach is attractive because it provides a single point of responsibility and communication from the project owner's perspective. However, it may create a higher risk of poor design quality when the contractor is responsible for the design as well. To increase their profit, the D&B contractor may not choose the best design that demonstrates the lowest life-cycle cost. These problems lead to an increasing use of the design, build and operate (DBO) delivery model.

Under a DBO arrangement, the contractor is responsible for design, construction and operation of the facility for an agreed period. In this case, the DBO contractor has a financial incentive to adopt a design which can reduce the operating and maintenance cost as he has to operate the facility after construction. The contractor's knowledge in project design and construction allows him to develop a tailored operational program that anticipates and addresses potential problems. With the combination of design, construction and operation, there will be a greater continuity of private sector involvement in government projects (The National Council for Public-Private Partnerships, 2013).

As shown in **Figure 3.4**, the DBO contractor enters into a primary contract with the project owner to design, construct and operate the facility over a specified period. The DBO contractor then enters into back-to-back contracts with design firms to produce design; construction companies to build; and a facility management company to operate the facility. The project owner pays the design and construction service in a lump sum and the operating service on an indexed base price basis (so as to adjust inflation or deflation) (Palmer, 2000). User fees generated from the facility will also be treated as a source of finance during the operation phase. To ensure the quality of operating service, the project owner will conduct reviews on a regular basis, e.g. biannually, to assess the facility performance, operating costs and further funding arrangement.

Some local examples of design, build and operate projects

In Hong Kong, the DBO method has been used by government departments such as the Environmental Protection Department to encourage private sector participation in the environmental infrastructure. Recently, the biggest sports venue in Hong Kong, the Kai Tak Sports Park, was also awarded by DBO. Some notable DBO projects are listed in **Table 3.3**.

Figure 3.4 Contractual relationships in the design, build and operate method.

Table 3.3 Examples of Local Projects Delivered by the Design, Build and Operate Approach.

Projects	*Project Owner*	*Project Description*
Kai Tak Sports Park[1]	Home Affairs Bureau	Design, build and operate for a 25-year contract of a sports park which consists of a covered sports avenue as the main axis, connecting all key facilities of the sports park such as a 50,000-seat main stadium, a 10,000-seat indoor sports centre, a 5,000-seat public sports ground, sports and health centres and retail areas.
Organic Waste Treatment Facilities, Phase 1[2]	Environmental Protection Department	Design, build and operate the food waste treatment facilities phase 1 (to receive and treat food waste from commercial and industrial sectors) for a 15-year operation period.
Shuen Wan Landfill Restoration Project[3]	Environmental Protection Department	To restore the Shuen Wan landfill site and develop it into a golf driving range. The project included design, construction, operation, restoration, aftercare and management of the Shuen Wan golf driving range for a 30-year period.
Waste Electrical and Electronic Equipment Treatment and Recycling Facility[2]	Environmental Protection Department	Design, build and operate a waste electrical and electronic equipment (WEEE) treatment and recycling facility for a tentative period of 10–15 years.

Sources: [1] Home Affairs Bureau, HKSAR, 2019
[2] Environmental Protection Department, 2013
[3] Surman and Kebergang, 2003

Advantages and disadvantages

Advantages:

- Inherits all benefits associated with design and build method.
- Consideration of life-cycle costing in design is more guaranteed. This is important especially when the operating and maintenance cost of the facility is very significant.
- Contractor's knowledge of design and construction can be fully utilised to establish a long-term operation and maintenance plan.

Disadvantages:

- Higher tendering cost.
- DBO contractor may try to minimise the residual value of the facility at the end of the operation contract period by deferring maintenance (LaBonde, 2010).

- Prices are difficult to negotiate due to high complexity and long time frame of the DBO contracts.
- Project owner will not have complete control over all operational and maintenance issues if some are not included as key performance indicators (KPIs) in the DBO contract.

> **Points to note**
>
> Since the DBO model requires the contractor to operate the facility for a number of years, its application is more popular in government or non-profit making projects. If contract terms are well-drafted, a DBO contract can provide incentive for strong management and substantial risk transfer (Palmer, 2000). In the case of the environmental infrastructure projects, the DBO method allows the government to monitor and regularly review the DBO contractor's performance during the operation period. Non-compliance with any KPIs listed in the contract will allow the government to deduct points and the contractor will not receive full contract payment.
>
> In DBO, the title to the facility remains with the project owner.

Design, build, finance and operate

The distinct characteristic of the design, build, finance and operate (DBFO) model is that the contractor (a DBFO Consortium) is responsible for financing the project, though the extent of financial responsibility may vary considerably in different DBFO arrangements (The National Council for Public-Private Partnerships, 2013). This method is usually used in public projects like tunnels and toll highways as one of the viable methods to privatise public projects. To make this model feasible and attractive, the project has to generate sufficient revenue over the contract period to finance the design, construction and operating expenditures. The project originator, which is usually the government, eventually becomes a service buyer. For example, in a toll highway DBFO project, the government buys the maintained highway service rather than constructing and operating the highway.

As shown in **Figure 3.5**, the DBFO consortium has to secure private funds to pay for the design and construction of the project.

In Hong Kong, the government has delivered various infrastructural projects by build-operate-transfer (BOT) (see **Table 3.4**). In the local BOT projects, the

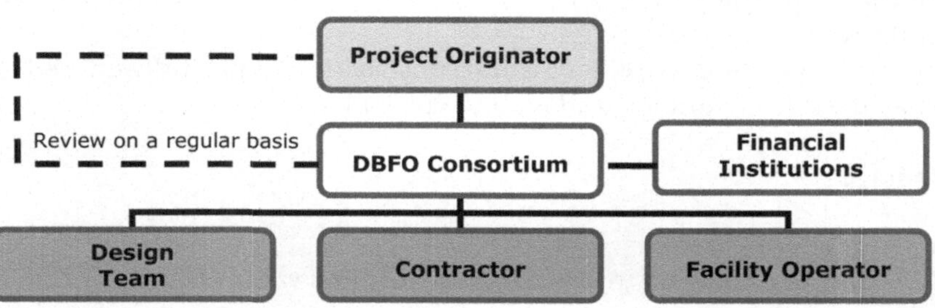

Figure 3.5 Contractual relationships in the design, build, finance and operate method.

Table 3.4 Examples of Local Projects[1] Delivered by Build-Operate-Transfer.

Projects	*Franchise Period*	*Project Description*
Western Harbour Tunnel	30 years, up to 2023	Financing, design, construction, commissioning and operation of a 1,975 m tunnel between West Kowloon and Sai Ying Pun.
Tate's Cairn Tunnel	30 years, up to 2018	Financing, design, construction, commissioning and operation of a tunnel linking Diamond Hill and Shatin.
Route 3 (Tai Lam Tunnel)	30 years, up to 2025	Financing, design, construction, commissioning and operation of a tunnel extending under the Tai Lam Country Park from Ting Kau in the south to Ho Pui in the North.

Source: [1] Legislative Council, HKSAR, 2010.

government enters into a franchise agreement with a private sector company, usually a consortium, to finance, develop and operate a facility over a franchise period. During the franchise period, the BOT company owns the facility and receives revenue from its operations. After the period, the facility will be transferred at no cost to the government. Considering that design, construction, finance and operation responsibilities all rested on the BOT company, local BOT arrangement is the same as DBFO.

Reminder

In fact, BOT and DBFO are the same in nature – to procure a consortium to design, construct, finance, maintain and operate the facility for the project originator (Miller, 2000). Although BOT usually applies in government projects, there are applications in the private sector when the owners have limited funds and resources to develop the facility (Menheere and Pollalis, 1996).

Advantages and disadvantages

Advantages:

- Inherits all benefits associated with the design, build and operate method.
- The project originator's funding requirement is substantially reduced. This is a concern when public funds are heavily constrained.
- Serves as an efficient way to privatise a public service.

Disadvantages:

- The drawbacks associated with DBO approach generally apply here.
- DBFO is comparatively more expensive than DBO as the cost of finance is higher (the interest rates charged by commercial banks are generally higher than the cost of government funds).

Points to note

Public-private partnership (PPP) is often discussed with design, build, finance and operate (DBFO). What is PPP in principle?

The PPP model is used by the government to recruit a private sector entity to have their skills and assets shared when delivering a service/facility for public use in addition to sharing risks and potential rewards. PPP emphasises the long-term contractual relationship to deliver the service by the public and private sectors jointly. PPP differs from other forms of private sector involvement such as outsourcing or privatisation. In outsourcing, the government solicits the private sector through shorter-term service contracts. In privatisation, the government transfers the entity to the private sector in perpetuity and the government only acts as a regulator during the period (Efficiency Unit, HKSAR, 2008). There are many PPP cases in Hong Kong, such as Hongkong Disneyland and AsiaWorld-Expo. The government's participation in PPP usually involves the contribution of land and initial project finance. The sharing of risks and revenue between the government and the private entity has to be clearly defined in the agreement to avoid disputes.

PPP focuses on the collaboration between public body and private sector to deliver the facility/service, and it can take many forms like D&B, BOT, DBO, DBFO, etc. (The National Council for Public-Private Partnerships, 2013). According to the Reference Guide on Selection of Procurement Approach and Project Delivery Techniques prepared by the HKSAR government (Environment, Transport and Works Bureau, 2004a), DBFO is described as an umbrella term for various arrangements including private finance initiative (PFI), public-private partnership (PPP) and build-operate-transfer (BOT). However, this interpretation deviates from most overseas literature and there are two points that are worth clarifying here:

- PFI or PPP involves the private sector to collaborate in public projects. However, DBFO and BOT can be applied to any project in principle, although usually in government projects. There are applications of DBFO/BOT in NGO and non-profit projects as well.
- PPP can exist in many forms which may not have design, construction and operation responsibilities transferred to one private company (Menheere and Pollalis, 1996).

Reminder

Private finance initiative (PFI) is the term widely used in the United Kingdom and Australia, which refers to a procurement method where a private sector firm takes on the responsibility for constructing, operating and maintaining a public service over a prescribed concession period. In the United Kingdom, the term is often used interchangeably with public-private partnership.

Allocation of responsibilities and risks

Through formation of a contract, a project owner can transfer some of the project responsibilities to other parties. While allocating responsibilities to others, the associated risks are also transferred. Each project delivery method inherits a different pattern of responsibility allocation that can be summarised in **Figure 3.6**.

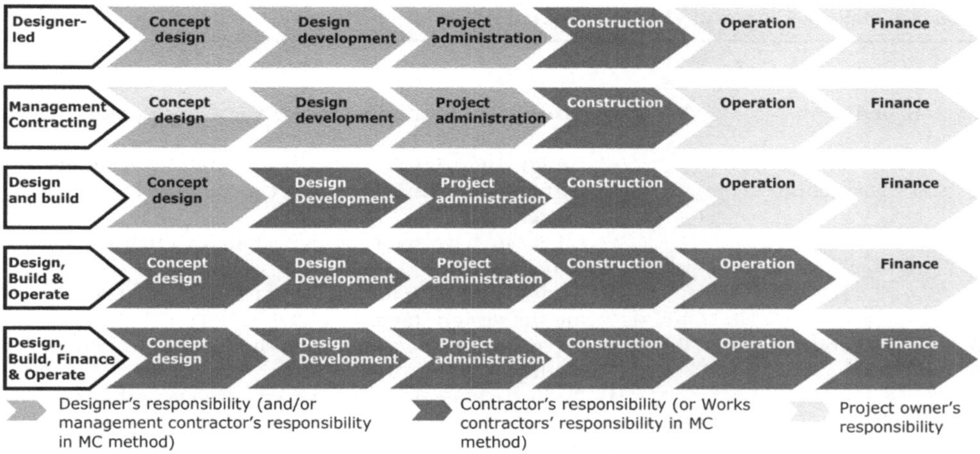

Figure 3.6 Allocation of responsibility between project owner, designer and contractor by project phase and delivery method.

The illustration in **Figure 3.6** is a generic one and may be adjusted slightly under different circumstances to suit specific needs. For example, in some DBFO contracts, the project owner may jointly finance the project with the DBFO consortium.

> **Points to note**
>
> An appropriate choice of delivery method helps the project owner to transfer project risks to other parties, but it does not relieve the project owner from the related costs. If the risk is allocated to the contractor, the contract price is expected to be higher. Sometimes, the project owner may choose to accept a certain proportion of risk rather than transfer it entirely to the contractor, especially if the project owner has greater control over the occurrence of such a risk item.

Selection of project delivery options

Each project has a unique set of characteristics including project type, scale and construction method, as well as requirements on budget, schedule and risk allocation. These project characteristics and requirements lead to project risks of varying degrees and different natures (Georgia State Financing and Investment Commission, 2003). While considering the acceptance of risk, a project owner has to analyse several factors including schedule, project scope, project owner's capability, contract arrangement and legal/funding concerns. These factors and their possible constraints that may affect the choice of delivery option are listed in **Table 3.5**.

Selecting an appropriate delivery method is critical to the achievement of project objectives and risk transfer. For example, if there is a schedule constraint, design and build or management contracting can be chosen to allow the construction phase to overlap with the design phase. Or, if the employer has the ability to oversee design and prefers a single point of responsibility, design and build will be an appropriate method.

Table 3.5 Major Factors and Constraints That Affect the Decision on Delivery Method.

	Factors	*Possible Constraints*
1	Schedule	• Insufficient time to complete the design prior to construction commencement. • Design and construction phases have to be overlapped.
2	Project scope	• Difficult to define the project scope clearly during the tender stage. • Likely to have many design changes during the construction stage. • Desirable to have design input from the contractor during the design stage.
3	Project owner's capability	• Project owner doesn't have in-house resources to verify quality of work. • Project owner doesn't have experience with a specific delivery method. • Project owner doesn't have the expertise to operate / maintain the facility.
4	Contract arrangement	• Project owner prefers to oversee project design. • Project owner wants to have direct control over the contractor's work. • Project owner prefers a single point of contact.
5	Legal / funding concerns	• Public accountability is a concern. • Required to justify the use of funds.

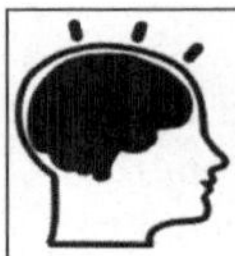

Question 3.1

Put a tick where the objectives are achievable under the specified project delivery option or a cross to those that cannot be achieved.

Project Objectives	*Designer-led*	*Management Contracting*	*Design and Build*
Allow full project scope to be determined later as the site work proceeds			
Price certainty before commencement of work			
Facilitate early completion			
Provide high quality of design and construction			
Have full control of design			
Transfer most risk to contractor			
Enable contractor's input during design stage			
Allow single contractual link with project owner			
Have impartial designer's advice during construction			
Enable highly complex design or advanced technology to be adopted			
Enhance collaboration between designers and contractor			

Suggested answers

Question 3.1

Project Objectives	*Designer-led*	*MC*	*D&B*
Allow full project scope to be determined later as the site work proceeds	×	√	×
Price certainty before commencement of work	√	×	√
Facilitate early completion	×	√	√
Provide high quality of design and construction	√	√	×
Have full control of design	√	√	×
Transfer most risk to contractor	×	×	√
Enable contractor's input during design stage	×	√	√
Allow single contractual link with project owner	×	×	√
Have impartial designer's advice during construction	√	√	×
Enable highly complex design or advanced technology to be adopted	×	√	×
Enhance collaboration between designers and contractor	×	×	√

4 Payment mechanisms

Principles of payment mechanisms

In a construction contract, the fundamental obligation of a contractor is to build the facility, whereas that of a project owner is to pay the contractor for the works. Payment method is of critical concern to both contracting parties as it directly relates to the level of risk taken by them. There are two basic methods of making the payment by a project owner: fixed price and cost reimbursement.

Fixed price vs cost reimbursement

Fixed price items refer to

> items paid for on the basis of a predetermined estimate of the cost of the work, including an allowance for the risk.... The estimated price is paid by the client, irrespective of the cost incurred by the builder.
>
> (Aqua Group, 2007, p.58).

Cost reimbursement items refer to the items paid in accordance with the actual cost of work (Murdoch and Hughes, 2008). Although the contract may have laid down specific rules or methods to calculate the cost, the amount to be paid by the project owner depends on how much the contractor spends on the work.

These two approaches allocate different cost estimation and fluctuation risks to the contracting parties. In a contract containing fixed price items, the contractor agrees to carry out the work at a price he has estimated during the tender stage. Any risk of error or underpricing has to be borne by the contractor. In a contract containing cost reimbursement items, neither the project owner nor the contractor knows the exact amount to be paid or received when they sign the contract. Any unforeseeable expenditure by the contractor is directly absorbed by the project owner. This includes additional costs due to contractor's problems (e.g. poor management, low productivity, high wastage etc.) or environmental problems such as shortages of labour/material and inflation.

Application of fixed price and cost reimbursement payment methods

Although the fixed price payment method favours the project owner, contract administrators do not apply fixed price items blindly to the whole contract. If the information for tender pricing is inadequate or uncertain, the tendering contactors will set the price higher to cover the potential risk. If the eventualities never occur, the project owner will pay more. Therefore, to strike the best balance between the risk transfer and the associated cost, contracts often contain a combination of fixed price items and cost reimbursement items for clearly defined work and incomplete designed work respectively.

Based on the predominant payment method set out in the contract, we can classify the contracts into three main types: lump sum, remeasurement and cost reimbursement contracts.

Type of contracts

Lump sum contract

Lump sum contracts are characterised by a single fixed price i.e. contract sum being set for the entire contract before construction work is started. The contractor agrees to complete the work in return for the lump sum of money, which the project owner agrees to pay upon satisfactory completion of work.

There are two types of lump sum contracts, lump sum with quantities and lump sum without quantities. For the lump sum with quantities type of contract, bills of quantities are prepared by the QS consultant with reference to the tender drawings and specifications. The bills of quantities contain every item of work required and the tenderers are invited to price against each of the items (as in **Figure 4.1**) to arrive at a tender sum.

For lump sum without quantities types of contract, bills of quantities are not provided. Tenderers are given drawings and specifications to estimate a tender sum. Detailed schedules of rates and quantities are usually prepared by each tenderer to show the build-up of his/her tender sum. Alternatively, to enhance pricing consistency across all tenderers, a schedule of rates and quantities may be provided by the QS consultant to the tenderers for pricing.

Reminder

The term 'schedule of rates and quantities' is used for the itemised breakdown in lump sum without quantities contracts. This is to differentiate it from the standard BQ in a lump sum with quantities contract.

Question 4.1

True or False?

i If we find 'quantities' of work in a lump sum tender, it is surely a lump sum with quantities contract.

ii The format and contents of a BQ and a schedule of rates and quantities will be identical if they are prepared for the same project because the scope of work is the same.

					CONCRETE WORKS
	Description	Quantity	Unit	Rate	HK$
	<u>IN-SITU CONCRETE</u>			Tenderer enters a price for each item	Item total = Quantity × Rate
	<u>Reinforced concrete 35/30</u>				
	Beds				
A	100mm thick	30	m^3		
	Suspended slabs; horizontal				
B	250mm thick	50	m^3		
C	300mm thick	20	m^3		
D	375mm thick				

Figure 4.1 Excerpt of a BQ page for tendering.

Advantages and disadvantages

	Advantages	*Disadvantages*
Lump sum contracts in general	• A tried and tested method that the market is familiar with. • Firm price is available to both contracting parties before work commencement. • Tender sums are available for tender evaluation. • Provide incentive to a contractor to manage his work effectively to maximise profit.	• Drawings and specification have to be completed before inviting tenderers.

Question 4.2

In a lump sum contract, will the contract sum be adjusted when there is a variation?

Question 4.3

Under the following conditions, if there is an error found in a quantity of the schedule of rates and quantities (SOR), will the contract sum be adjusted?

i SOR is prepared by the tenderer
ii SOR is prepared by the consultant QS

	Advantages	*Disadvantages*
Lump sum with quantities contracts	• Bills of quantities provide a good basis for tendering and valuation of variations. • Tenderers do not need to take-off their own quantities. This saves the tenderers' resources and lowers the risk of measurement error, which will be reflected in the tender prices.	• Preparation of bills of quantities takes time. • The project owner is ultimately responsible for any errors in the bills of quantities.
Lump sum without quantities contracts	• The project owner does not need to bear the risk of measurement error. Such risk is transferred to the contractor.	• The format of schedule of rates prepared by tenderers is not standardised, which makes tender analysis difficult. • The contractor must allow a higher price for any uncertainty not clearly shown but implied on drawings or described in specification, which will then be paid by the project owner irrespective of whether or not the eventualities occur. • Difficult to price and agree variations if there is no detailed breakdown of the tender sum.

Application

Lump sum contract is the most popular form used in the industry as it provides a firm price to a project owner. This type of contract can be applied in most projects as long as the scope and design are certain at the tender stage.

Remeasurement contract

In remeasurement contracts, approximate bills of quantities are given to the tenderers for pricing. It is noteworthy that the quantities in the bills of quantities are approximate only and there is no firm contract sum. Here, the bill of quantities constitutes a schedule of rates only (RICS, 2013). Both the project owner and the contractor know that the quantities are subject to remeasurement upon completion of work. The interim payments, including variations ordered during construction period, are valued at the contract rates.

Points to note

When using a remeasurement contract, a schedule of rates may be used instead of approximate bills of quantities for tenderers to price. The advantage of using a schedule of rates is to save time and effort without preparing the approximate quantities. It is particularly helpful when an ad-hoc project has to commence immediately. However, any possible bulk purchase discount will not be offered by the tenderers. Further, in the absence of approximate quantities, tenders received will be more difficult to compare.

Remeasurement contracts are suitable for projects where there is a level of uncertainty in terms of the exact scope and quantity of work. For instance, excavating soil and excavating underground concrete/masonry work are two items that cost differently. If the information about the underground condition of a site formation project is limited but the project owner insists on using a lump sum without quantities contract, the contractor may allow a higher price to cover the potential risk of concrete/masonry work excavation. The project owner may pay more if there is no such item involved. If the project owner chooses a remeasurement contract, the contractor will allow competitive prices for excavating the soil, underground concrete and underground masonry. As the project work proceeds, no payment will be made for underground concrete excavation if the remeasured quantity is nil.

Points to note

When dealing with variations, we often come across with situations where no contract rates are applicable to rate the additional work. This can happen in lump sum contracts as well as remeasurement contracts. In this case, the contractor will build up new rates (also called star rates) for the new work items. The QS consultant will then assess the rates accordingly. Details of build-up rates will be elaborated in **Chapter 21**.

Advantages and disadvantages

Advantages:

- A complete design is not required at the tender stage and therefore, an early project start is possible.
- Project scope and quantities are easily adjustable.

Disadvantages:

- Final cost is not known until project completion.
- More staff may be needed to monitor and measure the quantities of work.

Application

Remeasurement contracts are often applied in civil engineering projects and renovation projects where the exact quantities of work are difficult to define precisely before commencement of work.

Cost reimbursement contract

Unlike the previous two types of contract that rely on the contractor's pre-contract estimates to establish the payment amount, a cost reimbursement contract does not refer to any prior estimates for payment. The contractor is paid the actual cost plus a fee covering profit and overheads. The actual cost is ascertained by evidence of expenditure such as material invoices, labour hiring invoices and plant hire records submitted by the contractor, whereas the fee is usually calculated on a pre-agreed method which gives rise to various forms of cost reimbursement contracts.

Similar to remeasurement contracts, there is no definite contract sum when the contract is signed.

It is quite obvious that with this contract type, the employer bears all the risks including inflation, low productivity and problems of resource availability. On the contrary, the contractor receives full reimbursement for his cost and profit. There are different arrangements of cost reimbursement contracts to provide varying extents of control over contractor's expenses. They are cost plus percentage fee contracts, cost plus fixed fee contracts and guaranteed maximum price (GMP) contracts.

A cost plus percentage fee contract provides the least control of a contractor's expenses. Besides reimbursing the contractor for his project cost, the contractor can receive a percentage fee for his profit and overheads. There is no incentive for the contractor to save cost as the more he spends, the higher the fee. Therefore, this arrangement is rarely applied in practice. Nevertheless, it is ideal when there is insufficient time or information for the tendering process.

An improved approach is to set a fixed amount for the contractor's fee – cost plus fixed fee contract. However, this type of contract requires more information in order to have a fixed fee agreed between the parties during the tender stage. Although there is a slight incentive for the contractor to complete the work earlier to earn the fee, there is no incentive to minimise the waste as the fee is fixed.

When a cost reimbursement contract is used, many project owners choose a guaranteed maximum price (GMP) contract instead of the previous two. GMP is often regarded as a target cost contract where a guaranteed maximum price is estimated together with the share percentages and agreed between the project owner and the contractor at the tender stage. During the course of work, the contractor is paid according to his expenditure plus a percentage fee for overhead and profit, but the amount to be received by the contractor is limited to the cost ceiling agreed. In other words, any extra spending beyond the cost ceiling or guaranteed maximum price has to be borne by the contractor. However, any savings from cost underruns will be shared between the contractor and project owner as a reward. In this way, the project owner is safeguarded by a cost ceiling and the contractor is motivated to run his project efficiently below the cost ceiling so as to share the savings.

The setting of cost ceiling is the most critical step in GMP contracts. However, the contractor will inevitably try to set a higher figure to minimise his risk. This is why cost-plus contracts are difficult to establish and require mutual trust between the project owner and the contractor.

Points to note

Since the early 2010s, the Development Bureau has promoted the New Engineering Contract (NEC) and has drawn much attention to GMP or target cost contracts. The NEC adopted by the government so far is mainly Option C (Target contract with activity schedule) and Option D (Target contract with bill of quantities). In principle, both of them are GMP contracts to provide a higher incentive to contractors to complete the contract by having the 'pain/gain' sharing with the contractors. **Figure 4.2** illustrates three possible cases of 'final cost + fee' of the project. In case A, the project owner will not share the extra loss on top of the 'pain share'. In case B, the employer will share the loss but not 100%. So, the contractor cannot reimburse all his expenses. In case C, the two parties will share the gain (i.e. savings) according to the pre-agreed percentage.

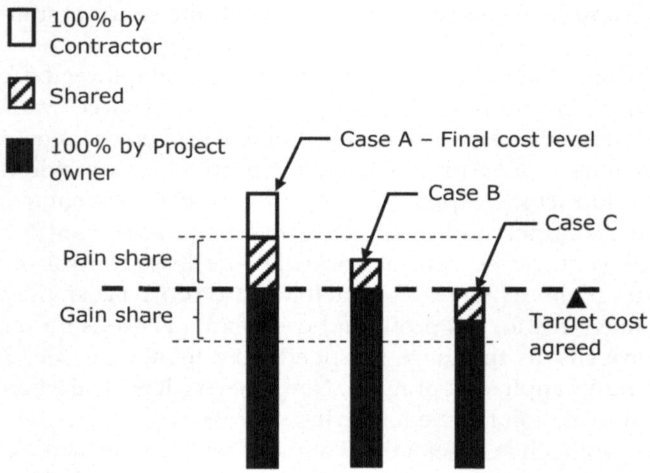

Figure 4.2 Pain/gain sharing between the project owner and the contractor.

Question 4.4

Below is the pain/gain share range stated in an NEC Option C (Target Cost) contract; with the target cost agreed at HK$100 million.

Band	*Share Range of Target Price*	*Employer's Share %*	*Contractor's Share %*
8	Over 120%	0%	100%
7	Over 115% up to 120%	50%	50%
6	Over 105% up to 115%	70%	30%
5	Over 100% up to 105%	90%	10%
4	Over 95% up to 100%	80%	20%
3	Over 85% up to 95%	70%	30%
2	Over 75% up to 85%	90%	10%
1	Under 75%	100%	0%

How much will be paid by the project owner if the final cost (including fee) is HK$84 million?

Advantages and disadvantages

Advantages:

- Work can be carried out immediately when specifications are ready.
- Project scope and quantities are highly adjustable.
- The project owner pays no premium to cover uncertainties – he only pays the actual cost plus fee.

Disadvantages:

- Unable to predict the final cost before project completion.
- More staff are required to check the actual cost spent by the contractor.

- The project owner takes all the risk on the contract unless a 'realistic' price is agreed in the GMP contract.
- Very little incentive for the contractor to carry out the work effectively (except for the GMP contract).

Application

In Hong Kong, a cost reimbursement contract is not popular due to the substantial risk taken by a project owner. Cost plus percentage fee or cost plus fixed fee contracts are ideal for emergency cases like restoration work after accidents or damages where the projects can be commenced quickly. GMP contracts have been used in private projects like One Island East in Quarry Bay and the Charter House in Central. As mentioned in the 'Points to Note', the Development Bureau has been strongly promoting NEC in recent years and a number of GMP/target cost contracts have been issued by the Drainage Services Department such as the Fuk Man Road Nullah in Sai Kung, the Happy Valley underground stormwater storage scheme and the Yuen Long and Kam Tin Sewerage. Early completion and/or cost saving was achieved (Drainage Services Department, HKSAR, 2019).

Selection of payment methods

The payment method of a contract influences a project owner's risk on the project cost and a contractor's incentive to finish the work effectively. **Figure 4.3** illustrates the relative magnitude of these two dimensions among different types of contracts.

Question 4.5

If both lump sum with quantities contract and lump sum without quantities contract are 'lump sum' in nature, why does the lump sum without quantities contract present less risk to a project owner?

Question 4.6

Can we have different payment methods applied in a single contract for different items?

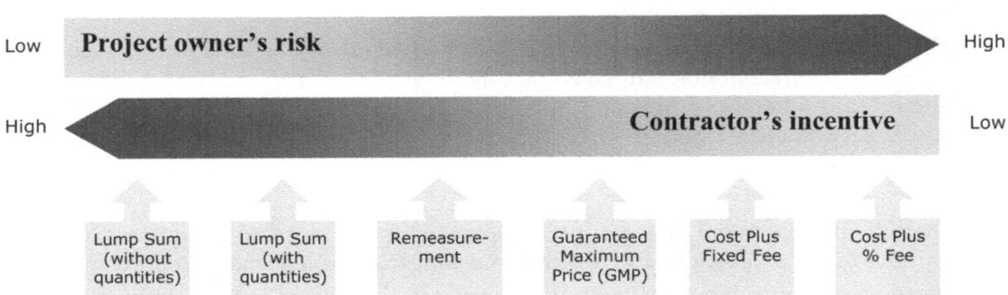

Figure 4.3 Project owner's risk and contractor's incentive in different payment methods.

Table 4.1 Major Factors Affecting the Choice of Payment Methods.

	Factors	*Considerations*
1	Project nature	• New, renovation or maintenance project?
2	Scope of work	• Can the project scope and size be clearly defined before work commencement? • Is there any uncertainty due to adoption of new technology?
3	Project schedule	• Is an early start required? • Are there any constraints in project duration?
4	Certainty of final cost	• Is a fixed contract sum preferred?
5	Changes during construction	• Is it likely to have a lot of design changes during construction?
6	Risk allocation	• Does the project owner prefer the lowest possible risk even if he needs to pay a premium to cover the risk?
7	Capabilities of project participants	• Are project participants familiar with the intended payment method?
8	Market condition	• Is the market buoyant with plenty of construction activities? • Is there any difficulty finding necessary resources for work? • Is price fluctuation likely to be significant over the project period?

When selecting a payment method for a project, careful analysis of the desired risk allocation and project factors should be made. **Table 4.1** lists the critical factors to be considered when deciding on the suitable payment method.

Suggested answers

Question 4.1

i False

In a lump sum without quantities contract, schedule of rates and quantities can be provided to/by the tenderers in order to work out the lump sum amount.

ii False

Unlike BQ, drafting of the schedule of rates and quantities is not bound by specific rules. Therefore, a BQ and a schedule of rates and quantities prepared for the same project can have very different layout, item classification and description.

Question 4.2

Although the project owner and the contractor both agree on the total price figure in a lump sum contract, it does not mean that the 'lump sum' figure cannot be adjusted. If there is any addition or omission involved in a variation, the contract sum will be adjusted accordingly.

Question 4.3

If there is an error found in a quantity of the schedule of rates and quantities (SOR), regardless of which party prepares the SOR, no adjustment will be made to the contract sum. In other words, no adjustment will be made to the contract sum in both (i) and (ii) conditions.

Question 4.4

The final cost (HK$84M) is equivalent to 84% of the target price. Therefore, according to the pain/gain share range agreed, the final cost payable to the contractor is:

Band	*Share Range of Target Price*	*Contractor's Share %*	*Payment by Project owner (HK$)*	
4	Over 95% up to 100%	20%	100M×5%×20%	= 1M
3	Over 85% up to 95%	30%	100M×10%×30%	= 3M
2	Over 75% up to 85%	10%	100M×1%×10%	= 0.1M
1	Under 75%	0%	-	
		Gain share:	4.1M	
		Final cost (incl. fee):	84.0M	
		Total payable to contractor:	88.1M (HK$)	

Although apparently the project owner pays an extra of HK$4.1M on top of the actual cost, he enjoys a cost saving of HK$11.9M compared with the agreed target price.

Question 4.5

The lump sum without quantities contract presents less risk to the project owner because any errors in the measurement of quantities for the preparation of tender is borne by the tenderers, not the project owner.

Question 4.6

We can apply different payment methods in a single contract for different items. For instance, in a lump sum with quantities contract, the entire contract is paid predominately by pre-agreed unit rates. However, if some items are foreseeable but the exact quantities are not determinable during the tender stage, we can apply provisional quantities to those items. Their quantities will be remeasured upon completion. In addition, we can allow prime cost sum items by which the main contractor can receive payment according to the amount expended (for the nominated subcontractors concerned) plus a pre-agreed percentage fee (details of the prime cost sum will be covered in **Chapter 7**).

Part 3

Preparing for tender

5 Tendering methods

Why tender?

Tendering is a process used by organisations to appoint a service provider to complete the desired work. On the one hand, project owners (i.e. the employers of construction contracts) always seek and select the most suitable contractors for their projects through tendering. On the other hand, contractors who hunt for projects will make an offer (or bid) to an invitation to tender sent by the project owner.

Tendering can involve different procedures depending on the method used. We can broadly classify tendering methods into four approaches based on the level of competition: open tendering, selective tendering, single tendering and negotiation.

Open tendering

This is a tendering method that allows all contractors who are interested in the project to submit their tenders. Notices of invitation are usually placed in mass media like newspapers, trade journals and websites. For government projects, notices of invitation are published in government gazettes and, if necessary, in the local press and on the internet. **Figure 5.1** below illustrates the main steps in open tendering.

Usually, information about the proposed project including type, scale and timeframe of work is listed in the notice. Any interested contractors can send a request to obtain the tender documents. In some projects, the contractor may need to pay a deposit to receive the tender documents. This may be used to cover the administrative cost or it may be refunded to the contractor upon submission of a bona fide tender.

Points to note

A bona fide tender refers to a tender that is prepared and submitted in good faith, without fraud.

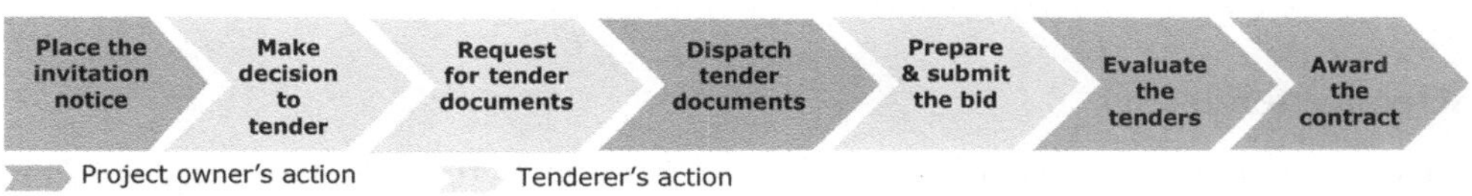

Figure 5.1 Main steps in open tendering.

The open tendering method allows any interested contractor, even a new or unknown company, to bid the job. Since the level of competition in open tendering is the highest, the opportunity cost to a contractor is also the highest (as the chance of winning is very low). Open tendering also requires substantial tender evaluation efforts as the number of tenders received can be huge. Nevertheless, government departments often use the open tendering method for the purpose of public accountability.

Advantages and disadvantages

Advantages:

- Maximum competition is available to the benefit of the project owner.
- No tender list or approved list of tenderers is used and therefore no favouritism.
- Prevents bidders from colluding.
- Tenderers are not invited and therefore all tenderers should be genuinely interested.
- Allows unknown contractors to tender for the work and break into the market.

Points to note

Collusion, which often appears as bid-rigging, 'occurs when, without the knowledge of the person calling for bids or requesting a tender, two or more competitors agree they will not compete with each other for tenders, allowing one of the cartel members to "win" the tender' (Competition Commission (HK), 2019).

Question 5.1

Can tenderers cooperate to rig the bid if there is no express term in the tender document prohibiting them to do so?

Disadvantages:

- Cost of tendering is expensive to project owners. The administrative work required to handle a potentially large number of tender bids is huge.
- More time is required for the project owner to evaluate the tenders received.
- From the contractor's perspective, the chance of success is the lowest due to keen competition. The cost of abortive tendering has to be absorbed by the contractors.
- An incompetent or unsuitable contractor may be chosen.

Selective tendering

Selective tendering is the most popular tendering method in Hong Kong. This method is used when the project owner has a list of contractors and the tender is only opened to the selected companies.

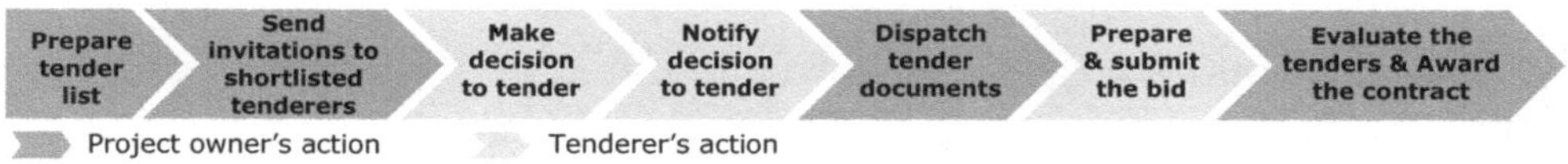

Figure 5.2 Main steps in selective tendering.

Figure 5.2 presents the main steps involved in selective tendering. First, the project owner or his representative will shortlist a number of contractors to put on a 'tender list'. Preparation of the tender list is critical as the contract is likely to be awarded to one of the tenderers within the list.

Most government departments and private developers maintain a long list of contractors but it is uneconomical to invite all contractors in the company's list to tender for each job. To lower the tendering cost while maintaining sufficient competition, only a small number of contractors are invited to tender. The number of tenderers should be restricted to six (Aqua Group, 2003; National Joint Consultative Committee for Building (Great Britain) et al., 1997). The tenderers should be shortlisted objectively based on their company background, past performance and the proposed project requirements. An initial enquiry containing some basic information of the project will be sent to the shortlisted contractors to see if they are interested to tender. Detailed tender documentation will be sent to the interested tenderers for preparation of the bid. At least four weeks should be allowed for the tender period if selective tendering is used.

In some complex projects such as those using an alternative project delivery method, a non-standard contract or a special construction technique, prequalification can be used to shortlist a small number of competent contractors who are interested in the proposed project to undergo the competitive bidding process (Environment, Transport and Works Bureau, 2004b).

Points to note

Prequalification is a formal process of evaluating the competency of companies so as to prepare a list of tenderers who are qualified to tender for the proposed project.

The prequalification process involves three basic steps: invitation to prequalification, evaluation of qualification submissions and notification of results. In the case of government projects, notices of invitation to prequalify are published in government gazettes and local newspapers, as well as on the internet. The notice should contain basic project information such as the nature and quantity of work, estimated commencement date and completion date. Companies that are interested to apply for prequalification may be asked to submit general information (e.g. past projects information, company annual turnover etc.) as well as technical details like their views on design/specifications or contingency plan for the project. To maintain fairness, objective prequalification criteria should be set. In many countries like the United Kingdom and the United States, prequalification questionnaires are used to collect general information from the applicants to facilitate objective evaluation. After evaluation, the project owner should inform each applicant of the prequalification result.

Question 5.2

True or False?

i Prequalification is to shortlist tenderers to submit a tender. Therefore, it must be included in the selective tendering process.
ii Prequalification is the same as selecting contractors to the project owner's approved list of contractors.

Single-stage vs two-stage selective tendering

Selective tendering can be delivered in one of the two ways: Single-stage selective tendering or two-stage selective tendering (Finch, 2011). Both methods are selective tendering by their nature as they involve invitations to tender from a shortlist of companies that satisfy the selection criteria. The selective tendering model described in **Figure 5.2** is single-stage selective tendering which is the most popular method adopted in Hong Kong. For two-stage selective tendering the contractor will be appointed in the design stage to help develop the best design and construction solution for the project. In the first stage, selected tenderers are invited to submit tenders based on minimal information. Their tenders may include construction programme, method statement, pricing of head office and site overheads, profits, and schedules of rates for some work packages. The first stage concludes with the appointment of a preferred contractor based on a pre-construction service agreement or a consultancy agreement. The second stage is typically a negotiation between the project owner and the previously appointed contractor. Based on the prices and allowances agreed in stage one, a lump sum for the project will be agreed as the scheme is developed. Two-stage selective tendering can be applied in D&B projects so as to allow contractor's involvement at the design stage.

Points to note

Apparently, most projects involve two stages in the tender process: competitive tendering followed by negotiation. Although a certain amount of negotiation is often inevitable, such as resolving items qualified by the contractor in the tender or agreeing on a discount to the work, the negotiation process does not automatically constitute a 'second stage' of tendering. In most single-stage selective tenders, negotiation exists but

> the discussion and negotiation of residual issues should not be such as in any way to undermine the integrity of the process
>
> (JCT, 2002, p. 12).

Advantages and disadvantages

	Advantages	*Disadvantages*
Selective tendering in general	• As the tenderers are known, the risk of awarding the contract to an unqualified contractor is lower. • Due to a smaller number of bidders involved, the time and cost required to evaluate the tenders is less. • The opportunity cost to tenderers is lower due to the relatively low competition.	• New bidders who may offer more innovative ideas are excluded. • Prices may be higher when compared with open tender. • There is a higher possibility of collusion if the firms get to know the probable list of tenderers in the area. The only way to rectify this is to change the shortlist regularly, either entirely or partly.

Two-stage selective tendering	• Early appointment of contractor leads to shorter project duration. • Contractor's input in design helps to improve buildability of the design. • Collaboration between contractor and design team is enhanced.	• The preferred contractor's role in design development strengthens its negotiation position in the second-stage tender which may result in a difficult bargain on the project owner's side.

Question 5.3

Can I use single-stage selective tendering to appoint a D&B contractor? If yes, what is the difference of using single-stage and two-stage selective tendering to appoint a D&B contractor?

Single tendering

In the single tendering method, only one tenderer is invited to submit a tender for a proposed project. Normally, this is not a recommended method except for the following circumstances:

- in extreme urgency, such as restoration of works after unforeseeable events or accidents;
- when copyrights or patented works are involved;
- when specialised service or equipment must be procured to meet certain compatibility or specification requirement; or
- when lease terms for occupation require the work to be carried out by a particular company.

The steps involved in single tendering are basically the same as selective tendering. The only difference is that a single company, instead of a few companies, is invited to tender.

Advantages and disadvantages

Advantages:

- Shorter tendering period as the administrative work involved in tender examination/evaluation is much reduced.
- Capability of the contractor is secured.
- Abortive tendering cost to the contractor is minimal.

Disadvantages:

- No competition is available and there is a risk of a higher price.
- The project owner finds it difficult to check whether the bid received is competitive or not.
- It is difficult to satisfy the test of public accountability when public funds are involved.

Negotiation

If open tendering represents an extreme tendering method that maximises the extent of competition, negotiation characterises another extreme tendering method that gives no competition. When the negotiation method is used, a single contractor is selected based on his past performance, experience and working relationship. As shown in **Figure 5.3**, having confirmed the company's interest to tender for the project, design information and pricing documents such as a bill of approximate quantities will be given to the contractor for pricing. Any disagreed prices and terms will be subject to negotiation between the project owner and the contractor. Negotiation will proceed with the contractor until an agreement is reached and the contract is signed.

Sometimes, situations may arise when the project owner wants to place an extension contract with a contractor. Imagine if a proposed commercial building contract is awarded to Contractor A. After contract commencement, a footbridge is confirmed to be constructed between the new commercial building and a nearby existing building. If Contractor A performs satisfactorily in the commercial building project, it is desirable to have the same team to build the footbridge. There will be synergistic benefits in site planning and resources organisation. It is also likely to bring time and cost savings to the project owner. However, considering the scope of work in the commercial building contract, it may not be suitable to issue a variation order to Contractor A to construct the footbridge. An extension contract awarded to Contractor A will be a more reasonable arrangement.

Since the extension contract is often smaller in scale and the contractor is engaging in an existing contract, it is uneconomical to adopt competitive tendering for the extension work. The existing contract can serve as a good basis for the project owner and contractor to negotiate the terms and prices for the extension contract.

Advantages and disadvantages

Advantages:

- Allows early selection of the contractor before design is fully completed.
- Capability of the contractor is secured.
- Abortive tendering cost to the contractor is minimal.
- Continuity of work can be allowed in case of extension/continuation contract.
- Facilitates the development of team spirit between the consultants and the contractor, which can help to eliminate avoidable claims during the contract period.

Disadvantages:

- Inherits all disadvantages associated with single tendering.
- Loss of time if negotiations are unsuccessful.

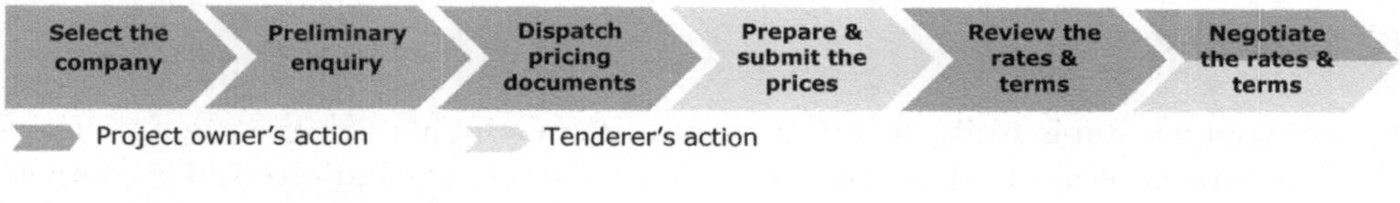

Figure 5.3 Major steps in negotiated tenders.

Suggested answers

Question 5.1

Under the Competition Ordinance, Cap. 619, it is illegal for the tenderers to rig the bid. In other words, tenderers are not allowed to do so with or without any express prohibition in the tender document.

Question 5.2

i False

Prequalification can be considered as an objective method to shortlist contractors to enter the tender list. The project owner can shortlist tenderers by selecting from his approved list of contractors or seeking advice from the consultants. Therefore, prequalification is not compulsorily required in all selective tendering process.

ii False

Prequalification is different from admitting a contractor to an approved list. An approved list of contractors held by a project owner is a permanent list applicable to multiple contracts but the prequalified contractors selected in a prequalification exercise are only applicable to the concerned project (Environment, Transport and Works Bureau, 2004b).

Question 5.3

Yes, we can appoint a D&B contractor by single-stage selective tendering method. In other words, we can use either single-stage or two-stage selective tendering to appoint a D&B contractor. The Code of procedure for selective tendering for design and build (National Joint Consultative Committee for Building, 1985) suggested many variations in the two-stage tendering routes for D&B, although not all of them are common in Hong Kong. In summary, the main difference between the single-stage and two-stage approaches is that if we use the single-stage approach, the D&B contractor will be appointed once, for the design and construction of the facility. If we use the two-stage approach, the contractor will be appointed twice. The first appointment is for the pre-construction design service and the second appointment is for the construction service. Under the two-stage selective tendering model, the project owner can defer the award of construction contract until he is satisfied with the contractor's performance in the stage one package.

6 Tendering procedures

Tendering procedures in general

The primary aim of tendering is to allow a project owner to find the right contractor to carry out the work. Different tendering methods and their general features have been illustrated in **Chapter 5** already. Although the detailed tendering procedures vary with different tendering methods, five basic stages are involved:

- preparation of tender documents,
- invitation to tenderers,
- issuance of tender documents,
- preparation of tender bid, and
- evaluation of tenders.

To illustrate the steps and tasks carried out by different parties in the tender process, competitive tendering is used as a reference.

Tendering procedures of single-stage selective tendering

Most of the competitive contracts in Hong Kong are awarded by single-stage selective tender. To provide a full picture of the tender process, detailed stages of selective tendering from production of tender documents to contract award are portrayed in **Figure 6.1**. The dominating party in each step is highlighted in the diagram to explain the interactions between the parties. For simplicity, the QS consultant is assumed to be the project owner's adviser in the tender process.

Points to note

As shown in **Figure 6.1**, the project owner or the QS consultant participates in the entire tender process from start to end. However, the contractors are involved in the middle stage only. This part of tender process is often regarded as the estimating process undergone by a contractor to prepare the bid.

Step 1: Produce tender documents

Having decided the procurement strategy and tendering method (for details refer to **Chapters 3 to 5**) for the project, the QS consultant has to prepare the tender

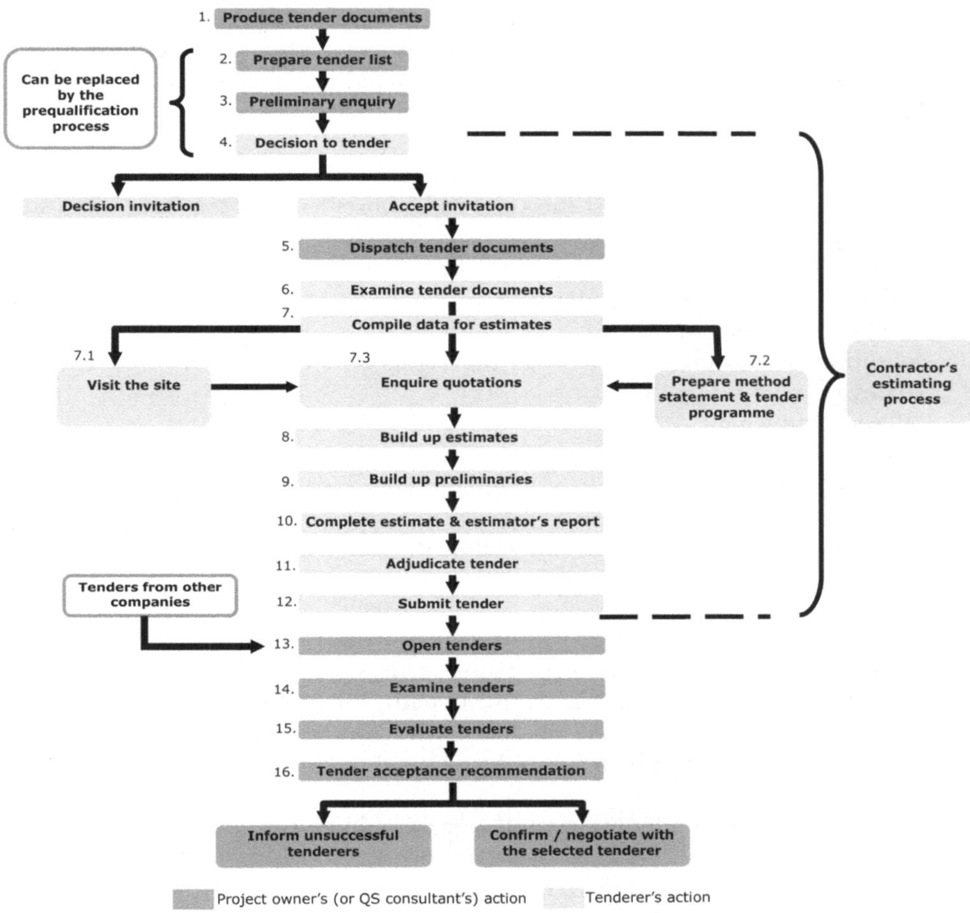

Figure 6.1 Single-stage selective tendering procedures.

documents. Tender documents are the information to be delivered to each tenderer for preparation of the tender bid. The 'Practice Notes for QSs – Tendering' (HKIS et al., 2012) has laid down guidelines for QS consultants to conduct the tender process.

The QS consultant has to make every effort to produce the tender documents in a timely and cost effective manner. The QS team leader has to hold regular meetings with the measurers or assistants to:

- review the production schedule for the tender documents;
- ensure that the use of a standard method of measurement and the BQ format is consistently applied;
- check that all drawings and specifications are coherent and adequate for measurement;
- verify that all works are properly measured; and
- ensure that specific requirements of the project owner are included in the preliminaries.

A bulk check on the quantities should always be done after the bills of quantities have been completed. Before issuance to the tenderers, the QS consultant should ensure that a complete set of tender documents including the following sections are produced:

- Instructions to Tenderers/Notes to Tenderers.
- Conditions of Tender that list the rules to be observed by a tenderer when submitting a tender.
- Form of Tender to be completed and signed by the tenderer. It includes the tenderer's details and the tender sum.
- Articles of Agreement.
- Conditions of Contract that covers the rights and obligations of contracting parties.
- Full set of drawings.
- Specifications.
- Bills of Quantities or Schedule of Rates and Quantities where applicable.

Further details of the tender document will be discussed in **Chapter 7**.

Step 2: Prepare tender list

To ensure adequate competition in the tender process, the project owner can select a list of contractors to form the tender list. Usually, the architect, engineer and QS consultant will give advice to select capable contractors to enter the tender list. The number of tenderers should be restricted to six and one or two extra names can be held in the reserve.

When compiling the tender list, several areas should be considered:

- Size of company: It should match the project size.
- Nature of work engaged by the company: The company should have experience relevant to the proposed project.
- Past projects completed by the company: The completed projects provide a good portfolio to tell the company's standard of workmanship, technical expertise and knowledge.
- Reputation of the company: Check the reputation of the company to meet deadlines, supervise quality and safety of works, rectify defects and settle final accounts.
- Financial stability of the company: It should be credit-worthy and capable of financing the workload.

Alternatively, the preparation of the tender list can be replaced by the prequalification process described in **Chapter 5** to select a small number of contractors to tender for the project.

Step 3: Preliminary enquiry

The objective of the preliminary enquiry is to ensure that a required number of tenderers are returned, and if any contractor declines the invitation, a substitute can be drawn from the reserve list to make up the required number. A preliminary enquiry letter that provides the general information of the project will be sent to all tenderers.

Usually one to three weeks are allowed for the contractors to decide whether they wish to tender or not. The following information should be included in the preliminary enquiry letter:

- Project owner and designers
- Location of the site
- Nature of work to be carried out
- Conditions of contract used as well as any significant special conditions
- Anticipated contract commencement and completion dates
- Any special features or requirements, e.g. environmental control measures, contractor's design requirement, bonds or guarantees.

Step 4: Decision to tender

Although many contractors are reluctant to decline an invitation to tender, it is uneconomical to submit bids on all the projects that are out for bid. Estimating is a costly process and contractors should spend their estimating resources on projects that offer the highest chance of success. That is why contractors should appraise the project carefully before committing substantial resources to the estimating process.

Based on the information provided in the preliminary enquiry letter, the contractor should consider the following issues:

- Experience: Is the company experienced in this type of work? Is the company technically capable of undertaking the work?
- Resources: Does the company have the necessary supervisory staff, plant and capital to finish the project?
- Location: Is the project located within/outside Hong Kong? Are there any implications on the project risks and overhead requirements?
- Workload: Does the estimating department have sufficient manpower to prepare the bid? Is the tender period enough to prepare a sensible bid?
- Anticipated risk: Are there too many risks in the project?
- Competition: What is the level of competition? How many contractors will be invited to tender? What will be the cost of preparing the tender?
- Project team: Is there any experience of working with the project owner, designers and QS consultant? Does the project owner have the financial resources available to pay for the work?

After assessing the above factors, a decision to tender or not is made. The contractor should then send a reply to the project owner or his agent.

Step 5: Dispatch tender documents

Having received all replies from the contractors, the QS consultant will dispatch the tender documents. Nowadays, to save paper, many tender documents are prepared in electronic format, including the bills of quantities, drawings and specifications. The e-document can be delivered to the tenderers easily through an online portal.

The QS consultant has to work out a tendering period which allows reasonable time for the tenderers to prepare the bid. Normally, a minimum of three weeks is required for the tendering period but a longer period may be required depending on the size, nature and complexity of the project; the standard of tender documents and the extent of special tender requirements (e.g. requirement to submit alternative designs or proposals).

During the tender period, tender addenda are often issued to the tenderers. To avoid future disputes, the QS consultant should set out the changes to the tender documents clearly and tenderers should confirm the receipt of addenda by returning the signed acknowledgement slip. In order to allow sufficient time for the tenderers to incorporate the addenda information into their estimates, tender addenda should be issued not later than one week before the tender closing date (HKIS et al., 2012).

Reminder

Tender addenda refer to the supplementary tender information issued to the tenderers after the formal dispatch of tender documents, usually for the purpose of rectifying errors, providing additional information or clarifying ambiguities in the tender documents.

Step 6: Examine tender documents

The tender documents are usually passed to the estimator for pricing. The estimator should check the documents to make sure that the tender documents are complete and inform the QS consultant of any discrepancies immediately. A preliminary report is usually prepared to summarise the general features of the project for top management review.

Today, many tenders require the tenderers to submit a method statement, tender programme and maybe even an alternative design proposal together with the priced BQ/priced schedule of rates. A tender team comprising the estimator, project manager and planning engineer is necessary to work out a feasible and realistic bid. The team members will examine the tender documents to highlight anything unusual in the specifications or drawings, any items that are easily missed and any peculiar contract terms. The estimator will usually act as the leader to coordinate the views and comments of the team members to prepare the tender.

Step 7: Compile data for estimates

Before the estimator can estimate the price for each item in the bills of quantities, a lot of data about the project have to be collected. These data are not merely derived from the tender documents but require a thorough analysis of how the work will be carried out, the type of mechanical plant and specialists required, the possible risks to be encountered, the latest market prices on labour and materials and so forth.

Three primary steps are involved to collect the necessary information for pricing, including:

- visit the site
- prepare a method statement and tender programme
- acquire quotations from subcontractors and suppliers

Step 7.1: Visit the site

During the tender period, the project owner may arrange a site visit for the tenderers to ascertain the site condition and any likely restrictions which may affect the execution of work. The site visit is best to be arranged after the tender documents have been reviewed by the tenderers. With the knowledge of the project gained from their review, the estimating team can focus on the specific areas of concern or interest during the site visit. Normally, the following information will be gathered from the site visit:

- Topography of the site, e.g. site profile, density and condition of vegetation, surface water, overhead obstruction etc.
- Proximity, condition and use of adjacent buildings and structures.
- Accessibility of site, conditions of traffic and transportation facilities for labour and materials.
- Need for temporary roads.
- Obvious utility location and type of service.
- Local restrictions like parking restrictions, noise control and restrictions on the use of mechanical plant.

Many companies adopt a standardised site visit report to ensure that all critical items are not overlooked. Photographs should be taken on the site visit as the condition of the site may change in the time between the bid and the project start. The condition of adjacent buildings/structures should also be photographed. The photographs can act as useful reference to determine whether damage on an adjacent structure occurred before or after the project began. Only by collecting reliable and detailed data can the contractor estimate the likely risks and allocate appropriate sums to the bid to cover these risks.

Step 7.2: Prepare method statement and tender programme

A method statement is a written document that details how the works are to be planned, procured, constructed and handed over. Today, method statements are frequently requested as part of the tender submission, which allows the project owner to gain an insight into the way the contractor operates. Even if the method statement is not required in the tender submission, tenderers are recommended to consider how the works will be carried out before pricing the tender as the project cost can vary significantly with different construction methods. For instance, the use of different formwork systems, such as traditional timber form or mechanised system formwork, can have very different financial implications. That is why a thorough consideration of the construction method is essential to pricing. Some companies may adopt their patented technologies in the proposed project. By reviewing the method and sequence of work, the potential benefits, risks and constraints can be evaluated more precisely. This can help the estimating team to work out a realistic and competitive bid.

A tender programme refers to a schedule outlining the sequence of construction activities from the start to completion of a project. During the tender stage, the contractor will draw up a tender programme. The tender programme can show how the contractor intends to sequence the work. Normally, a tender programme will not form part of the contract document, but it can be a useful reference for the project owner to understand the contractor's plan of work. Therefore, contractors are often required to include a programme as part of their tender submission.

Regardless of the requirement to submit a tender programme with the tender, information about the durations for different sections of work is essential for estimating the cost of time-related preliminary items such as:

- site management cost
- expenses associated with site accommodation, electricity, water and the like
- rental charges for mechanical plant and equipment
- rental charges for temporary works

Moreover, based on the tender programme, the risk associated with inclement weather, such as executing external work or concreting during summer, can be more accurately estimated. Therefore, a tender programme should be sufficiently detailed to serve the above purposes.

While the method of work should be coherent with the sequence and duration of work sections, the method statement and tender programme should be developed concurrently by the estimating team with the same care and thoroughness.

Step 7.3: Acquire quotations

The majority of general contractors subcontract most or all of the sub-trades. To build up prices for the BQ items, the contractor must first obtain up-to-date quotations from his/her subcontractors and suppliers. Based on the tender documents and the method statement established earlier, the project can be sub-divided into packages for quotations enquiry. The estimator can select suitable subcontractors or suppliers from the company database (or the suppliers/subcontractors list if named suppliers/subcontractors are specified in the tender documents) and invite them to submit quotations for the tender. To avoid receiving insufficient quotes, the general contractor should contact the subcontractors/suppliers to ascertain their willingness to submit a quotation before invitation.

Enquiries to suppliers

Enquiries to suppliers are made to obtain quotations for materials. Besides the specification and quantities of the required materials, some important information should also be included in the enquiry for quotation:

- project title and location
- contract period and if possible, indicative delivery dates

- delivery requirement, such as ex works; free on board (FOB) origin; cost, insurance and freight (CIF); and delivered duty paid (DDP)
- any special requirement like provision of design service, testing or warranty by the supplier
- any restrictions on access to site

Points to note

Suppliers often used different trading terms to describe the responsibility of buyer and seller. The following interpretations are based on the trade rules developed by the International Chamber of Commerce (Law, 2016).

Ex works: the seller only needs to make the goods available at a specified location, usually the supplier's factory, and the buyer is responsible for the customs, taxes, insurance and all transportation costs from supplier's factory to buyer's site.

FOB origin: FOB refers to the time when the seller is no longer responsible for the goods and the title of goods transfers to the buyer. Therefore, FOB origin means the seller is responsible for the risk up to loading the goods onto the carrier at the departure port. FOB destination will require the buyer to take responsibility from the seller's specified location or destination.

CIF: The seller is responsible for the transportation cost and insurance coverage up to the arrival port. However, the title of goods is transferred to the buyer once the goods are loaded to the carrier. The supplier is responsible to clear customs in the country of export whereas the buyer must clear customs in the country of import and arrange local transportation to the site from the port.

DDP: The seller is responsible for all the risk and delivery costs until the goods are transferred to the agreed destination in the buyer's country (usually the arrival port).

Question 6.1

A contractor orders 100 boxes of tiles from a Spanish supplier. The contractor places this order requiring FOB destination to a construction site in North Point, Hong Kong. The contractor agrees to pay the local charges to deliver the tiles from Kwai Tsing Container Terminals to the site. The supplier accepted the order and dispatched the goods. Unfortunately, during the transit to the site, the truck crashes in an accident and most of the tiles are damaged. Who is responsible for the loss? The contractor or the supplier?

Enquiries to subcontractors

Before sending an enquiry to subcontractors, the estimator should be clear about the nature of subcontractor service required, which can be one of the three types:

- Labour-and-materials subcontractor: for conventional supply and installation work packages.
- Labour-and-plant subcontractor: for hiring a mechanical plant together with the operator.

- Labour-only subcontractor: for providing labour-only service like concreting. Any materials required to be procured separately by the general contractor.

The estimator must carefully extract the relevant pages of the bills of quantities and drawings to each invited subcontractor for quotation submission. Irrelevant items on the extracted BQ pages should be crossed out to avoid confusion. In addition, the estimator should include the following information in the enquiry letter:

- project title and location
- name of employer, designers and consultants
- general description of work
- contract period and, if necessary, an outline programme (especially when subcontractors are required to submit their own tender programme to the general contractor)
- conditions of main contract and amendments
- conditions of subcontract and amendments
- any special subcontract conditions/requirements
- site plant and facilities to be provided by the main contractor
- services and attendance to be provided by the main contractor
- location where full contract details and drawings can be inspected

Reminder

The request for quotation procedures are in fact very similar to the selective tendering procedures used by the project owner to choose a main contractor. Here, the general contractor has to compile a list of subcontractors/suppliers, prepare and dispatch the relevant parts of tender documents to them, invite them to submit quotations, collect quotations from them and finally choose the preferred quotation (usually the lowest one) for pricing. The purpose here is to obtain the market information or the lowest prices for bid estimation, not to find a company to do the work. Therefore, the general contractor does not need to inform the subcontractors/suppliers whether or not their prices are used for the bid estimate.

Assuming that the tender period available to the general contractor is five weeks, it will only allow three weeks or less to the subcontractors or suppliers to prepare their quotations. This is to leave sufficient time for the general contractor to evaluate the quotations for pricing the bid. In view of the limited time available to the subcontractors and suppliers, the general contractor should make every effort to facilitate these companies in the quotation preparation process by giving precise and complete information and by bringing peculiar requirements to their attention. Estimators should treat the subcontractors and suppliers as partners in order to complete the tender estimate satisfactorily within the limited time frame.

Step 8: Build up estimates

With all the quotations received, the estimator will select the best quote (usually from the lowest quotation) to build up the rate for each BQ item. Information system can provide great assistance to compare subcontractors' or suppliers' quotations on a sub-trade level. Many estimating software or simply a spreadsheet can easily cross-compare the rates and check for arithmetical errors.

Step 9: Build up preliminaries

Besides building up unit rates for all trade work items, the preliminaries section of the bills of quantities needs pricing as well. Preliminaries, commonly known as project overheads, are

> the site cost of administering a project and providing general plant, site staff, facilities and site-based services and other items not included in all-in rates
> (Chartered Institute of Building, 2009).

Site management staffing, site accommodation, hoardings, scaffolding, protection of works, drawings production, insurance and so forth are examples of preliminaries. To avoid missing some essential preliminary requirements, it is recommended to price the preliminaries at a later stage when the estimator fully understands the tender documents. Further, the development of the resourced programme, organisation chart, site layout and method statement should precede the pricing of preliminaries, to identify the type of facilities required for each activity and the periods of requirement.

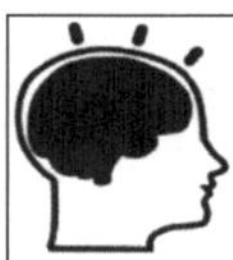

Question 6.2

Which of the following is/are not preliminaries?

- wastage of steel bars
- removal of debris from site
- water for car washing bay
- temporary support for formwork

Step 10: Complete the estimate and estimator's report

Once all the unit rates are calculated, the BQ can be extended and summed up to give the base estimate. Although the use of a spreadsheet helps us to calculate the total figures effectively, it is relatively difficult to identify mistakes especially when formulae are used. A mathematical check must be carried out to ensure that the base estimate is free from errors. At the same time, the estimator must find some ways to bulk check the pricing. Different companies may adopt different approaches, e.g. by comparing the costs of labour and material in each trade to see if it is reasonable.

Even the BQ is priced and checked, it is not yet ready for submission. Top management has to review the base estimate on a strategic level to decide the profit and mark-up. The estimator will provide analysis and supporting documents to help management understand the full technical and commercial requirements of the project. Many contractors have developed a standard format of estimator's report, which usually covers the following information:

Project review:

- brief description of the project
- method of construction
- resourced programme
- list of problems or risks associated with the project but not properly covered by the BQ items
- non-standard contract conditions or amendment to contract conditions
- assumptions made in the estimate preparation

Estimate analysis:

- summary of trade/bill totals
- analysis of subcontractors, materials and labour cost at trade level
- assessment of projected cash flow and profitability (based on likely profit mark-up)

Market analysis:

- review of the relevant past projects developed by the project owner
- any pertinent information of market conditions, industrial trend and competition.

Step 11: Adjudicate tender

The conversion of an estimate into a tender bid involves a lot of commercial considerations and it is the responsibility of management. Based on the estimator's report, management will review the costs in the base estimate and decide the appropriate bidding strategy. Factors like market share, current and future workload, relationship with the client and client's reputation can affect the decision on bidding strategy. Having decided the bidding strategy, management will see if any adjustment to the all-in-rates used in the base estimate is required for market trends such as inflation, interest rate and the like. The use of computer-aided estimating can allow the adjustments to be easily performed with their effects on the total price quickly calculated.

Then, an appropriate amount for head office overheads and profits has to be considered. Usually, a percentage will be decided and applied to the BQ items to cover such costs. This process relies heavily on the experience and professional judgement of the management. If the percentage is too high, the chance of winning the job will be low. The alternative is to apply a low percentage so as to win the contract but bear a high risk. Management has to decide tactically to strike a good balance between competitiveness and profitability.

When the trade work items, preliminaries, profits and head office overheads are all settled, the amounts can be entered into the BQ. Having added the provisional sums, prime cost sums and contingencies, the tender figure is computed. All the adjustments made during tender adjudication must be carefully entered into the final BQ. Bulk checks should be made by the estimator on the bill totals and summary total to confirm the same figures are decided in the adjudication.

Step 12: Submit tender

The priced BQ, together with the required submissions such as tender programme, method statement and organisation chart should be put in a sealed envelope and submitted to the designated location before the tender closing date. For submission using electronic files, all files should be digitally signed by the tenderer. Organisational e-cert issued for electronic authentication or signature must be valid as at the tender closing date.

Step 13: Open tenders

The tenders received should only be opened by the tender board or authorised tender opening parties, which usually include the QS consultant, project owner's representative and the architect. Before contract award, communications with tenderers should be restricted.

After opening of tenders, the QS consultant should check whether the form of tender is duly signed and the conditions of tender are complied with. Any non-compliance should be dealt with in accordance with the conditions of tender.

Step 14: Examine tenders

Before tender evaluation, the QS consultant should check for any errors in the tenders. If errors are found, QS consultant has to judge whether they are:

- errors that do not affect the completeness of the tender such as arithmetical errors, typographical errors or inconsistent pricing; or
- non-compliances where the tenderer fails to provide the essential information as set out in the tender document such as missing the technical submission.

Most of the conditions of tender should contain procedures to deal with the above situations. In general, post-tender contacts with the tenderers for the sake of error clarification/correction should be kept to minimum. If such contacts are necessary, clear records should be made and no advantage should be given to the contacted tenderer over the other tenderers.

After checking for errors, the QS consultant should examine the tender bid price and the item unit rates to see if they are realistic. Although the contracts may not be awarded to the lowest bidders in many cases, tender price is still the determining factor affecting the award decision. Low tender offers are certainly more favourable in competitive bidding. However, if the tender price is unreasonably low, the tenderer may have difficulty completing the project satisfactorily or have a higher risk of claims. While high tender offer will be less competitive, having only a few rates being unexceptionally high may not affect the overall competitiveness of the bid but will give an advantage to the contractor in future claims. All these cases, if found, should be highlighted in the tender report to bring to the project owner's attention.

Step 15: Evaluate tenders

While tender evaluation can be applied in different tendering approaches, all of them share the same primary function: to prioritise the tenders based on a set of objective criteria so as to choose the most suitable one. In general, there are two main types of evaluation system: lowest price method and price quality method. Some private firms prefer lowest price method as it is relatively straightforward by evaluating tenders based on the tender price only. This is usually used for small projects as the lowest-bid contractor may not offer the best service. Using the price quality method, the QS consultant can set weightings on the relative importance of price against other non-price criteria such as past experience, track record, safety performance, design capability, management system and so forth.

For projects that involve public funds, fairness and objectivity in the tender evaluation process is a major concern. In government departments, two methods are used in the tender evaluation for their projects: the formula method and the marking scheme method (Environment, Transport and Works Bureau, HKSAR, 2009). Both methods are price quality approaches by nature, and include price and other attributes such as contractor's past performance as evaluation criteria. These criteria, together with their assigned weightings, must be included in the conditions of tender.

When the project is complex, a tender interview may be held to allow the project owner and consultants to gain a better understanding of the tenderers' proposal. Performance of the tenderer in the tender interview usually contributes to the tender evaluation as well.

Points to note

The formula method uses the tender price and the tenderer's past performance as the evaluation criteria. The following formula is used to calculate the score of each tenderer:

Score of tenderer A = 60 × (Lowest tender price ÷ Tender price of A) + 40 × (A's performance rating ÷ highest performance rating among tenderers)

The performance rating means the rating held in the government database (Contractors' Performance Index System) on the tender closing date.

In the standard marking scheme designed for the marking scheme method, tender price is still the primary factor affecting a tenderer's score. In addition to tender price, four other attributes, namely tenderer's experience, tenderer's past performance, tenderer's technical resources and tenderer's technical proposal, are assessed to arrive at a 'technical score'. Clear performance-based assessment criteria are laid down in the Technical Circular (Works) No. 8/2004 (Environment, Transport and Works Bureau, 2009) to illustrate the assessment of the four attributes and will not be detailed here. After scoring the mark for each attribute, the following formula is used to calculate the overall score of a tenderer:

Score of tenderer A = 60 × (Lowest tender price ÷ Tender price of A) + 40 × (A's technical score ÷ highest technical score among tenderers)

It is noteworthy that the standard marking scheme involves assessment of the tenderers' attributes rather than being based on a rating held in the government database. That is why an assessment panel has to be formed with at least two members plus a chairman (Environment, Transport and Works Bureau, 2009).

Step 16: Tender acceptance recommendation

With the use of a proper evaluation system, the 'best' tender can be identified objectively. The QS consultant can prepare the tender report to make a recommendation.

The tender report should include a clear recommendation for the appointment of the successful tenderer. Such recommendation should be supported by the evaluation analysis and review of the tender sum breakdown. Any substantial irregularities in the pricing of the recommended tender should also be highlighted in the report. Where the highest score tenderer or lowest price tenderer is not recommended (under the price quality evaluation method or lowest price method), the reasons must be clearly stated. The followings are suggested for inclusion in the tender report (HKIS et al., 2012):

- summary of tender received
- copies of forms of tender and qualifying letters
- summary of the actions taken following receipt of tenders (e.g. tender interview, negotiations etc.)
- results of arithmetical and technical checks and actions taken as a result of such findings
- reconciliation of tenders against estimates, cost limits, target cost etc.
- conclusion and recommendation

Other information can also be included, such as:

- any tenderer who has withdrawn his tender
- if the number of tenders received is small, an explanation for the poor response rate
- claims history of the recommended tenderer

In some cases it may be in the project owner's interest to negotiate some details with the chosen tenderer before the award of a contract. For instance, when the recommended tender includes:

- a counter-proposal that does not substantially deviate from the tender requirements but is disadvantageous to the project owner; or
- errors and discrepancies in the bills of quantities

these matters should be resolved before signing the contract. Negotiation may be conducted by exchange of correspondences and/or meetings.

Upon approval of the recommendation by the project owner, the letter of acceptance can be sent to the successful tenderer. Unsuccessful tenderers should also be notified in writing.

Good practice for contractor selection

The *Code of Practice for the Selection of Main Contractors* (Construction Industry Board, U.K., 1997) illustrates the principles of good practice on the selection of contractors. The key guidelines are summarised below.

- The tender list should be compiled systematically and the number of tenderers should be limited to six. Further names can be held in reserve.
- Sufficient time should be allowed for the preparation and evaluation of tenders.

- Standard forms of contract from recognised bodies with minimal amendments should be used where available.
- Information provided to all tenderers should be the same and sufficient for preparation of the bid.
- Clear tendering procedures should be followed to ensure fair and transparent competition.
- Tenders should be evaluated and accepted on quality as well as price.
- Qualified tenders should be discouraged. A contractor should be asked to withdraw the qualifications if there are significant qualifications or else to face rejection.
- Information in the tender documents should be treated with strict confidentiality.
- In cases where the project owner's budget is exceeded, negotiation with the lowest tenderer can be carried out to look for savings.

Question 6.3

Why should qualified tenders be discouraged? Are they not representing 'tenders with good quality'?

Tender performance review

After the tender process has been completed, it is recommended that both the project owner and the tenderers should carry out their own tender performance analysis. For project owners, review of the tender performance allows them to keep track of the contractors' bidding performance so as to maintain a more comprehensive list of approved contractors. For the contractors, tendering is a costly process and should be effectively monitored. Different aspects of the company's tender performance can be analysed, for instance:

- cumulative ratio of tenders to contracts won over a period
- value of contracts won over a period
- type of contracts won over a period
- profit margins of the contracts won

The above analysis is useful to provide indicators for the management to devise bidding strategy for future projects.

Suggested answers

Question 6.1

Since the supplier agreed to deliver FOB destination to the contractor's site, he is responsible for any damage during the whole transit until the tiles reached the site. Although the contractor has agreed to pay the local delivery charges, the supplier is still the one who owns the tiles while they are in transit. The supplier is responsible for the damage and he can only claim compensation from his insurance company.

Question 6.2

'Removal of debris from site' and 'water for car washing bay' belong to preliminaries. 'Wastage of steel bars' and 'temporary support for formwork' are related to specific trade work and should be priced in the respective trade work items.

Question 6.3

'Qualified tender' refers to a tender that contains statements added by the tenderer to limit his liabilities if he is awarded the contract. Typical examples of qualifications in a tender include:

- tender bid excludes the provision of insurance, bond, shop drawings or the like
- tenderer counter-proposes to use some material not specified in the tender document
- tenderer requires a deposit payment upon contract award
- tenderer proposes different payment or retention terms

7 Tender documentation

Composition of tender documents

The tender documents are typically composed of several major parts as shown in **Table 7.1**. When preparing a contract, items 3 to 8, together with the letter of acceptance and other relevant correspondences issued after tender distribution must be included in the contract documents.

Notes to tenderers

The notes to tenderers aim at giving some general guidelines that may help tenderers to prepare the tender. The document is also used to highlight some points for attention like the communication or enquiry method, tender briefing and site visit arrangement etc. A sample of notes to tenderers is shown in **Figure 7.1**.

Conditions of tender

These are the general and tender-specific rules that govern the content of a tender, its submission, and the evaluation and contract award process. Many consultants or large project owners maintain standardised general conditions of tender and special conditions of tender. General conditions of tender include the rules and terms that can be generally applied to most projects. Any alterations to a general condition of tender will be effected by a special condition of tender, such as the alternative design required from tenderers, electronic tender submission details and so forth.

Some common issues to be included in the general conditions of tender such as tender submission details, tender validity and essential information required to be submitted are shown in **Figure 7.2**. Excerpts of tender conditions are included for better illustration.

Form of tender

The form of tender is a covering document prepared by the QS consultant and signed by the tenderer to indicate that it understands the tender and accepts the various terms, conditions and other requirements of participating in the tender exercise.

Table 7.1 Composition of Tender Documentation.

	Components	*Nature*	*Prepared by*
1.	Notes to Tenderers	For tendering	QS Consultant
2.	Conditions of Tender	For tendering	QS Consultant
3.	Form of Tender	For tendering	QS Consultant
4.	Articles of Agreement	General	QS Consultant
5.	Conditions of Contract	General	QS Consultant
6.	Drawings	Describes the scope of work	Designers
7.	Specifications	Describes the scope of work	Designers
8.	Bills of Quantities (or alternatives such as Provisional Bills of Quantities or Schedule of Quantities and Rates)	For pricing	QS Consultant (may be prepared by tenderers in case of Schedule of Quantities and Rates)

NOTES TO TENDERERS

Tenderers shall read this Tender Document carefully prior to submitting their tenders. Any tender which fails to comply with the requirements contained herein may render the tender invalid.

1. Introduction

Tenders are invited for the construction of xxx located at xx road, xx, Hong Kong. The Employer is xxx Company.

2. Tender Document

This Tender Document identified as xxx consists complete sets of:

(a) Form of Tender;
(b) Conditions of Tender;
(c) Form of Contract;
(d) Conditions of Contract;
(e) Bills of Quantities;
(f) Schedules 1 – xxx;
(g) Annex A – Reply Slip for Site Visit;
(h) Annex B - Marking Scheme

3. Communication

All questions relating to the Tender and related documents should be submitted in writing to the point of contact below. All questions received in relation to the Tender, together with the answers, will be made available to all Tenderers, to ensure equality of information. The source of the questions will not be divulged.

Mr xxx, xxx Company, xx road, xx, Hong Kong

Tel: 1234-5678 Email: xxx

4. Tender Briefing Session and Site Visit

A tender briefing session will be held on 2 Jan 20xx at 10:00 a.m. at the xx office at xx road, xx, Hong Kong. Tenderers are strongly advised to attend the tender briefing session and site visit in order to fully acquaint themselves with the contract requirements and to determine the scale and costs of the Service to be provided.

5. About These Notes

These Notes do not form part of the Tender Document or the Contract. In the event of any conflict between these Notes and the Tender Document, the Tender Document shall prevail.

Figure 7.1 Sample notes to tenderers.

GENERAL CONDITIONS OF TENDER

GCT 1 Definitions
GCT 2 Tender documents issued
GCT 3 Submission of tender
GCT 4 Validity and acceptance of tender
GCT 4 Submission of essential requirements
GCT 5 Financial information
GCT 6 Unauthorised alterations
GCT 7 Discrepancies in the documents
GCT 8 Clarification of documents
GCT 9 Errors in tender submission
GCT 10 Correction rules for tender errors
GCT 11 Unreasonably low bids
GCT 12 Tender clarifications
GCT 13 Tender addenda
GCT 14 Tender evaluation
GCT 15 Tender negotiation
GCT 16 Anti-collusion
GCT 17 Ethical commitment

Excerpt 1:
Tenderer is required to submit the following information. Failure to submit shall render his tender invalid:
- Master construction programme in MS Project format (latest version)
- Method statement outlining the proposal for carrying out the works including those for the following special issues as separate sections:
  - Temporary works erection;
  - Access route for personnel, materials and debris;
  - Noise and nuisance control
- Organisation chart and C.V. of key staff of the tender
- Project safety plan and project quality plan

Excerpt 2:
The offer or acceptance of an advantage or other inducement by any person with a view to influencing the placing of the tender may be an offence under the Hong Kong Prevention of Bribery Ordinance, but in any event such action will result in the rejection of the tender.

Figure 7.2 Example of general conditions of tender items and excerpts of conditions.

It represents the offer made by the tenderer to the employer. As shown in **Figure 7.3**, the form of tender should contain the following information:

- project title
- description of works
- tender price: lump sum for the works (where applicable)
- validity of offer
- time for completion of the whole works (to be entered by the tenderer)
- tenderer information (company name, name of signatory and his/her signature)
- name of witness and his/her signature
- date

Question 7.1

What is the nature of the following documents in the context of 'offer and acceptance'?

- letter of invitation to submit a tender (issued by the employer)
- tender bid from the contractor
- letter of acceptance issued by the employer

FORM OF TENDER

For the construction of xxx at xxx Road, Hong Kong

Notes:

(1) If a tender is being submitted by a partnership or an unincorporated body, the names and residential addresses of all partners should be given in the spaces provided⁺.

(2) In all cases, the tenderer must give the number and date of the business registration certificate here.

Number: ______________________________.

Expiry Date: ______________________________.

To: Architect

1. I/We, having inspected the Site, examined the Samples, Drawings, Conditions of Contract and Specifications for the construction of xxx do hereby offer to execute, design as required, complete the whole of the Works in accordance with the tender documents for the sum of Hong Kong Dollars ______ ______________________(HK$______________________) or such sum as may be ascertained in accordance with the relevant Conditions of Contract.
2. I/We undertake if my/our Tender is accepted to possess the Site within seven days of receipt of the Architect's order to commence and to complete and deliver the Works comprised in the Contract within ____________________ calendar days* from and including the Commencement Date. The Commencement Date is the seventh day after the date receiving the Architect's order to commence.
3. I/We agree to abide by this Tender for a period of three months commencing from and including the day following the date stipulated receiving it and it shall remain binding upon me/us and may be accepted at any time before the expiration of that period.
4. I/We declare that the Master Copy as described in Clause xxx of the Conditions of Tender will form part of the Contract Documents if my/our offer is accepted unless it is superseded by the relevant documents issued after the date fixed for receiving this tender.
5. Unless and until a formal Agreement is prepared and executed this Tender together with your written acceptance thereof, subject to the provision of Clause xxx hereof, shall constitute a binding contract between us.
6. I/We understand that it is your intention not to create any contractual relations in this invitation to tender until the award of the Contract and you are not to accept the lowest or any tender.

Name in Block Letters ______________________________

Signature ______________________________

in the capacity of ______________________________

duly authorized to sign tenders for and on behalf of ______________________________

Registered address of firm ______________________________

Telephone ______________________ Facsimile No ______________________

Date ______________________________

Signature of Witness ______________________________

Name in Block Letters ______________________________

Address ______________________________

Occupation ______________________________

Date ______________________________

If a tender is being made by a partnership or an unincorporated body, the names and residential addresses of all partners shall be given in the space provided below.

Name of Partners	**Residential Addresses of Partners**

Figure 7.3 Sample form of tender.

Points to note

Putting a form of tender into the tender document ensures that all tenders are prepared in accordance with the tender documents and that the tender offers are comparable. Once the employer accepts the tender offer, the form of tender creates a contract on the terms specified with an understanding that a form of contract will be entered into (Powell-Smith, 2000). Therefore, tenderers must not alter or delete the contents of the Form of Tender other than as indicated; and any such actions may cause the tender to be disqualified.

Articles of agreement

The articles of agreement is the most important part in the contract document as it spells out the core statement of obligations between the two contracting parties – the contractor to complete the works in accordance with the contract documents and the employer to pay the contractor as specified in the contract. If standard form of contract is used, the articles of agreement is usually included in the standard form. Some information must be included in this document to constitute a clear contract:

- title of the project work including the location
- names of the employer and the contractor
- contract sum
- date of the Articles being signed
- signatures of the two parties
- signatures of the witnesses

Reminder

The statement in the articles of agreement that spells out the obligations of the two parties describes the 'consideration' of the contract.

For better illustration, a simplified version of the articles of agreement is shown in **Figure 7.4.**

Question 7.2

Why are articles of agreement found in a tender document? Should the tenderer complete it when he returns the tender bid?

Conditions of contract

Conditions of contract are the terms or requirements as agreed by the two contracting parties. These terms define the contractual obligations and rights of the employer and the contractor. Typical terms to be included in a contract are listed in **Table 7.2**. In many places including Hong Kong, standard forms of contract have been developed to take into account the possible events that may occur

ARTICLES OF AGREEMENT

This Contract is made on the ____________ day of ____________________________________

BETWEEN: xx Company (hereinafter called "the Employer");

AND: ____

(hereinafter called "the Contractor");

WHEREAS the Employer is desirous of having ____
____ (hereinafter called "the Works") at ___
____

executed in accordance with the General Conditions of Contract and the Special Conditions of Contract, the Tender and the acceptance thereof by the Employer, the Bills of Quantities or Schedule of Rates, the Drawings and the Specification;

WHEREAS the Contractor has agreed to execute the Works subject to the terms and conditions hereinafter contained;

THE PARTIES AGREE AS FOLLOWS:

1. All words and expressions in this agreement shall have the same meanings as are respectively assigned to them in the General Conditions of Contract and the Special Conditions of Contract hereinafter referred to.
2. For the consideration hereinafter contained, the Contractor shall execute the Works to the satisfaction of the Architect in accordance with the General Conditions of Contract and the Special Conditions of Contract, the Tender and the acceptance thereof by the Employer, the Bills of Quantities or Schedule of Rates, the Drawings and the Specification.
3. The Contractor shall execute the Works within the period stipulated in the Contract or within such further time as may be determined by the Architect in accordance with the provisions of the Contract.
4. The Employer shall pay to the Contractor the sum of HK$______________ (the "Contract Sum") or such other sum that becomes payable at the times and in the manner specified in the Contract.

IN WITNESS WHEREOF, the parties have executed this Agreement:

By the Employer:

______________________________	______________________________
(signature of signatory for the Employer)	(signature of witness)
______________________________	______________________________
(name and office held by signatory for the Employer)	(name and occupation of witness)

By the Contractor:

______________________________	______________________________
(signature of signatory for the Contractor)	(signature of witness)
______________________________	______________________________
(name and office held by signatory for the Contractor)	(name and occupation of witness)

Figure 7.4 Articles of agreement – a simplified version.

during the construction period. Major standard forms of contract currently in use locally are:

For government projects:

- General Conditions of Contract for Building Works, 1999 Edition (HKSAR)
- General Conditions of Contract for Civil Engineering Works 1999 Edition (HKSAR)
- General Conditions of Contract for E & M Engineering Works, 1999 Edition (HKSAR)

Table 7.2 Typical Items in the Conditions of Contract (based on the HKIS Standard Form of Building Contract, With Quantities, 2005).

Cl.	*Contract Terms*	*Cl.*	*Contract Terms*
1.	Interpretation and definitions	21.	Insurance against injury to persons
2.	Contractor's obligations	22.	Insurance of the works
3.	Master programme	23.	Possession, commencement and completion
4.	Architect's instructions	24.	Damages for non-completion
5.	Documents forming the contract and other documents	25.	Extension of time
6.	Statutory obligations	26.	Delay recovery measures
7.	Setting out the works	27.	Direct loss and/or expense
8.	Materials, goods, workmanship and work	28.	Notice of claims for additional payment
9.	Intellectual property rights	29.	Nominated subcontractors and nominated suppliers
10.	Contractor's site management team	30.	Persons engaged by employer
11.	Access for the architect to the works	31.	Facilities for statutory undertakers and utility companies
12.	Architect's representative	32.	Certificates and payments
13.	Variations, provisional quantities, provisional items and provisional sums	33.	Surety bond
14.	Contract bills	34.	Antiquities
15.	Contract sum	35.	Determination by employer
16.	Materials and goods on or off-site	36.	Determination by contractor
17.	Substantial completion and defects liability	37.	Determination by employer or contractor
18.	Partial possession by employer	38.	Fluctuations
19.	Assignment and sub-letting	39.	Notices, certificates and other communications
20.	Injury to persons and property and indemnity to employer	40.	Recovery of money due to the employer
		41.	Settlement of disputes

For private projects:

- Agreement and Schedule of Conditions of Building Contract for use in the Hong Kong Special Administrative Region, With Quantities, 2005 Edition (HKIA, HKICM and HKIS)
- Agreement and Schedule of Conditions of Building Contract for use in the Hong Kong Special Administrative Region, Without Quantities, 2006 Edition (HKIA, HKICM and HKIS)

The main benefit of using a standard form is that both contracting parties are familiar with the content, thus saving time and cost in tendering. The parties' confidence in contract arrangement is also enhanced (Ramsey, 2007).

To cater for the specific needs of individual employers, additions, deletions, substitutions or revisions may be applied to the standard form clauses, which sometimes create ambiguity or disputes when the alterations are poorly drafted. If a standard form is used with amendments, the contract conditions should be classified into general conditions of contract and special conditions of contract. Special conditions can be used to supplement the amendments, additions, deletions and amplifications to the general conditions of contract.

The appendix section of a standard form enables the parties to insert provisions that vary from job to job, such as contract commencement and completion dates, amount of liquidated damages, percentage of retention to be withheld, length of defects liability period and so forth.

Drawings

Drawings are provided in almost all projects to illustrate the scope and details of work. From the contractual perspective, there are drawings which form part of the contract and drawings which do not form part of the contract. The first type represents most of the drawings like building plans, sections, details and the like that are given to the tenderers during the tender period. These drawings will become part of the contract documents. Some drawings are not included with the tender document, but are made available for inspection by the tenderers during the tender period. These drawings include existing utility layouts and site investigation plans that only provide information for tenderers' reference. The nature of the second type of drawings should be highlighted clearly to the tenderers to avoid possible misunderstanding or future claims.

Specifications

Specifications are prepared by an architect or design engineer to provide technical information related to the quality of materials and workmanship. There are two main classes of specification: general specification and particular specification. General specification covers general requirements of operations and materials. However, different projects may need a varying degree of modifications to the general specifications to suit the project characteristics, e.g. long-span structural members required for a stadium, a higher standard of fire curtain for a theatre, a better level of workmanship in the finishing work for a landmark facility etc. Any specific modifications to the general specification will be stated in the particular specification.

To economise the design and production effort, many large organisations such as the government have implemented standard design in various components e.g. precast units, reinforcement details, finishing work and the like. Also, the standard specification can be used with the standard designs. In government projects, the general specification is not sent to the tenderers but is downloadable from the website to save paper. The incorporation of the general specification into the contract is brought into effect by reference in the particular specification.

Bills of quantities

Based on the drawings and specifications, QS consultants prepare bills of quantities that set out the quantities and descriptions of the project work for the tenderers to price. This document facilitates the comparison of competitive bids. During the contract period, the BQ can provide a basis for valuation of variations and payments of completed work. Unlike the other parts of the tender documents, there are widely accepted rules for the measurement and presentation of the bills of quantities. The Hong Kong Standard Method of Measurement of Building Works 4th Edition Revised 2018 (HKSMM4 Rev 2018) issued by the HKIS is extensively used for measuring local building works. There are other standard rules developed for measuring civil engineering works but they are beyond the scope of this book. More illustration on the formats and measurement rules for preparation of BQ will be given in **Part 4** of this book.

As a quick review, the bills of quantities can be divided into four main sections:

- preambles
- preliminaries
- measured work
- provisional sum and prime cost sum

Preambles

The preambles are non-measurable items that do not cost any money themselves but may affect the cost of work sections to which they apply. The preambles section contains the general preambles that describe the method of measurement used in preparing the bills of quantities. Information that is necessary for pricing the bills of quantities but not covered elsewhere should also be included in the general preambles. Normally, the standard method of measurement applied (e.g. HKSMM4 Rev 2018) is stated clearly here. However, we do not need to reproduce the entire SMM. The incorporation of the SMM into the contract is effected by reference in the general preambles and specifications. Any deviations from the standard rules given in the SMM are highlighted in the particular preambles so that tenderers can make appropriate allowance when pricing.

Preliminaries

The preliminaries bill lists the site overheads requirements and contract conditions, such as description of work, form of contract used, site supervision and site facilities. As mentioned in **Chapter 6**, tenderers have to estimate the likely cost needed to fulfil the requirements. More details on the coverage and estimating method for preliminary items will be discussed in **Chapter 23**.

Measured work

Where the contract type is lump sum with quantities, the quantities of work for the entire project will be measured and listed in this section for pricing. **Figure 7.5** shows a sample page of the bills of quantities. According to HKSMM4 Rev 2018, measured items are grouped in trade sections such as excavation, concrete works, waterproofing and so forth.

Provisional quantities

Sometimes, an item or a section of work cannot be accurately measured at the time the tender documents are issued. Provisional quantity, which is only an estimated quantity of work, can be used under this circumstance. When the term 'provisional' is used, it means that the described work will be subject to remeasurement after execution. The value of this item or this part of work in the bills of quantities will be deducted from the contract sum and the value of work executed will be added. For instance, in **Figure 7.5**, the exact number of signage

Bill No. 4 Carpenter and Joiner

Ref	Description	Qty	Unit	Rate	HK$	c
	<u>SUNDRIES</u> <u>Timber Bench</u> Timber bench; comprise 20mm thick plywood; finished with plastic laminated sheet on all exposed surface; include all necessary ironmongeries, fixing accessories, sealant; all as detailed in Specification and Drawing no. abc-001 to abc-002					
A	overall size 1500 x 350 x 400 mm high	2	nr			
	<u>Countertop</u> Timber countertop to Room G01 and Room G02 on G/F; including supporting frames; all associated ironmongery and accessories; all as detailed in Specification and Drawing no. abc-003 to abc-004					
B	overall size 2000 x 500 mm x 1120 mm high	2	nr			
	<u>Signage (**ALL PROVISIONAL**)</u> Acrylic signage with silk screen letters, numerals, symbols, characters and punctuations including fixing and painting; all as detailed in Specification and Drawing nos. abc-005 to abc-006					
C	Ceiling mounted; overall size 800 x 150 x 6 mm thick; all as detailed in Drawing no. abc-005 (Signage code A)	50	nr			
D	Signage to walls; overall size 800 x 150 x 6 mm thick; all as detailed in Drawing no. abc-006 (Signage code B)	15	nr			

Figure 7.5 A sample page of bills of quantities.

required is not confirmed at the time of tender although the general design is ready. Provisional quantities can thus be used for the signage so that the actual quantities will be remeasured after execution and the contractor will be paid according to the remeasured quantity.

Prime cost sums and provisional sums

Both provisional sums and prime cost (P.C.) sums are included in the bills of quantities as lump sum items to cover some works of a project that cannot be measured in full detail at the time the tender documents are prepared.

Prime cost sum is a sum of money provided for a work or service to be executed by a nominated subcontractor or nominated supplier. Such sum is estimated by the QS consultant and deemed to be exclusive of any profit required and attendance to be provided by the main contractor. As illustrated in **Figure 7.6**, the contractor has to insert its required profit and attendance when pricing the P.C. Sums.

Frequently, the P.C. sum (e.g. HK$1.7 million for the electrical works in **Figure 7.6**) is inserted into the bills of quantities before quotations have been received from the nominated subcontractors. For this reason, the P.C. sum is 'provisional' in nature. The contractor's profit and attendance calculated on the $1.7 million P.C. sum must be adjusted when the final sum on electrical works is spent.

Bill No. 10 Prime Cost Sums and Provisional Sums

Ref	Description	HK$	c
	Notes: The Prime Cost Sums shall be expended as directed by the Architect or deducted in whole or in part if not expended. PRIME COST SUMS		
A	Air-Conditioning and Mechanical Ventilation System Provide a Prime Cost Sum of Hong Kong Dollar: Two Million Six Hundred Thousand Only (HK$2,600,000.00) for Air-Conditioning and Mechanical Ventilation System • Allow for profit .... % • Allow for attendance .... $	2,600,000.	00
B	Electrical Works Provide a Prime Cost Sum of Hong Kong Dollar: One Million Seven Hundred Thousand Only (HK$1,700,000.00) for Electrical Works • Allow for profit .... % • Allow for attendance .... $	1,700,000.	00

Figure 7.6 A sample page of bills of quantities – prime cost sums.

Reminder

Nominated subcontractor is a subcontractor who is selected unilaterally by the employer to carry out a portion of project work. The main contractor has to employ the subcontractor with very limited grounds for objection, and the main contractor has to accept the subcontractor's quotation as the subcontract sum.

Provisional sum is the amount of money allowed for the work that is required but cannot be sufficiently defined or measured at the time of tender. As in **Figure 7.7**, the landscaping works is not yet designed at the time of tender but it is foreseeable that landscaping is required for the project. As a result, an estimated amount of HK$200,000 is inserted in the bills of quantities by the QS consultant.

Points to note

Provisional quantities, prime cost sums and provisional sums in a contract are carried out upon architect's instructions. Although all these items require subsequent adjustment of the amount of money in the bills of quantities, they should not be used indiscriminately. The choice is basically depended on the extent of information available at the time of tender. If the nature and the type of work is detailed enough but not quantifiable at the time tender documents are issued, provisional quantities can be used. If both the nature and quantity of work cannot be determined, prime cost sum or provisional sum should be more applicable. If the use of a nominated subcontractor is confirmed at the time of tender, prime cost sum should be used instead of provisional sum.

If the work cannot be clearly defined, then why should we include a provisional item in the tender document? The reason is to allow an indicative price for the work concerned in the BQ. In this way, tenderers do not need to make allowance for the risk associated with the provisional item. Furthermore, the employer can have a more realistic estimate of project sum for budgeting. Each bid collected from the tenderers will be prepared based on the same data.

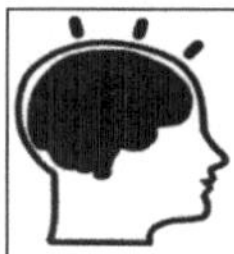

Question 7.3

An existing commercial arcade is going to carry out alteration and addition (A&A) works. The project uses a lump sum with quantities contract and the consultant is now preparing the tender. The project owner does not want to employ any nominated subcontractors.

Below are a few terms that the project owner wants to add to the tender document. Suggest in which part of the tender document they should be put: (1) conditions of tender, (2) conditions of contract, (3) measured items in the BQ, (4) preliminaries, (5) provisional quantities, (6) provisional sums or (7) P.C. sums.

i Submit the list of A&A projects undertaken by the tendering company in the past five years.
ii Take down and set aside the dilapidated timber doors (single leaf) according to Architect's Instruction.
iii Make allowance for the green roof construction but the design has to be confirmed during the construction stage.
iv Make allowance for night work to carry out the noisy operations in order to avoid nuisance to the existing tenants.

Bill No. 11 Provisional Sums

Ref	Description	HK$	c
	PROVISIONAL SUMS		
	Notes: Where Provisional Sums are included for work to be carried out by the Nominated Subcontractor which has not been specified in detail at the time of tendering, these sums shall be deducted from the Contract and the work carried out shall be measured and valued in accordance with the Conditions of Contract and the value added to the Contract Sum.		
	The Provisional Sums shall be expended in part or in whole as directed by the Architect or wholly deducted from the Contract Sum if not required:		
A	Provide a Provisional Sum of Hong Kong Dollar: Two Hundred Thousand Only (HK$200,000.00) for Landscaping Works	200,000.	00
B	Provide a Provisional Sum of Hong Kong Dollar: Five Hundred Thousand Only (HK$500,000.00) for Children Play Equipments	500,000.	00

Figure 7.7 A sample page of bills of quantities – provisional sums.

Provisional/approximate bills of quantities

In general, bills of quantities are prepared by the QS consultants for tenderers to price. These are 'firm' bills of quantities with the quantities of work accurately measured by the QS consultants. No adjustment to the quantities will be made unless there are variations during the contract period. However, a BQ can be 'provisional' if it is used for a remeasurement contract where all the items will be subject to remeasurement. Except for the 'provisional' nature of all the quantities in the BQ, the layout of a provisional BQ is the same as that of a firm BQ.

Schedule of rates (or schedule of quantities and rates)

If the contract used is a lump sum without quantities form, no bills of quantities will be provided to the tenderers. Instead, a schedule of rates (or a schedule of quantities and rates) will be prepared by the QS consultant for tendering purpose.

Alternatively, the tenderers may be asked to prepare their cost breakdown of the tender bid in the form of a schedule of quantities and rates. The layout of a schedule of quantities and rates is very similar to a BQ, but the errors in the schedule will not constitute any variations to the contract sum.

Points to note

Theoretically, the layout of a schedule of quantities and rates prepared by a tenderer should be very similar to a BQ. Nevertheless, the tenderers are not bound to follow the HKSMM rule when they prepare the schedule. Therefore, in practice, the schedule prepared by different tenderers can be very different. This will bring difficulty to the QS consultants when they compare the tenders.

Suggested answers

Question 7.1

- Letter of invitation to submit a tender (issued by the employer): Invitation to treat
- Tender bid from the contractor: Offer
- Letter of acceptance issued by the employer: Acceptance of offer

Question 7.2

The articles of agreement included in the tender document allow the tenderers to understand the future contract to be signed (if the tender is accepted). In this way, the tenderers can estimate their likely commitment and risks.

Since the contract has not been awarded yet, the tenderer does not need to complete/sign the articles of agreement when returning the tender.

Question 7.3

i In the conditions of tender.
ii In the BQ as a provisional quantity item.
iii In the BQ as a provisional sum item. P.C. sum will not be used because the project owner does not want to employ nominated subcontractors.
iv This is not a trade-specific item and should be added to the preliminaries bill of the BQ. Tenderers can price the item accordingly.

8 Approximate estimating techniques in the pre-contract stage

The purpose of approximate estimating

The construction cost is of primary interest to both project owners and contractors. Project owners need an early estimate of the probable construction cost to assess the feasibility of a project proposal and the cash requirement during the project period. In a large public project, the budget estimate is required at an early stage to secure sufficient funds, which have to be approved by the Legislative Council. In addition to project owners, the consultant firms and the contractors also devote many resources to improve the accuracy of cost estimating. Consultants, especially the QS consultants, have to utilise their cost database to advise the project owners on the probable project cost based on the available design information. Contractors have to estimate the construction cost required to complete the project as described in the tender documents. However, cost estimating has never been simple.

A project begins with an inception, which is often a vague idea during the strategic definition stage. At that time, only a crude estimate of costs can be worked out based on the maximum amount that the project owner is willing to pay. When the design develops, a preliminary estimate can be prepared by the QS consultant. The project cost estimate continues to evolve through stages of refinement until the tender drawings and specifications are fully completed.

Points to note

The pre-contract stage refers to the project period before the award of the construction contract. According to the RIBA plan of work 2020, the pre-contract period covers five phases, namely the strategic definition stage, preparation and brief stage, concept design stage, spatial coordination stage and technical design stage (Sinclair, 2020).

Since the amount of information available for estimating is limited in the early pre-contract stages, it is impossible to prepare an accurate estimate of the project cost. Therefore, we describe the early stage (or pre-contract) estimates as approximate estimates.

The major uses of these approximate estimates can be summarised as follows:

- conduct feasibility studies of design alternatives
- set project budget or cost limit

- maintain financial control of projects
- produce comparative studies

Overview of approximate estimating techniques

Several approximate estimating techniques have been widely used by QS consultants over the years, including:

Single-rate approximate estimating methods:

- functional unit method
- floor area method

Multiple-rate approximate estimating methods:

- elemental cost plan
- approximate quantities (detailed design cost plan)
- pre-tender estimate

These approximate estimating techniques can be used in various stages of the pre-contract period. Based on the practice notes issued by the Hong Kong Institute of Surveyors (2016), the application of approximate estimating techniques is summarised in **Figure 8.1**. Needless to say, the level of estimating accuracy increases as the design information develops. When moving towards the preparation of BQ, the approximate estimate will get closer to the contract sum to be agreed with the contractor. However, one should understand that the contract sum can be different from the pre-tender estimate for many reasons, such as the effect of competition in the market and the quality of past data.

Once the complete set of bills of quantities are ready, contractors and QS consultants can use the analytical estimating technique to estimate prices for work in a more accurate manner. Further details will be discussed in **Part 5**.

Points to note

Analytical estimating is to estimate the unit rate of a work item by a thorough consideration of the resources and the respective amounts needed (Brook, 2017).

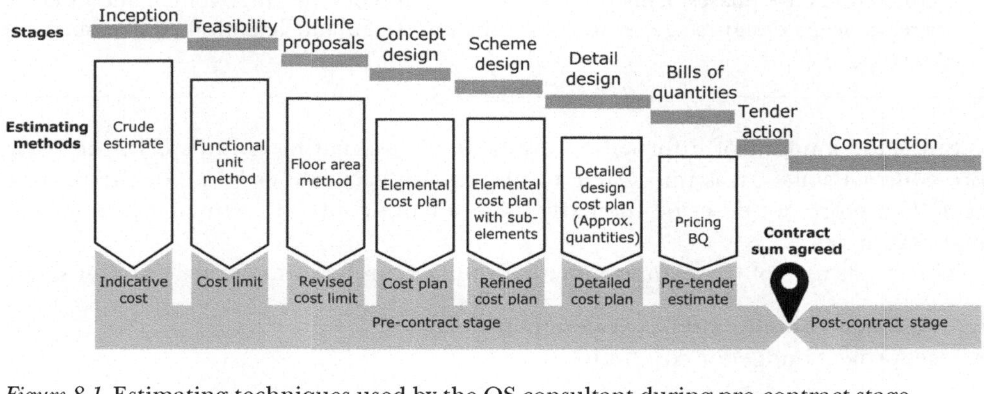

Figure 8.1 Estimating techniques used by the QS consultant during pre-contract stage.

Points to note

Approximate estimating techniques are useful tools for QS consultants to evaluate designs and control the budget for the project owners during the pre-contract stage. It is noteworthy that no matter how close the pre-contract estimate and the contract sum are, the final account sum is not automatically tied within the budget. The final account sum can deviate from the original contract sum for many reasons such as design variations or unanticipated events happened during the post-contract stage. QS consultants have other cost control measures for use in the post-contract stage but they are beyond the scope of this book.

Reminder

Post-contract stage (not contract stage) refers to the stage after contract award until contract close-out. Contract stage is the period (after tendering) when the contract is developing and often involves negotiation between the two parties.

The use of past project cost data for approximate estimating

In response to the limitation on time and information availability, approximate estimating in early stages relies heavily on the use of cost analysis data. Usually, when preparing the budget for a new project, past project(s) of a similar type and size is/are chosen as reference for estimating. The cost analysis figures of the past project(s) will be used to predict the cost of the new project. If any differences between the referenced project and the new project are encountered, cost adjustments will be made.

Points to note

Cost analysis is a full appraisal of costs involved in past projects to obtain reliable cost data so as to assist cost estimating for future projects. Cost analysis covers a wide range of studies including analyses of the total project cost, trade sections of BQ, labour and material costs, single rates and composite rates.

Cost adjustments are made to respond to a variety of circumstances, such as:

- differences in quality standards, e.g. different finishing material used
- changes in construction methods, e.g. precast vs in-situ concrete members
- differences in design components, e.g. some items found in the new project but not the referenced project or vice versa
- location factor, e.g. a remote site with limited accessibility will involve higher project costs
- differences in site condition, e.g. differences in adjacent building conditions, existing utilities, site profile etc.
- changes in price level

Except for the price level adjustment which can be made by referring to cost indices, most of the cost adjustments (such as adjustments for location factor, specification level and construction method) are rather subjective and rely on the experience of the QS consultants. That is why the selection of a comparable project is essential at the outset.

Points to note

Often, QS consultants and large project owners analyse cost trends to build their in-house database of cost indices. Two main types of cost trend analysis are maintained: (1) market price changes of some principle construction resources such as daily rates of labour, unit rates of timber, concrete etc. These rates can be computed into a composite index to represent the building cost change; and (2) tender price changes for different types of projects such as residential, office, hotels etc.

Single-rate approximate estimating

Functional unit method

The functional unit method estimates the project cost based on the functional unit of a project. For example, the functional unit of a carpark building is the carpark space, the functional unit of a school is the school place and so on. The following formula is used to estimate the project cost by functional unit method:

$$C = f \times n$$

where
C = estimated total cost of the project
f = cost per functional unit for the project (e.g. \$/hospital bed, \$/hotel room, \$/carpark space)
n = number of units to be provided in the project

At the inception stage, many projects like schools, carparks, hospitals, hotels and hostels are designed based on a target number of functional units to be provided. For instance, a project owner wants to construct a hospital that provides 400 beds. By analysing the unit costs (\$/bed) of a number of past hospital projects, the unit cost per bed (f) is HK\$20 million. Therefore, the estimated cost of the proposed hospital is:

$$C = \text{HK\$20 million} \times 400 = \text{HK\$8 billion}$$

If the unit cost per bed is derived from the hospital projects in 2016, a cost adjustment is necessary.

Assuming the cost index in 2016 = 150 and the current index = 155

C in money-of-the-day (MOD) = HK\$8 billion × 155/150 = HK\$8.27 billion

Although theoretically the unit cost can be adjusted for differences in design, form of construction or site condition between the projects under comparison, it is difficult to apply as the project is still in its infancy. Nevertheless, if recent comparable data is available, this method can provide a quick indicative estimate without much detailed design information required. Therefore, it is often used to establish the first estimate in the feasibility stage.

Question 8.1

A consultant is analysing the tender costs to calculate the tender price indices. The followings are the average tender costs for village houses collected over the past five years:

Year	*Cost per m²*
2019	25,500
2018	26,100
2017	26,350
2016	26,500
2015	26,350

Taking 2015 as the base year and the cost index as 100, calculate the cost index for each year.

Advantages and disadvantages

Advantages:

- A quick and simple method to establish a budget (or cost limit) for design direction or funding request.
- Requires minimal amount of information.

Disadvantages:

- A rough estimate with low level of accuracy.
- Difficult to make cost adjustments for design, form of construction and site condition.
- Not applicable to many building types where estimation of the number of functional units is impracticable or unsuitable, such as an open-plan office building, factory or warehouse.
- More applicable to project owners who maintain extensive records of building project prices of the same nature.

Floor area method

When development plans evolve and the total floor area can be calculated, project cost can be estimated by the floor area method, using the formula below:

$$C = s \times a$$

where
C = estimated total cost of the project
s = cost per unit area of similar projects ($\$/m^2$)
a = construction floor area of the proposed building(s)

In Hong Kong, large QS consultants have maintained building cost per unit area information for public access.

It is customary for QS consultants to use the construction floor area (CFA) to estimate the construction cost of a building. CFA is the covered areas fulfilling the functional requirements measured at each floor to the outer surface of the external walls (or in the absence of such walls, the external perimeter) of the building including all lift shafts, stairwells, mechanical plant rooms, refuse rooms and the like. Lightwells and voids are excluded from the area but no deductions are made for internal walls.

Points to note

The CFA used in cost estimating by a local QS is different from the gross floor area (GFA) used by architects for planning purpose. GFA is the area contained within the outer surface of the external walls of the building measured at each floor level (including any floor below the level of the ground), together with the area of balconies and the thickness of external walls. The calculation of GFA can exclude floor space occupied solely by machinery or equipment such as plant rooms, pump rooms, riser ducts etc. GFA concession may also be granted for private carparking spaces, water features in communal gardens etc. (Buildings Department, HKSAR, 2011). In general, the CFA used by QS consultants is larger than the GFA of the same building used by architects.

In the early design stage when floor plans have not yet developed, some QS consultants may estimate the approximate CFA by applying a conversion factor to the GFA.

Approximate CFA = GFA × conversion factor α

(e.g. $\alpha = 1.05–1.10$ for residential buildings; 1.10–1.20 for offices)

It is also noteworthy that the CFA applied by most local QS in cost planning is the area measured within the outer surface of the external wall, which is different from the U.K. practice where the gross internal floor area (GIA) is used (measured between the internal surfaces of the external walls). When referring to past project cost data, it is always advisable to check the definition of the floor area concerned.

Question 8.2

Is the area under a canopy regarded as covered area and included as the construction floor area?

The example below illustrates an estimate of the construction cost for a hypothetical shopping centre in Hong Kong, based on the information shown in **Figure 8.2**.

Proposed shopping centre

Area per floor (G/F to 3/F):

$= \text{Length} \times \text{width} - 4 \text{ corners}$
$= (5 + 10 + 5) \times (5 + 15 + 5) - 4 \times 5 \times 5$
$= 400 \text{ m}^2$

Basement area:

$$10 \times 15 = 150 \text{ m}^2$$

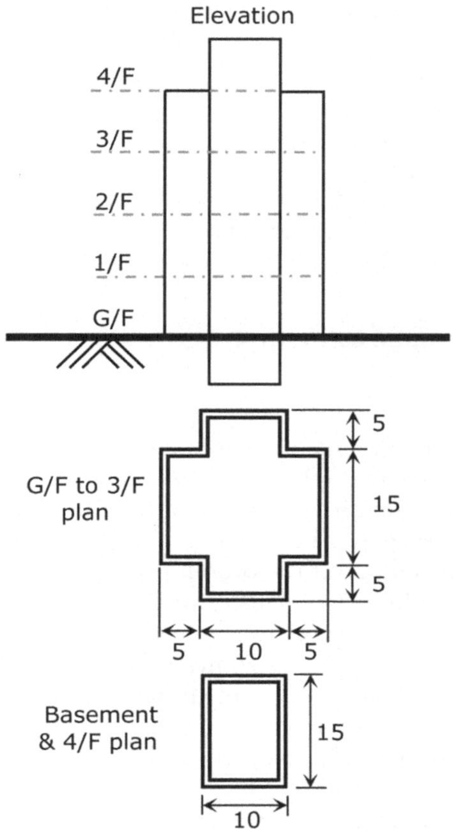

Figure 8.2 Layout of a hypothetical shopping centre project.

4/F area:

$$10 \times 15 = 150 \text{ m}^2$$

Current cost/m^2 (s) for shopping centre = HK$32,000

Total construction cost (C) for the shopping centre:

$$\begin{aligned} C &= s \times a \\ &= \text{HK\$ } 32{,}000 \times (400 \times 4 + 150 \times 2) \\ &= \text{HK\$ } 60{,}800{,}000 \end{aligned}$$

Advantages and disadvantages:

Advantages:

- Easy to calculate.
- Easy to understand as the cost is related to floor area.

- More accurate than functional unit method.
- Easily accessible cost data is available; no skill is required to assess the prices/rates.
- Applicable to all project types.

Disadvantages:

- Does not take into account the dissimilarity in building shape or height, which also have an impact to construction cost.
- Similar to the functional unit method, it is difficult to make cost adjustments.

Multiple-rate approximate estimating

Elemental cost plan

When more design information such as floor plans and sections are available, a more accurate estimate that uses multiple rates can be computed. A cost plan is an estimate of project cost presented in an elemental format. By breaking down the construction cost forecast into a list of elemental costs, the project owner and designer can have a good understanding of the overall project budget and how much of the budget has been allocated to each element of the project. The cost plan somehow equates the design requirements with the money spent. If every element in the building is allocated with an appropriate amount of money, a balanced cost framework should be seen. With the elemental cost plan, the design team can check whether a balanced design is achieved or not.

An elemental cost plan typically covers the following standard elements:

- Site preparation: including demolition, site formation and site investigation
- Foundation and substructure
- Carcase: including the structural frame, doors and windows
- Finishing
- Fixture and fittings
- Building services
- Miscellaneous work: such as external work, landscaping and utilities connection
- Preliminaries (% allowance)
- Contingencies (% allowance)

Points to note

Referring to the list of elements in a typical cost plan, 'contingencies' can be found as one of the elements. 'Contingencies' are sums allowed in the estimate to cover project risks such as design changes, pricing inaccuracies, inflation from preparation of cost plan to receipt of tender and other unforeseeable events which may occur during the construction period. Although the probability of occurrence of each risk item cannot be predicted precisely, the overall financial impact of the risks should be estimated and allowed in the budget to avoid cost overrun. In this respect, contingencies are similar to reserves allowed in the project owner's account. Contingencies are usually calculated by a certain percentage of the total project estimate.

The project estimate is derived by adding the elemental cost of each element together.

Based on the availability of information, there are three approaches to calculate the elemental cost:

- by floor area
- by proportion
- by elemental unit quantity

By floor area

This method is used when information is limited but sufficient for the preparation of a cost plan. The equation for calculating the elemental cost of each element is:

$$E_i = e_i \times a$$

where
E_i = estimated elemental cost of element i in the project
e_i = unit rate for element i (\$/m^2)
a = construction floor area (or GFA) of the proposed building(s)

Estimated total cost of the proposed project is:

$$C = \sum_{i=1}^{n} E_i$$

$$= \sum_{i=1}^{n} (e_i \times a)$$

where
C = estimated total cost of the project
n = number of elements

The unit rates are obtained by a cost analysis of similar previous projects with adjustment for price fluctuation. Where appropriate, cost adjustments can be applied for other factors. For instance, the external wall finish designed for a proposed residential project is of a higher grade than the type used in the referenced project, which is about 15% more expensive. The 15% cost difference can be added to the unit rate for the external wall element accordingly. Similar adjustments can be applied to specific elemental unit costs to take into account the differences between the referenced project and the proposed project.

This is a popular method used in Hong Kong as it requires less data and is easy for most project owners to understand.

By proportion

This method is used when ratios obtained from design data can improve the accuracy of elemental cost targets. For instance, the pile no. per m^2 ratio of a proposed project (r) is available from the engineer. Based on this information, the elemental cost of foundation can be estimated by the following equation:

$$E_i = e_i \times \frac{r}{r_o} \times a$$

where

E_i = estimated elemental cost of foundation in the proposed project
r = pile no. per m^2 ratio of the proposed building(s)
r_o = pile no. per m^2 ratio of the past project
e_i = unit rate for foundation referred to the past project (\$/m^2)
a = construction floor area (or GFA) of the proposed building(s)

By elemental quantity

This method should be considered first if information is available as it produces a more accurate estimate. The elemental cost for each element is calculated by the following formula:

$$E_i = u_i \times q_i$$

where

E_i = estimated elemental cost of element i in the project
u_i = elemental unit rate for element i (e.g. \$ per door, \$ per m^2 of external wall)
q_i = quantity of element i in the project (e.g. no. of door, external wall area)

$$C = \sum_{i=1}^{n} E_i = \sum_{i=1}^{n} (u_i \times q_i)$$

where

C = estimated total cost of the project
n = number of elements

The elemental unit rate is obtained from the cost analysis of past similar projects and the element unit quantity is estimated from the drawings of the proposed project.

Expanding elements to the sub-elements level

When more design information such as typical details is available, the elemental cost can be broken down to sub-elemental costs such as **Table 8.1** below:

The selection and number of elements or sub-elements used in a cost plan is dependent on the information available and the design of the new project. For instance, a commercial project with building automation may include a sub-element of building automation system under the 'Specialist engineering services' element whereas an industrial project may not have such item.

Reminder

The HKIS practice notes for the pre-contract estimates and cost plan are not the standardised rules or methods for pre-contract estimates, but only some general guidelines for better practice. Therefore, the pre-contract estimates or cost plans produced by different consultants for the same stage of design can be quite different.

Table 8.1 Elemental Cost Plan of a Commercial Development.

		Estimated Total Cost HK$	*HK$/m²*
1	**Foundation and substructure**		
1.1	Foundations and diaphragm wall	104,595,100	3,977
1.2	Substructure and basement	68,774,500	2,615
2	**Superstructure**		
2.1	Structural frame and slabs	67,275,400	2,558
2.2	Roof	3,234,900	123
2.3	Stairs and ramps	2,919,300	111
2.4	External walls, wall finishes, windows etc.	68,932,300	2,621
2.5	Internal walls and partitions	13,228,900	503
2.6	Doors	6,732,800	256
2.7	Ironmongery	2,104,000	80
3	**Finishes**		
3.1	Internal wall finishes	22,302,400	848
3.2	Floor finishes	9,073,500	345
3.3	Ceiling finishes	6,312,000	240
4	**Fittings and sundries**		
4.1	Metal work and sundries	4,523,600	172
4.2	Furniture and fitting	13,991,600	532
5	**Sanitary fittings**	12,150,600	462
6	**Specialist engineering services**		
6.1	HVAC systems	75,822,900	2,883
6.2	Electrical and lighting installations	27,404,600	1,042
6.3	Emergency diesel generator	2,314,400	88
6.4	Fire services installations	21,907,900	833
6.5	Plumbing and drainage system	24,669,400	938
6.6	Building automation system	8,231,900	313
6.7	Communication, security and control systems	3,708,300	141
6.8	Lift and escalator installations	43,289,800	1,646
7	**Builder's work in connection with services**	9,599,500	365
8	**Main contractor's profit and attendance on building services**	8,231,900	313
9	**External works**	2,051,400	78
10	**Preliminaries**	42,185,200	1,604
11	**Contingencies**	33,427,300	1,271
	Total:	708,995,400	26,956
	say	709 million	

Note: GFA = 26,300m²

Advantages and disadvantages

Advantages:

- More accurate than functional unit method and floor area method.
- Provides information on the cost limit for each element, which enables more effective budget control.
- Allows designers to check whether the money allocated to each element can efficiently fulfil the project owner's requirements.

Disadvantages:

- Time consuming to prepare.
- A large database is required to provide comparable cost data.
- Sufficient project information is required.

Approximate quantities method

As design proceeds with detailed information becoming available, approximate quantities of work can be measured from the drawings. A list of approximate quantities resembles a simplified bill of quantities, containing major cost-significant items that can be estimated. The extent to which the building components are grouped into one description depends on the ability to price the work. The approximate quantities are then priced at the composite rates which are obtained from a cost analysis of previous, similar projects or from published cost data.

$$Q_i = m_i \times y_i$$

where
Q_i = estimated cost of composite item i in the project
m_i = composite rate of item i
y_i = approximate quantity of item i in the project

$$C = \sum_{i=1}^{n} Q_i = \sum_{i=1}^{n} (m_i \times y_i)$$

where
C = estimated total cost of the project
n = number of composite items

Some examples of approximate quantities items are shown in **Figure 8.3**.

Points to note

Referring to the example in **Figure 8.3**, all the works including the staircase, mosaic tiling, nosings, balustrades and handrails are described as one composite item and measured in meter run. A single rate (the composite rate) is applied that covers all the described work within the description together with any sundry items and labour that are not described but required to construct the staircase. This approach is very different from the measurement method as stipulated in the standard method of measurement, where each of the work items (such as concrete, reinforcement, formwork, screeding, tiling etc.) has to be measured separately. Further details will be illustrated in **Part 4**.

Question 8.3

Based on the example in **Figure 8.3**, if there are extra steel balustrades on one side of the staircase, can we include the cost of balustrades in the estimate for staircases?

Other than the pre-tender estimate, the approximate quantities method is considered to be the most reliable and accurate among the various approximate estimating methods, provided that information is available.

Ref		Qty	Unit cost HK$		Total Cost HK$
1.	Staircases Reinforced concrete staircases 1.10m wide with mosaic tile coverings to treads and risers, non-slip nosings and stainless steel tubular handrails.	m 70	3,800	266,000	1,540,200
	Ditto 1.30m wide	90	4,700	423,000	
	Ditto 1.40m wide	152	5,600	851,200	
				1,540,200	
2.	Internal doors Single hollow core door, plywood faced including metal frame, average quality hardware and paint finish.	Nr 100	4,200	420,000	722,400
	Single solid core flush door, plywood faced including timber door frame, architrave, mouldings, average quality hardware and paint finish.	48	6,300	302,400	
				722,400	

Figure 8.3 Examples of approximate quantities items.

Advantages and disadvantages

Advantages:

- More accurate and reliable than the elemental cost plans.
- Quicker than full measurement of quantities.
- Once the composite rates are built up, they can be used on a variety of projects and for different estimating needs.

Disadvantages:

- Requires more detail about the project.
- Time consuming to estimate the approximate quantities and compile the composite rates.
- There is a greater chance that some work may be missed and not priced, resulting in under-estimation.
- The data is less transferrable between companies as different methods of estimation might have been used.

Pre-tender estimate

When the drawings and specifications are completed, the bills of quantities can be prepared. QS consultants can work out the pre-tender estimate by pricing each BQ item. The unit rates applied to BQ items are primarily past project cost data with necessary adjustments. If relevant cost data is unavailable, quotations can be obtained from suppliers/contractors to complete the costing.

$$T_i = p_i \times t_i$$

where
T_i = estimated cost of item i in the project
p_i = unit rate of item i
t_i = quantity of item i in the project

$$C = \sum_{i=1}^{n} T_i = \sum_{i=1}^{n} (p_i \times t_i)$$

where

C = estimated total cost of the project

n = number of items

The pre-tender estimate will be compared with the preliminary estimate prepared at the very beginning and any significant deviation should be brought to the project owner's attention. The pre-tender estimate will also serve as a useful reference for evaluating the tenders received.

Choosing an approximate estimating method

With the various approximate estimating methods available, the following points should be considered when making the choice of method:

- availability of design information
- degree of accuracy required by the project owner
- available time to prepare the estimate
- availability and reliability of past cost data
- the use of BIM in pre-contract design

Suggested answers

Question 8.1

To calculate the cost index of each year, take the cost of the year to calculate and divide it by the cost of the base year, then multiply by 100.

Year	*Cost per m²*	*Cost index*
2019	25,500	97
2018	26,100	99
2017	26,350	100
2016	26,500	101
2015	26,350	100

Question 8.2

Conventionally, areas covered by canopies or other external wall-mounted features such as projected planters are excluded from the calculation of construction floor area. Definitions of the floor areas should always be checked before using the published cost data.

Question 8.3

Yes, the cost of balustrades can be included in the staircases item as well. In Hong Kong, there is no standard method to standardise how to measure approximate quantities. Therefore, the approach being used to group the building components may vary between companies and individuals. The RICS rules (RICS, 2012) can be taken as guidance.

Part 4

Measuring quantities

9 Principles and rules of taking-off

What is quantity take-off?

Bills of quantities are usually prepared by QS consultants for contractors to price the tender bid. This document, set in a standardised format, consists of a complete list of items that give the description and quantities of work required in a project. To prepare this document, QS consultants have to measure the quantities of works from drawings and specifications in accordance with the rules laid down in the Hong Kong Standard Method of Measurement (HKSMM). We describe this process as 'taking-off'.

Reminder

The standard method of measurement (SMM) contains rules and guidelines on how to measure the various parts of construction work. It also includes coverage rules to identify the inclusions or exclusions of certain elements. Normally, preparation of the bills of quantities is based on the rules as stipulated in the SMM.

Quantity take-off is an essential skill as measurement of work is not only required during the tender stage for BQ preparation, but also during the contract period for materials ordering, preparation of interim valuations and final accounts. Both the QS consultant and the contractor's QS have to perform quantity take-off frequently throughout the project period.

Rules of taking-off

Measurement or take-off of quantities has to follow a set of predefined rules in order to produce a BQ/schedule that is understood by all parties. The rules covered below are widely used by QSs in other countries as well.

Layout of dimension paper

When preparing measurement of works manually, a dimension paper that is formatted as **Figure 9.1** will be used.

The A4-sized dimension paper is divided into two halves, each with four columns. Column 1 is the timesing and dotting-on column in which the factor of multiplication

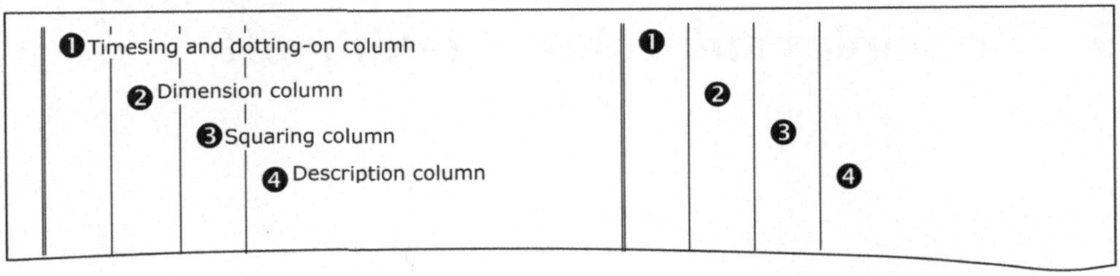

Figure 9.1 Format of dimension paper.

is entered when a multiple of a particular item is being measured. 'Dotting-on' allows the factors of multiplication to be added. Column 2 is the dimension column for recording the dimensions taken from drawings. Column 3 is the squaring column. Figures in column 1 and column 2 will be multiplied and recorded here for transfer to the bill. Column 4 is the description column where the description of measured items is entered.

Entering measurements in dimension paper

General format

It is always a good practice to write down the project name and referenced drawing number at the head of each page (as shown in **Figure 9.2**), and each page should be numbered consecutively.

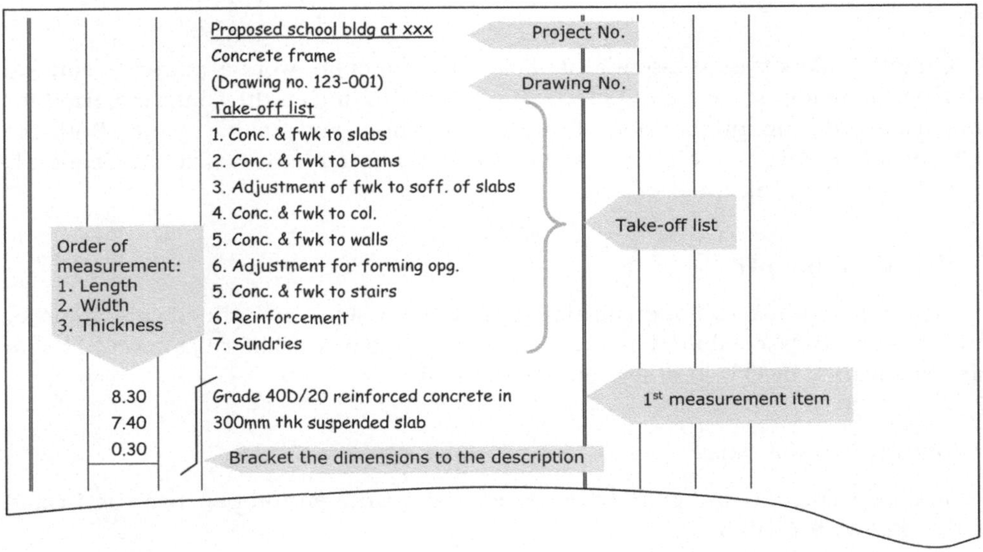

Figure 9.2 Entries to a dimension paper at the start of taking-off.

Take-off list

A take-off list should be given at the start of measurement to identify all items to be measured in a logical sequence. The take-off list can act as a checklist to avoid missing items. An example of take-off list for a concrete frame is shown in **Figure 9.2**.

Booking dimensions/measurements

The metric system is used for all measurements according to the HKSMM. Units of measurement used in taking-off include:

- Linear: m
- Area: m^2
- Volume: m^3
- Weight: kg
- Enumerated: nr (number) or set, as applicable
- Time: days or hours
- Itemised: item (for work items without a measured quantity, such as testing of waterproofing)

Dimensions must be entered in the order of length, breadth (or width) and depth (or height) as shown in **Figure 9.2**. All dimensions should be recorded in metre to two decimal places. A line should be drawn across the dimension column under each set of measurements (see **Figure 9.2**). By underlining each set of measurements, we can tell the unit of measurement ($m/m^2/m^3$) easily from the number of dimensions within the set.

Points to note

In **Figure 9.2**, the first measurement item consists of three dimensions (length, width and thickness). When measuring linear dimension or an enumerated item, only one value will be entered for the item concerned. When measuring an area, two values (such as length and width) will be booked. Further illustrations can be found in the worked examples of **Chapter 19**.

When entering the dimensions and description on the dimension paper, it is recommended to leave plenty of space in between items. This helps the readers to follow and, very often, allows space for missing items to be added back.

Question 9.1

How do you book the following in the dimension column?

i 16m long uPVC pipe
ii 16 number of timber doors

Question 9.2

The diagram shows the floor plan of a building, which contains a 200mm thick concrete wall and a 10m × 5m floor slab. The height of wall is 3m (measured from the floor surface). Take-off the concrete volume of the wall and the concrete volume of the floor slab on a dimension paper.

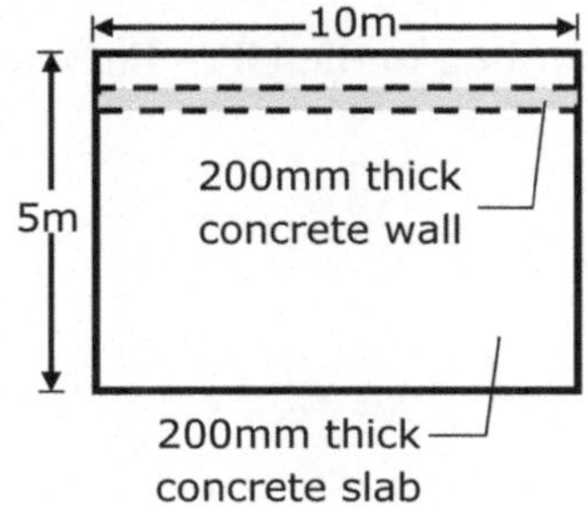

Figured dimensions vs scaled dimensions

Figured dimensions are the dimensions shown on drawings and scaled dimensions are the dimensions scaled off from drawings using a scale ruler. When taking-off, it is always preferred to use figured dimensions as the drawings are not always true to scale, especially when photocopied drawings are used.

Points to note

Some QSs may take-off quantities from 2D digital drawings on the computer screen. Before performing any on-screen measurement, the drawing must be checked to verify that it is drawn to scale and the listed scale is accurate.

Measured nett

When taking measurements from drawings, only the nett quantity should be measured. No adjustment will be made for the waste factor, shrinkage loss, bulking factor and the like. As shown in **Figure 9.3**, all voids such as lift shaft, window openings etc. should be deducted unless otherwise waived by the minimum deduction rules in the

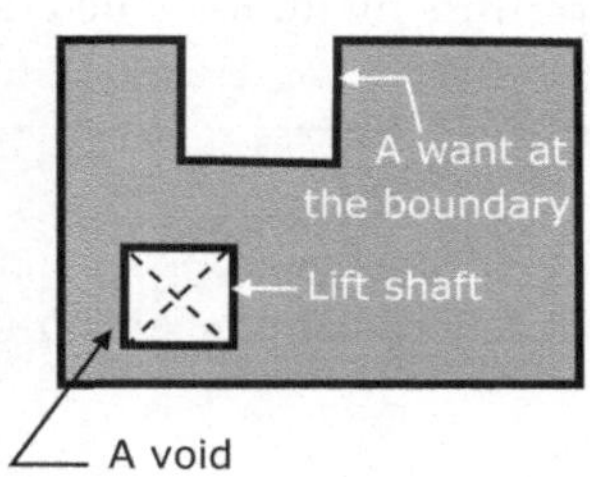

Figure 9.3 Void and want on a floor plan (only grey area is measured).

SMM. The minimum deductions for openings shall refer only to openings that are within the boundaries of the measured areas. Wants, which are at the boundaries of the measured areas, shall always be deducted irrespective of size.

Use of hyphen

In the item description, the use of hyphen between two dimensions shall mean a range of dimensions exceeding the first dimension stated but not exceeding the second. For instance, '200 – 300mm thick slab' means that the thickness of slab exceeds 200mm but does not exceed 300mm. Alternatively, if put in words, the range has to be written as 'exceeding 200mm but not exceeding 300mm thick'.

Annotating dimensions – signposting and waste calculation

One of the most important requirements of taking-off is to annotate the measurements in such a way so that others can understand. There are many chances when the measurements prepared by a QS have to be cross-checked by someone else, for instance, when settling the valuation of variations and final accounts. Signposting at the right-hand side of the dimension column (as shown in **Figure 9.4**) is useful to indicate a measured item's location on the drawings. Signposts should be kept brief, usually a single word or a short phrase.

To record the measurements clearly on a dimension paper, waste calculations and explanatory notes can be entered into the description column, usually at the right-hand side of the column. As indicated in **Figure 9.4**, all steps taken to arrive at the dimensions should be written in the waste calculation. The annotations in the waste calculation enable the readers to know where the dimensions have been taken.

Timesing

Timesing is used where dimensions are repeated. The measurement of the footing in **Figure 9.4** demonstrates the timesing technique. Four footings of the same

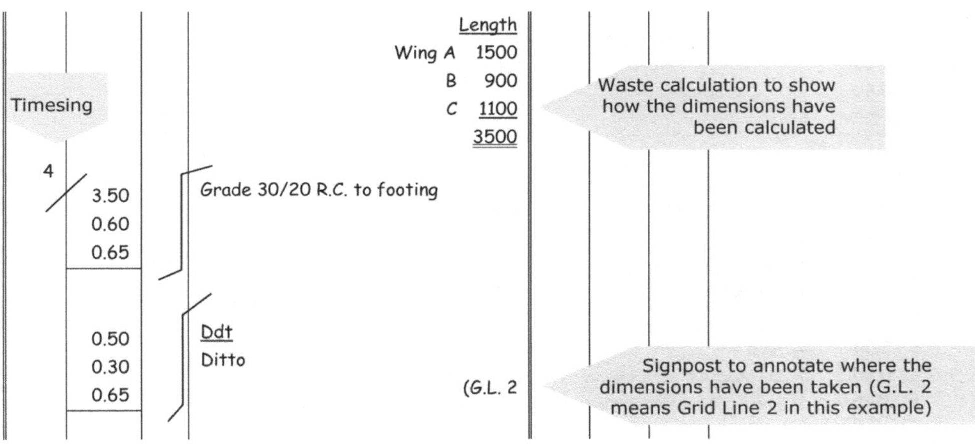

Figure 9.4 Example to illustrate basic dimensions entry.

length are to be concreted and therefore the dimensions are multiplied by four in the timesing column.

Dotting-on

If two factors of multiplication are to be added, dotting-on can be used. As shown in **Figure 9.5**, the first measurement entry means: (3 + 2) × 10.00m × 1.30m × 0.70m. Although the dotting-on practice is customary, care should be taken when using dotting-on to avoid others mistaking the dot for a decimal point.

> **Reminder**
>
> When inserting the dot for dotting-on, make sure that the dot is noticeably large enough.

Deductions and anding-on

Sometimes, we prefer to take an overall measurement and deduct the part which is not required, for instance, deduct the void after an overall volume measurement. In this case, we can enter the dimensions to be deducted in the dimension column under the heading 'Ddt', which means 'deduct' as in **Figure 9.4**.

All dimensions should be suitably bracketed with the corresponding description. Quite often, several sets of dimensions correspond to the same description and the bracket is useful to indicate their grouping (as in **Figure 9.5**). In some situations,

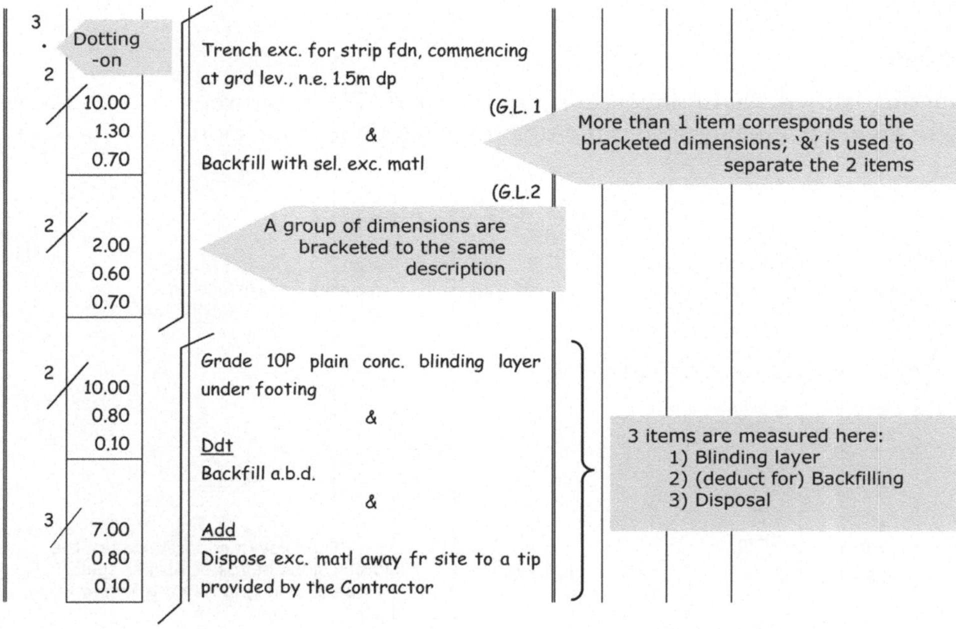

Figure 9.5 Example to illustrate bracketing and anding-on.

more than one item shares the same set of dimensions. To avoid repetition, 'and-ing-on' (indicated as an '&' symbol in **Figure 9.5**) can be used.

Question 9.3

Take-off the carpet area for the floor plan below.

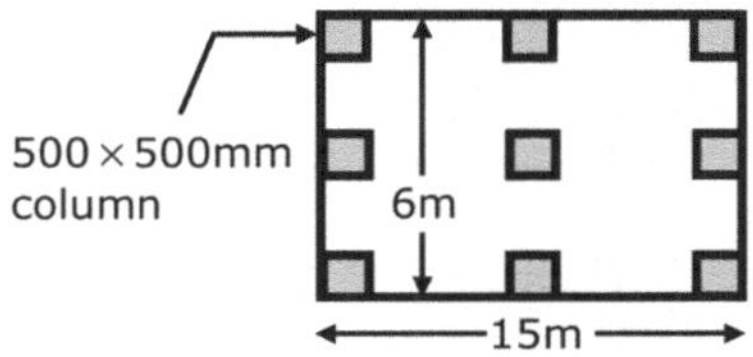

Question 9.4

With reference to **Figure 9.5**, what are the total volumes of the blinding layer and the disposal items?

Writing descriptions

To save long descriptions in measurement, abbreviations, which have been listed separately in the Appendix, are often used. Information which is 'deemed to be included' in the item as stated in the SMM is not required to be mentioned specifically. For instance, formwork measurement is deemed to include for falsework and re-propping, and it is therefore not necessary to mention in the description of formwork items.

Use of schedules

Sometimes, when measuring a number of items of similar characteristics such as doors, windows, ironmongeries, reinforcement and finishes, schedules can effectively assist the taking-off process. This is particularly the case when there are a large number of different items but with similar characteristics to be measured. It will be more efficient and tidier to measure the quantities in a schedule rather than on a dimension paper. The format of a schedule is highly flexible and expandable depending on the variety and nature of items to be measured. A sample schedule for measuring finishes is shown in **Figure 9.6**.

Taking-off procedures

To maintain clarity and accuracy in taking-off, the following procedures are recommended, irrespective of whether computer-aided measurement is used or not:

1 Identify the subject of measurement

Taking-off can be done for difference purposes such as BQ preparation, materials procurement, valuation of variations under architect's instruction

Location	No. of floors	Floor						Skirting			
		Factor	L	W	Type	m²	Remark	Factor	L	Type	m
G/F											
Plant Rm	1	1	2.50	1.30	F1	3.25		1	7.60	S1	7.60
Column	1	-2	0.30	0.50	F1	-0.30		2	1.60	S1	3.20
Door	1	-1	0.75	0.00	F1	0.00		-1	0.75	S1	-0.75

Location	Skirting	Ceiling						Wall					
	Remark	Factor	L	W	Type	m²	Remark	Factor	Girth	H	Type	m²	Remark
G/F													
Plant Rm		1	2.50	1.30	C1	3.25		1	7.60	2.80	W1	21.28	
Column		-2	0.30	0.50	C1	-0.30		2	1.60	2.80	W1	8.96	
Door		-1	0.75	0.00	C1	0.00		-1	0.75	1.80	W1	-1.35	

Figure 9.6 Example of schedule for taking-off finishes.

or interim payments and final account measurement. The same area or work item may be measured at different stages for different purposes. To avoid confusion, the subject of measurement should be stated clearly at the beginning of measurement.

2 Compile all relevant drawings and documents
All relevant drawings and specification should be obtained before starting the measurement task. In cases like measurement for architect's instructions, correspondence such as site memos, daily reports etc. should be studied. The taker-off should check if there are other instructions/revisions stated in the correspondence and all the information should be obtained for measurement.

3 Check any discrepancies or missing information
Takers-off should make sure that all necessary information is ready. If there is any discrepancy between the documents or missing information or ambiguities, a query list with all the questions should be prepared and passed to the architect for his/her clarification and confirmation. A typical query sheet consists of two columns: the left-hand column is for writing queries and the right-hand column is for consultants to put their answers.

4 Read through relevant SMM and preambles
The relevant part of the SMM should be read through to identify the measurement rules applied to the task. Where necessary, preambles should be noted to identify the amendments applied to the standard measurement rules.

5 Plan the take-off sequence
Prepare a take-off list and follow the list to measure.

6 Colour or mark on the measured items on drawings
To make sure that there are no missing or duplicated items in the measurements, it is a good practice to colour or check off the measured item on the drawings once the item has been measured.

7 Bulk check the dimensions
Bulk checking ensures the accuracy of measurement. Usually, items that involve large quantities or substantial cost will be checked. The taker-off should bulk check the detailed dimensions to see if they align with the overall length and width of the building as shown on the layout plan. If measurement is for the purpose of BQ preparation, bulk check should be carried out by another QS who is not responsible for measuring the respective parts of work.

Suggested answers

Question 9.1

i Length should be booked with two decimal places: 16.00
ii Item count should be entered as a whole number without decimal place: 16 (Note that only dimensions are entered with two decimal places in the dimension column.)

Question 9.2

Note the order of dimensions entered for walls and slabs.

Walls:		10.00 0.20 3.00	Order: - Length - Thickness - Height
Slab:		10.00 5.00 0.20	Order: - Length - Width - Thickness

Question 9.3

	15.00 6.00	Carpet; fixed to floor by adhesive	*Overall area is measured first.*
9/	0.50 0.50	<u>Ddt</u> Ditto (column	*The column areas are then deducted. Altogether 9 columns.*

Question 9.4

Volume of blinding layer: $2 \times 10m \times 0.8m \times 0.1m + 3 \times 7m \times 0.8m \times 0.1m = 3.28m^3$
Volume of disposal: $3.28m^3$
Note that both sets of dimensions apply to each item within the same bracket.

10 Length, area and volume measurement

Measuring areas and volumes of different shapes is one of the major steps in taking-off. A list of formulae for measuring areas and volumes of triangles, rectangles, circles etc. are included in the Appendix. Most of these formulae should be familiar to readers and are not explained here.

Girths and mean girth

When taking-off, measurement of the perimeter and the centreline (often described as mean girth in measurement) of a building or a room is often required. Mean girth is useful in calculating building quantities such as concrete wall volume, wall formwork area and finishing areas. Acquiring the skill to adjust a mean girth to and from a perimeter is particularly helpful in obtaining the required building quantities from the figured dimensions. A simple example is used to illustrate the technique of calculating a mean girth.

Assume we have a rectangular building and the plan is shown in **Figure 10.1**. The shaded area represents the external wall of the building. The mean girth of the wall is shown by the dotted line in the figure.

Depending on circumstances, the mean girth can be calculated from the external face or internal face of a wall. If the mean girth is derived from the total external length (i.e. the external girth) of a wall, a deduction for each external corner, which equals to the thickness of the wall as shown in **Figure 10.1**, has to be made. Alternatively, the mean girth can also be calculated by adding the four external corners to the total internal length (i.e. the internal girth) of the wall. **Table 10.1** summarises the mean girth calculation.

Even if the building plan is not rectangular in shape, the adjustment for external and internal corners can still be applied in the same manner. The building in **Figure 10.2** is planned with a set-back and a recess. The mean girth of the external wall can be calculated as if it were rectangular in shape.

As in **Figure 10.2** below, the external girth of the shaded wall at the set-back is in fact the same as if there were no set-back. At the recess, twice the depth of the recess (2a) will have to be added to the overall length.

To derive the mean girth from the external girth, the corners must be adjusted. In the current building plan, there are seven external corners and three internal corners (as shown in **Figure 10.3**). An external corner will balance the effect of an internal corner. Ultimately, only four external corners are required to be adjusted or deducted. Details of the mean girth calculation is shown in **Table 10.2**.

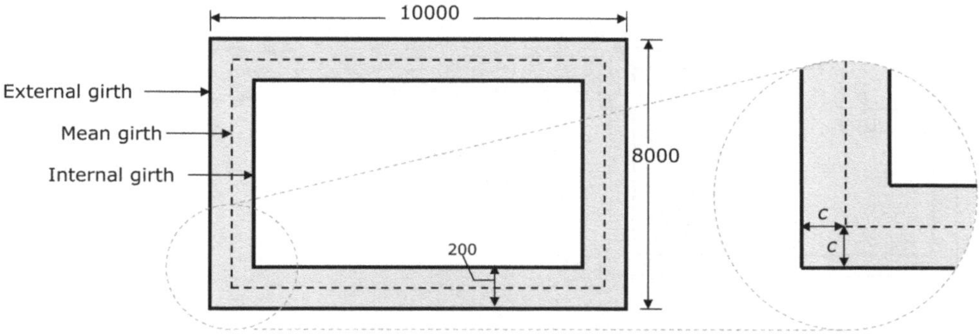

Figure 10.1 Plan of a rectangular building.

Table 10.1 Mean Girth Calculation for a Rectangular Building.

Measured from external face	Length	10,000
$G = 2(l + w) - 4 \times 2c$	Width	8,000
where	Sum of length and width	2/ 18,000
G = mean girth	External girth	36,000
l = length of building measured from external face	Less 4 corners 4/2/100	800
w = width of building measured from external face	Mean girth of the wall	35,200
c = separation between the mean girth and the external or internal girth		
Measured from internal face	Length	9,600
$G = 2(l' + w') + 4 \times 2c$	Width	7,600
where	Sum of length and width	2/ 17,200
G = mean girth	Internal girth	34,400
l' = length of building measured from internal face	Add 4 corners 4/2/100	800
w' = width of building measured from internal face	Mean girth of the wall	35,200
c = separation between the mean girth and the external or internal girth		

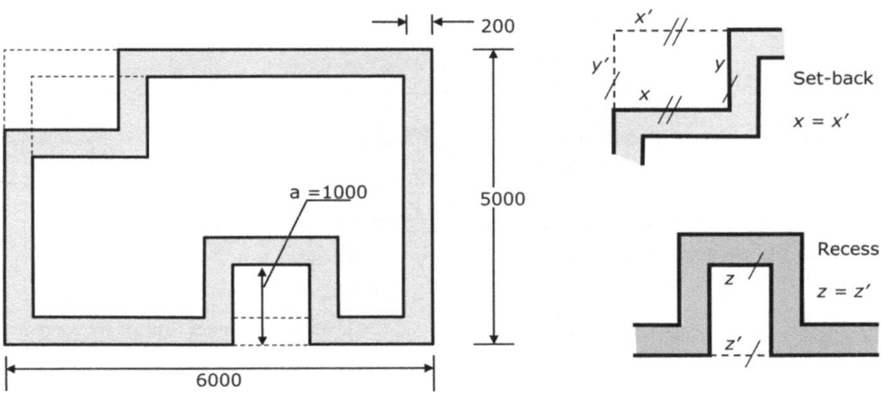

Figure 10.2 Plan of a non-rectangular building.

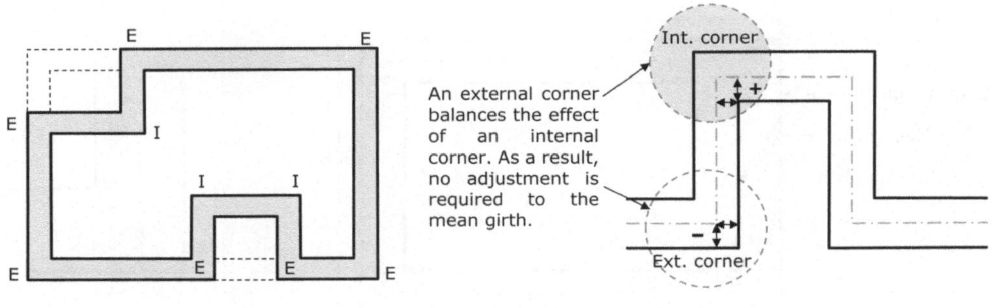

Figure 10.3 External and internal corners.

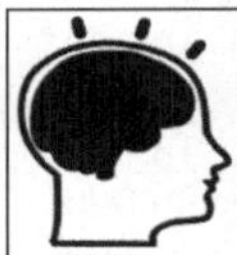

Question 10.1

i Measure the internal girth of the 300mm thick external wall.
ii Derive the mean girth and external girth from the internal girth.

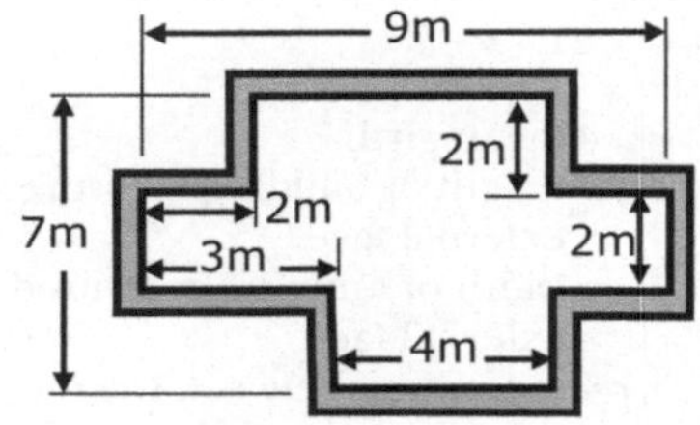

Question 10.2

i Measure the external girth of the 300mm thick external wall.
ii Derive the mean girth and internal girth from the external girth.

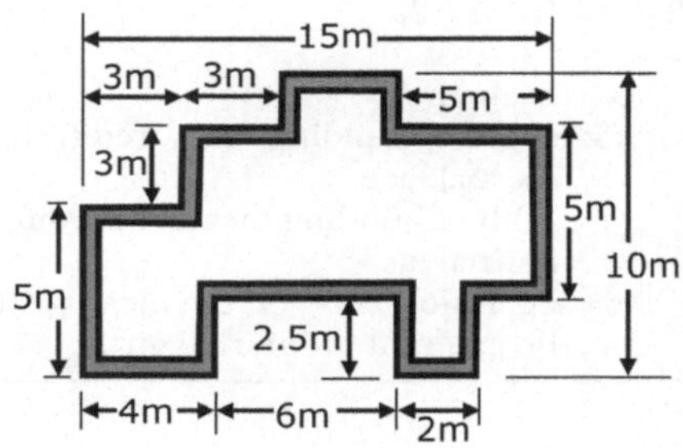

Question 10.3

i Measure the external/internal girth of the 300mm thick external wall.
ii Derive the other two girths from the answer to Question 10.3(i).

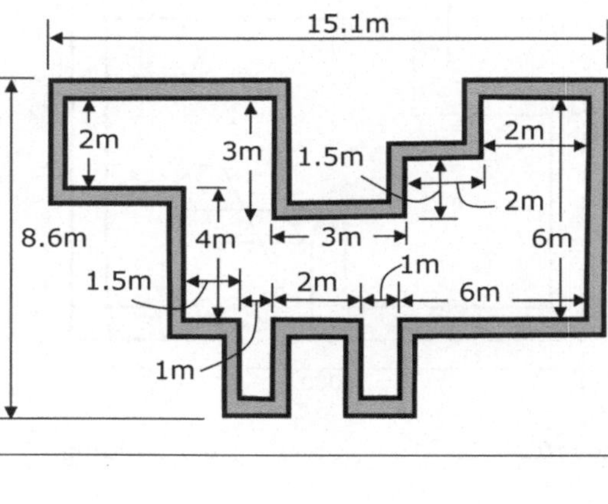

Table 10.2 Mean Girth Calculation for a Building with Set-back and Recess.

Measured from external face		*Measured from internal face*	
Length	6,000	Length	5,600
Width	5,000	Width	4,600
Sum of length and width	2/ 11,000	Sum of length and width	2/ 10,200
	22,000		20,400
Add 2a 2/1000	2,000	Add 2a 2/1000	2,000
External girth	24,000	Internal girth	22,400
Less 4 ext. corners 4/2/100	800	Add 4 ext. corners 4/2/100	800
Mean girth of external wall	23,200	Mean girth of external wall	23,200

Application of mean girth calculation

Mean girth is very useful in many calculations of excavation, concrete works, brick work and finishes. Using **Figure 10.1** as an illustration. If the external wall is in-situ concrete, the concrete volume and the formwork area can both be calculated easily with the mean girth as follows:

Volume of external wall concrete $= 2Gch$ where G = mean girth c = separation between the mean girth and the external / internal girth h = height	Explanatory notes: Volume of external wall concrete can be calculated by the simple formula: base area × height. The 'base area' of the concrete volume is $2Gc$
Area of formwork to concrete external wall $= 2Gh$ where G = mean girth h = height	Explanatory notes: The external wall formwork consists of two vertical boardings – the external and the internal boards. The sum of these two boarding areas is the formwork area we need to measure. The total area of the two boardings can be written as: $= 2h\,(l + w) + 2h\,(l' + w')$ $= h(G + 8c) + h(G - 8c)$ $= Gh + Gh$ $= 2Gh$ Instead of measuring the external boarding area and the internal boarding area separately to give the total formwork area, the use of mean girth saves us a lot of effort.

Question 10.4

Calculate the following by using the appropriate girth(s):

i concrete wall volume
ii nett formwork area to the concrete wall excluding the area where the canopy intersects

iii wallpaper area to the internal side of the wall
iv concrete volume of canopy
v painting area to the external side of the wall including the canopy

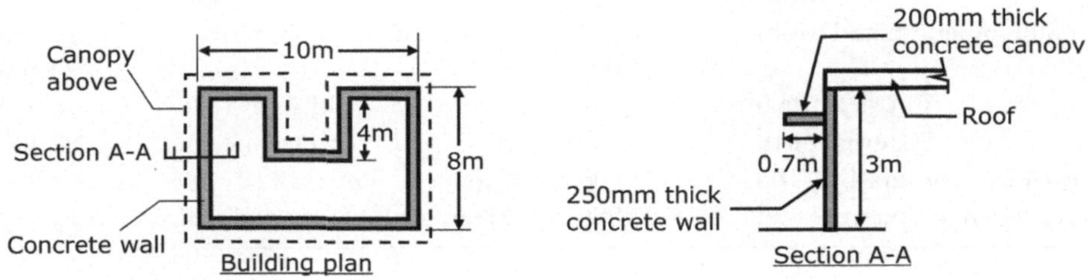

Measurement of irregular areas

When measuring irregular areas, one of the following methods can be used:

- By breaking down the area into a number of regular areas such as triangles and rectangles. The irregular boundary lines are dealt with by the give-and-take lines, as indicated in **Figure 10.4**.
- By Simpson's rule. As illustrated in **Figure 10.5**, the area is divided into a number of segments with equal spacing. Simpson's rule uses a suitably chosen parabolic shape which is based on three points to approximate the curved boundary of each segment. The smaller the intervals between each segment are divided, the greater the level of accuracy.

$$\text{Shaded area} = \frac{x}{3}[(y_0 + y_n + 4(y_1 + y_3 + \ldots + y_{n-1}) + 2(y_2 + y_4 + \ldots + y_{n-2})]$$

Where y_o, y_1, ... y_{n-1}, y_n are lengths of the segments
Note: n must be an even number.

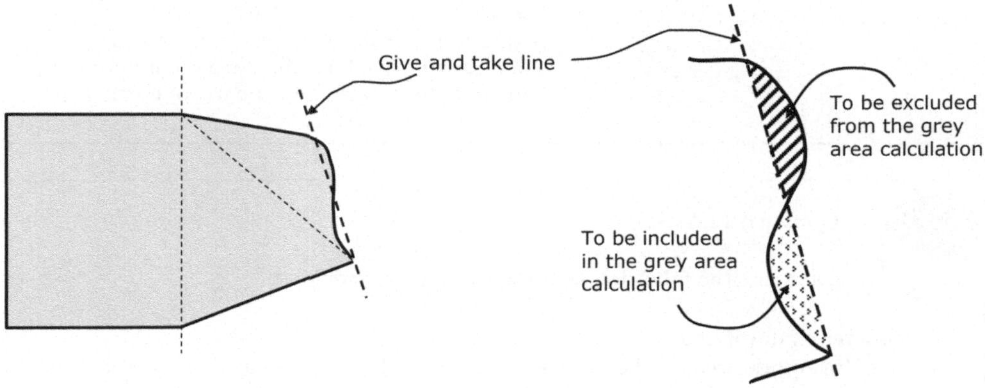

Figure 10.4 Dividing an irregular area for area calculation.

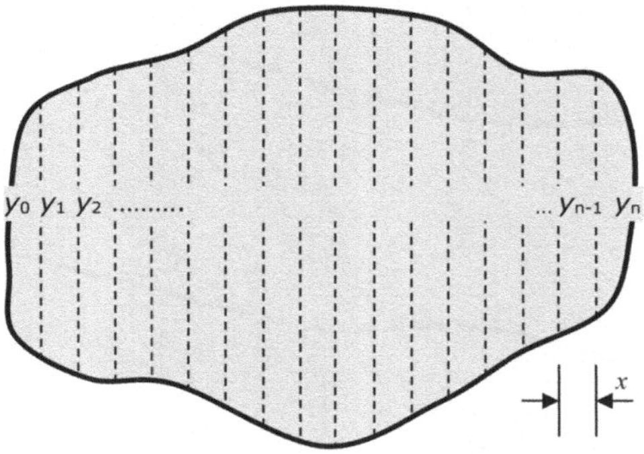

Figure 10.5 Area calculation by Simpson's rule.

- By the trapezoidal rule. Here, the trapezoidal rule approximates the curved boundary of each segment by a suitably chosen line based on two points. Similar to the Simpson's rule, the smaller the intervals between each segment are divided, the greater the level of accuracy. Further, there is no restriction on the number of intervals. In **Figure 10.5**, the shaded area can be calculated by the following formula:

$$\text{Shaded area} = x\left(\frac{y_0 + y_n}{2} + y_1 + y_2 + \ldots\ldots + y_{n-1}\right)$$

Where y_o, y_1, ... y_{n-1}, y_n are lengths of the segments

Measurement of irregular volumes

When measuring irregular volumes, one of the following methods can be used:

- By dividing the volume into a number of regular volumes.
- By Simpson's rule. This method is usually used to calculate the earthworks in connection with slope cutting and filling. Divide the volume by a series of equally spaced sections and obtain the areas of each cross-section. If the cross-sections are not available from the consultant engineer, the takers-off have to plot them on graph paper by themselves. Based on **Figure 10.6**, the volume can be calculated by the following formula:

$$\text{Irregular volume} = \frac{x}{3}(A_0 + 4A_1 + 2A_2 \ldots + 2A_{n-2} + 4A_{n-1} + A_n)$$

Where A_o, A_1, ... A_{n-1}, A_n are areas of the cross-sections

Note: n must be an even number, i.e. an odd number of cross-sections are required.

- By end-areas method. Similar to the Simpson's rule, the volume is divided into equally spaced sections but there is no restriction on the number of sections divided. The area of each cross-section has to be determined. According to **Figure 10.6**, the volume is equal to:

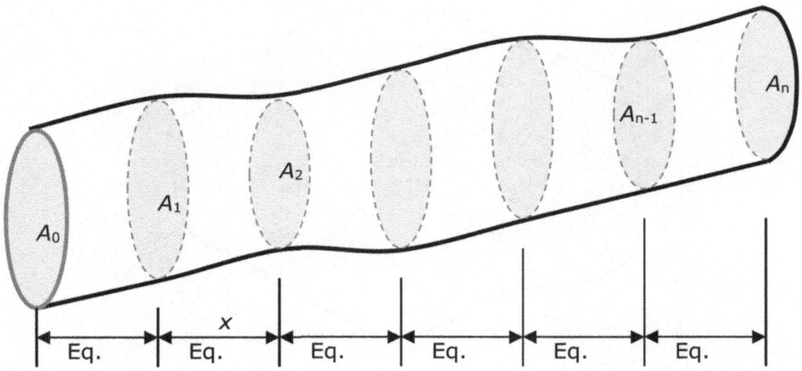

Figure 10.6 Volume calculation by Simpson's rule.

$$\text{Irregular volume} = x\left(\frac{A_0 + A_n}{2} + A_1 + A_2 + \ldots + A_{n-1}\right)$$

Where A_o, A_1, ... A_{n-1}, A_n are areas of the cross-sections

- By grid method. This method is often used to calculate the volume of cut and fill on a sloping site. Divide the site into a series of identical squares with the existing ground levels taken at the corners of each square. Usually, a 3 to 10 metre grid should suffice for building work purposes. Having the grid and levels ready, the average level over the entire gridded area can be found by the following formula:

$$\text{Average existing level} = \frac{\Sigma(L_1 + 2L_2 + 4L_4)}{\Sigma w}$$

where

L_1 = level at the extreme corner of the area
L_2 = level at intermediate point on the area boundary
L_4 = level at other intermediate point
w = weighting applied to level

Then, the average depth of cut/fill can be established. A simple example is shown below to illustrate the method.

Figure 10.7 shows the existing ground levels plotted on a 5 × 5m grid. The grid levels and corresponding weightings can be tabulated as follows:

Therefore, the average site level is = 62.7/24 m = 2.61m
Assuming the formation level is 2.00m, the average depth of excavation will be:

= 2.61m – 2.00m
= 0.61m

The volume of excavation will be:

(5.00m × 3) × (5.00m × 2) × 0.61m
= 91.5m^3

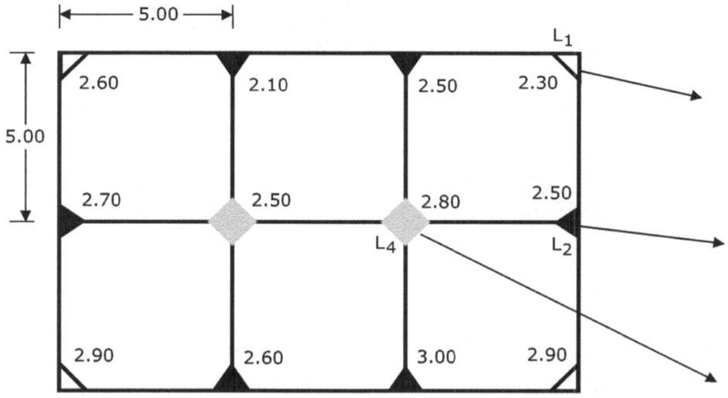

	Existing Level	Weighting	Weighted level
L1	2.6	1	2.6
	2.3	1	2.3
	2.9	1	2.9
	2.9	1	2.9
L2	2.1	2	4.2
	2.5	2	5.0
	2.7	2	5.4
	2.5	2	5.0
	2.6	2	5.2
	3.0	2	6.0
L4	2.5	4	10.0
	2.8	4	11.2
	Total:	24	62.7

Figure 10.7 A gridded area with existing ground levels.

Suggested answers

Question 10.1

Length		9,000	Explanatory notes:
Width		7,000	Except for the external length and width of the building, all given dimensions on the drawing are unnecessary in the calculation.
Sum of length and width	2/	16,000	
Internal girth		**32,000**	
Add 4 ext. corners 4/2/150		1,200	
Mean girth of the wall		33,200	
Add 4 ext. corners 4/2/150		1,200	Noticed that the external girth can be derived from the mean girth in the same way.
External girth of the wall		34,400	

Question 10.2

Length		15,000
Width		10,000
Sum of length and width	2/	25,000
		50,000
recess 2/2,500		5,000
External girth		**55,000**
Less 4 ext. corners 4/2/150		1,200
Mean girth of the wall		53,800
Less 4 ext. corners 4/2/150		1,200
Internal girth of the wall		52,600

Question 10.3

Length		15,100			
Width		8,600			
Sum of length and width	2/	23,700			
		47,400			
upper recess 2/3000		6,000			
lower recess 2/2000		4,000		Lower recess Length	
External girth		**57,400**			8,600
Less 4 ext. corners 4/2/150		1,200	Less:	6,000	
Mean girth of the wall		56,200		wall 2/300	6,600
Less 4 ext. corners 4/2/150		1,200			2,000
Internal girth of the wall		55,000			

Question 10.4

Wall girth calculations					
Length	10,000		=	$G_{wm} \times 0.25 \times 3.00$	m^3
Width	8,000		=	$43.00 \times 0.25 \times 3.00$	m^3
Sum of length and width 2/	18,000		=	32.25	m^3
	36,000				
recess 2/4,000	8,000	ii		Formwork area to concrete wall:	Unit
Wall external girth (G_{we})	**44,000**		=	$G_{wm} \times 3.00 \times 2 - G_{we} \times 0.20$	m^2
Less: 4 ext. corners 4/2/125	1,000		=	$43.00 \times 3.00 \times 2 - 44.00 \times 0.20$	m^2
Wall mean girth (G_{wm})	**43,000**		=	249.20	m^2
Less 4 ext. corners 4/2/125	1,000				
Wall internal girth (G_{wi})	**42,000**	iii		Wallpaper area to internal wall:	Unit
			=	$G_{wi} \times 3.00$	m^2
Canopy girth calculations			=	42.00×3.00	m^2
Wall external girth (G_{we})	44,000		=	126.00	m^2
Add: 4 ext. corners 4/2/350	2,800				
Canopy mean girth (G_{cm})	**46,800**	iv		Concrete volume of canopy:	Unit
Add: 4 ext. corners 4/2/350	2,800		=	$G_{cm} \times 0.70 \times 0.20$	m^3
Canopy external girth (G_{ce})	**49,600**		=	$46.80 \times 0.70 \times 0.20$	m^3
			=	6.55	m^3

Note: the header row of the second table also reads: i Concrete wall volume: Unit

v Painting area to the external side of wall and canopy: Unit

=	$G_{we} \times 3.00 - G_{we} \times 0.20 + G_{cm} \times 0.70 \times 2 + G_{ce} \times 0.20$	m^2
=	$44.00 \times 3.00 - 44.00 \times 0.20 + 46.8 \times 0.7 \times 2 + 49.6 \times 0.2$	m^2
=	$\underline{\underline{198.64}}$	m^2

Note that the area of concrete wall where the canopy intersects is not the same as the area of the vertical edge of the canopy. Although the height of both areas is 200mm, the girth length of them is different, which can be seen from the building plan.

11 Introduction of HKSMM4 Rev 2018

The standard method of measurement

The standard method of measurement (SMM) is a document that sets out the rules for measurement and description of construction works, aiming at providing a uniform basis of measurement for project participants.

Points to note

Although HKSMM is not a mandatory document for building measurement, it is normally adopted by QS to prepare BQ in Hong Kong.

One of the prime purposes of measurement is to prepare BQ for tendering. With an SMM, it is easier to convey to the tendering contractors what is and what is not measured in each BQ item. This enables more accurate pricing and reduces future disputes (Greenhalgh, 2013). Also, the administration and financial control of the project during the construction stage can be facilitated.

The Hong Kong Standard Method of Measurement 4th Edition Revised 2018 (HKSMM4 Rev 2018)

Before production of the first HKSMM, the U.K. Standard Method of Measurement had been followed by the industry on an arbitrary basis. In 1962, the first edition of the HKSMM for building works was written by representatives from surveyors and contractors. The first edition was soon revised in 1966 (HKSMM2), followed by the third revision in 1979 (HKSMM3). After more than thirty years, the HKSMM4 was published in 2005 by the Hong Kong Institute of Surveyors (HKIS). Besides revamping the layout from bullet-point form to tabulated form, the measurement rules for building services were included. Non-local and obsolete rules were also removed. The HKSMM4 was then further streamlined by the HKIS in 2018 to enhance consistency and simplicity e.g. timber doors are measured together with frames, linings, louvres, glazing, holdfasts and anchor bolts as a set.

Information provided by HKSMM4 Rev 2018

Use of the tabulated rules

The measurement rules in HKSMM4 Rev 2018 are set out in a tabulated format as **Figure 11.1**. In general, the whole SMM is divided into 22 work sections such as excavation, concrete works, wood works etc. Within each work section, the information to be provided, classification table, measurement rules and supplementary rules are tabulated.

As in **Figure 11.1**, a horizontal double line separates the SMM rules into two parts. All the rules shown above the double line are the basic rules apply to all work within the Section or Sub-section. The rules below the double line only apply to the particular group of items against which they are shown.

Information provided

This section ensures that all the relevant details about the work which the tenderers require for pricing are provided either in drawings, specifications or preambles as the case may be. For example, at the beginning of Section VI (b) Underpinning, the SMM states that

> a description of the work to be underpinned shall be given, stating the length, the depth of the underpinning and the limit of the length to be carried out in one operation.

Classification table

There are five columns in the classification table. Except for the fourth column which indicates the unit of measurement, each column represents the sub-division of the left-hand side column. For instance, in **Figure 11.1**, items in box 1 are sub-divided into items in box 5 which can be further sub-divided into items in box 9. All these items should be measured in U1 (measuring unit) and supplementary items in box 13 should be applied. Similarly, items in box 2 can also be sub-classified into items in box 5 and further into items in box 10, with unit in U2 and supplementary items in box 13 applied.

In **Figure 11.1**, a broken line is drawn between box U5 and U6. This means that both rules in U5 and U6 can be used. The rules are read from left to right and governed by the horizontal lines. Therefore, in this example, units U5 and U6 can be used for items in boxes 4, 8 and 12.

<table>
<tr><td colspan="5">Information Provided</td><td>Measurement rules</td><td>Definition rules</td><td>Coverage rules</td><td>Supplementary information</td></tr>
<tr><td colspan="5">1.
(a)
(b)</td><td>M.1</td><td>D.1</td><td>C.1</td><td>S.1</td></tr>
<tr><td colspan="5">Classification table</td><td>Measurement rules</td><td>Definition rules</td><td>Coverage rules</td><td>Supplementary information</td></tr>
<tr><td>1.</td><td rowspan="2">5.</td><td>9.</td><td>U1</td><td rowspan="2">13</td><td>M.2</td><td rowspan="4">D.2</td><td rowspan="2">C.2</td><td rowspan="4">S.1</td></tr>
<tr><td>2.</td><td>10.</td><td>U2</td><td>M.3</td></tr>
<tr><td rowspan="2">3.</td><td>6.</td><td rowspan="2">11.</td><td>U3</td><td>14.</td><td>M.4</td><td>C.3</td></tr>
<tr><td>7.</td><td>U4</td><td>15.</td><td>M.5</td><td>C.4</td></tr>
<tr><td>4.</td><td>8.</td><td>12.</td><td>U5
- - - - -
U6</td><td>16.</td><td>M.6</td><td>D.3</td><td>C.5</td><td>S.2</td></tr>
</table>

Figure 11.1 Layout of HKSMM4 Rev 2018.

Measurement, definition and coverage rules, and supplementary information

The other half of the table on the right-hand side provides guidance to users to measure the physical quantities and to describe the work item appropriately. The measurement rules column set out when and how the work shall be measured, e.g. the minimum sizes of voids to be ignored in deductions. The definition rules column gives definition of the extent and limits of the work represented by a word or expression used in the rules and the BQ. For example, in VI(a).3.1, definition rule D.1 explains that *The term planking and strutting means everything requisite to uphold the face of excavation other than shoring as required to be measured under item.* The coverage rules column highlights particular incidental work which shall be deemed to be included or excluded from the item. For example, in VI(a).3.1, coverage rule C.3(a) states that surface excavation is deemed to be included for excavating in any ground encountered. Such guidelines will affect the pricing of work. The supplementary information column covers rules governing the additional information which shall be given. For instance, the supplementary information S.7 in VI(a).10 states that the kind and quality of filling material has to be stated for hardcore or granular filling.

References to SMM

References to the classification tables should be given in the following format:

VI(a).3.4.1.1

VI Excavation *(a) Excavation*				
1st column in classification table	**2nd column in classification table**	**3rd column in classification table**	**Unit**	**5th column in classification table**
3. Excavating	4. Basements and similar excavations	1. In successive stages of 1.50m	m^3	1. Commencing levels stated

As shown in the above example, the fourth column from the left which contains units of measurement is not included in the reference. Therefore, item VI(a).3.4.1.1 refers to the rule: ‘Commencing levels stated’. If referred to item VI(a).3.4.1., the rule is: ‘In successive stages of 1.50m’. A digit 0 within a reference represents no entry in the column in which it appears. For instance:

VI Excavation *(a) Excavation*				
1st column in classification table	**2nd column in classification table**	**3rd column in classification table**	**Unit**	**5th column in classification table**
7. Backfilling to excavation	-	1. Arising from excavation	m^3	1. Selected, details stated.

If one wants to refer to the rule ‘selected, details stated’ with respect to backfilling, the reference should be written as VI(a).7.0.1.1.

The same referencing method has been used throughout this book.

Using the SMM in practice

As explained in **Chapter 9**, to complete a take-off task, we need to measure from the drawings each item of work by booking the respective quantity and item description into the dimension paper (or schedule). With reference to the classification table in the HKSMM, the taker-off can split the measurement task into quantifiable items.

Quantities of work

Based on the coverage rules and measurement unit stated in the SMM, dimensions of the work item are taken nett, and recorded into the dimension paper.

Item description

When drafting the description for an item, the SMM can be used as a framing guide. One descriptive feature is taken from each of the first three columns of the classification table, together with the relevant descriptions from the fourth column and any information required in the supplementary information column. Section and sub-section headings of the SMM are usually taken as the headings for the items in the BQ. Some examples are given below for illustration.

> **Points to note**
>
> Many project owners and QS consultants have developed standard phraseology of item descriptions to facilitate BQ preparation and cost estimating. Nevertheless, the use of SMM to build up item descriptions is not affected.

Example 1 – Excavate 150mm thick surface soil from the site.

Description can be written as:
<u>Excavation</u>
Surface excavation, average 150mm deep

Relevant SMM item: VI(a).3.1.1

VI Excavation
(a) Excavation

Classification Table				Sup. Info
3. Excavating	1. Surface excavation ≤ 0.20m in average depth	1. Average depth stated		-

Example 2 – Fix standard timber formwork to soffit of suspended slabs. Thickness of the suspended slabs is 250mm and the height of soffit of slab measured from structural floor surface is 4m.

Description can be written as:
<u>In-situ concrete</u>
<u>Formwork</u>

Sawn formwork to soffits of suspended slabs; 200–300mm thick; strutting 3.50–5.00m high

Relevant SMM item: VII(d).2.2.1.2

VII Concrete Works
(d) Formwork

Classification Table					Sup. Info
2. Soffits of slabs	2. Slab thickness > 200mm – thickness stated in further stages of 100mm	1. Horizontal	1. Height to soffit ≤ 3.50m 2. Thereinafter in 1.50m stages		S.1 Kind and quality of materials

Example 3 – Fix 32mm diameter high yield steel bars to concrete suspended slabs. The bar length measured from drawing is 14m long.

Description can be written as:
<u>In-situ concrete</u>
<u>Reinforcement</u>
High yield steel bars, 32mm diameter, 12.0–15.0m long in general reinforcement

Relevant SMM item: VII(c).2.1.2.1

VII Concrete Works
(c) Reinforcement

Classification Table					Sup. Info
2. Bar reinforcement	1. Nominal size stated	2. General reinforcement > 12m in length, given in stages of 3m	1. Round bars		S.1 Quality of steel stated

When material specification is unavailable

According to the SMM, most items are required to describe the material involved. However, when taking-off, we may find that some articles such as finishes, sanitary fittings, ironmongeries etc. are not yet decided at the tender stage. To enable tenderers to estimate a price for the item concerned, a prime cost (P.C.) rate, i.e. an amount of money allowed to purchase a specified item, will be incorporated as part of the item description (as shown in **Figure 11.2**).

As discussed in **Chapter 7**, the prime cost rates are for the material cost of that item delivered to site only and the contractor will estimate his price based on the

	Description	Quantity	Unit	Rate	HK$
	<u>IRONMONGERIES</u>				
	<u>Supply and fixing ironmongeries; all as detailed on drawings and specification</u>				
A	Lever handle (P.C. rate HK$300/No. supply only)	30	Nr		

Figure 11.2 Typical example of a P.C. rate in BQ.

prime cost as stated and allow in addition for waste, fixing, ancillary materials required for fixing such as mortar for bedding and jointing, profit and overheads. Once the material is confirmed in the construction stage, the contract rate will be adjusted according to the relevant stipulation in the preambles. Prime cost rate may also be described as for supply and fix/apply. In that case, the provisional allowance in the rate is for the supply and fixing cost charged by a specialist subcontractor, and the contractor should allow for the profit and overheads when estimating his price.

Further details on the pricing of P.C. rate items and subsequent adjustment will be discussed in **Chapter 22**.

Work not covered in SMM

Some items of work may not be covered by the HKSMM4 Rev 2018. In those cases, the measurement methods should be stated in the Preambles of the BQ or in the description of the items to which they apply. The same should be done when the measurement method adopted deviates from the method stated in the HKSMM.

12 Measurement of excavation

Introduction

Site clearance and excavation is usually the first construction operation in a building project. Unlike other work sections, tender drawings normally provide very little detail about site clearance and excavation because the work is largely related to a contractor's method statement. However, as laid down in the SMM rules, excavation work measurement is based on the existing site condition and the characteristics of the proposed underground features, rather than the construction method to be adopted by the contractor.

Major work items covered in Excavation Section VI of the HKSMM4 Rev 2018 are:

- Site clearance
- Excavation
- Earthwork support
- Disposal
- Backfilling
- Filling to make up levels/to form embankments or terrace
- Hardcore or granular filling
- Blinding to surfaces of hardcore or other filling materials
- Surface treatment

Measurement of other sundry items such as cutting off heads of piles and work associated with underpinning is not included here.

Site clearance

Essentially, before the starting of any construction work, the site has to be cleared. This includes clearing away vegetation and disposal of existing rubbish. Trees or bushes that are on the site have to be removed. Any stumps and roots have to be removed from the ground as well to give a clear ground as level as possible.

VI(a).2.1 Cutting down trees with girth > 0.6m

- Measured in number.
- Group by the girth of trees: 0.60–0.90m, and in further stages of 0.30m.

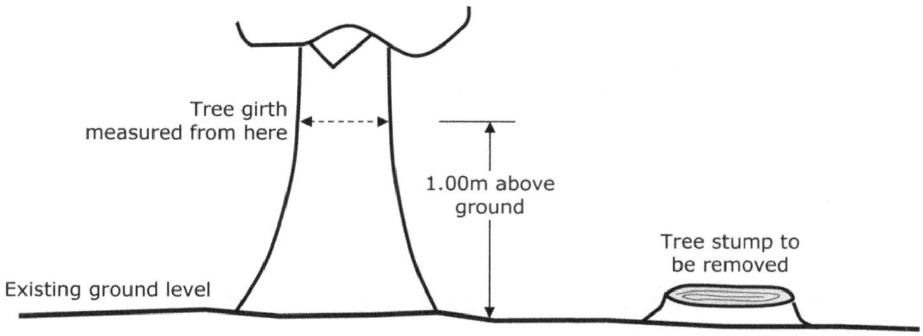

Figure 12.1 Measurement of tree girth.

- Tree girths are measured at the height of 1.00m above ground (VI(a).M.1), as shown in **Figure 12.1**.
- Grubbing of roots, disposal of materials and filling of voids in association with tree removal are deemed to be included (VI(a).C.1).

Points to note

'Deemed to be included', this phrase appears in the SMM frequently. When item A is deemed to be included in item B, A will not be measured in the taking-off. However, when pricing the work, the estimator has to allow the cost of doing item A in item B.

VI(a).2.2 Cutting down trees which are not in an open area

- If cutting down trees which are not in an open site area or are in an area without vehicular access, they shall be measured separately in number and details to be stated.
- Other measurement rules are the same as VI(a).2.1 above.

VI(a).2.3 Removal of existing tree stumps

- Where trees have been cut down by others, removal of existing tree stumps shall be measured in number.
- Grubbing of roots, disposal of materials and filling of voids in association are deemed to be included (VI(a).C.1).

VI(a).2.4 Site clearance of tall grass, shrubs, bushes and trees ≤ 0.6m girth

- Measured as an item.
- State the approximate site area.

Reminder

Removal of trees > 0.6m girth is measured in number;
Removal of trees ≤ 0.6m girth is measured as an item.

VI(a).2.5 Cutting turf to be preserved

- Turf is a surface layer of grass sod cut from the ground and preserved for future use. The method of preservation should be given in the specification.
- Measured in m^2.
- State the average depth, method of storage and preservation and location of stacking.

VI(a).2.7 Disposal of existing spoil heaps or rubbish dumps left by previous contractors

- Measured as an item.
- State the spoil or rubbish to be removed 'from site to a specified dumping area provided by the employer' or 'from site to a tip provided by the contractor'.
- Multi-handling to and from temporary spoil heaps, loading and transportation of spoil are deemed to be included (VI(a).C.2).

Excavation

Working space allowance

According to VI(a).M.2, working space has to be allowed in the measurement of excavation according to the following principles, with reference to **Figure 12.2**:

Allow 0.25m working space where:

Case 1 – the formwork does not exceed 0.60m high; or
Case 2 – the bottom of the formwork does not exceed 0.60m below the starting level of the excavation.

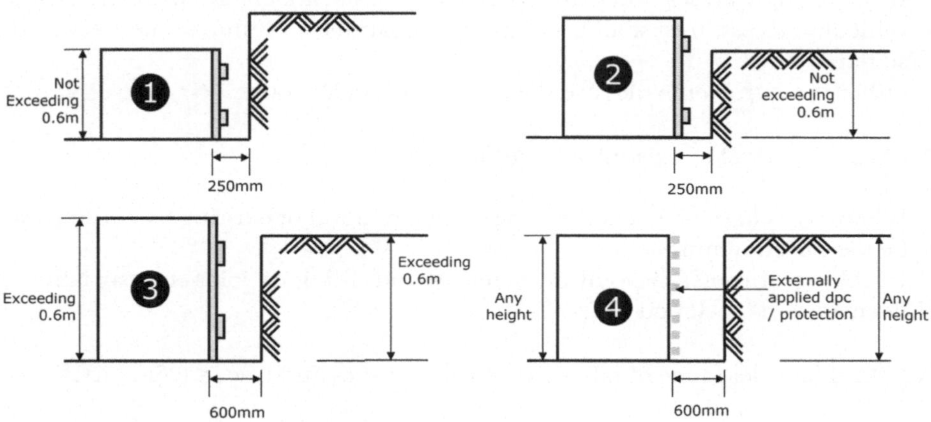

Figure 12.2 Allowance of working space in excavation.

Allow 0.60m working space where:

Case 3 – the formwork exceeds 0.60m high and the bottom of the formwork exceeds 0.60m below the starting level of the excavation; or
Case 4 – workers are required to operate from the outside such as to apply damp proof covering externally.

Points to note

Working space allowance should not be confused with open cut excavation. Open cut excavation is characterised by sloping the sides of excavation to a safe angle. When measuring open cut excavation, the volume of excavation should be calculated with reference to the cutting profile as shown in the engineering drawings.

No working space is allowed where:

- no formwork is required for the concrete work below ground (such as the blinding layer), and no work is required to be operated from outside after concreting.

Question 12.1

If we need to construct a foundation which consists of precast concrete pads with structural steel columns, do we need to allow working space for the foundation excavation?

The working space rule applies to excavations such as surface excavation, excavation to a reduced level, excavation for basement, trenches, column bases, isolated piers, manholes and the like.

Question 12.2

The plan shows a 650mm thick R.C. footing below wall. If the depth of trench excavation for the footing exceeds 600mm, how should we allow the working space for this trench excavation?

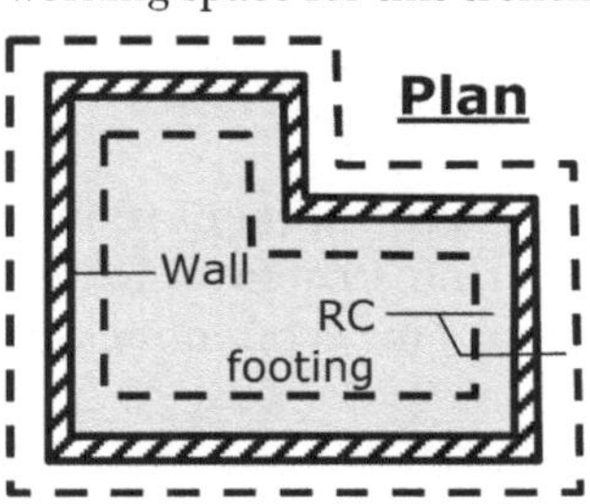

Question 12.3

Based on the working space rules in HKSMM4 Rev 2018, what is the working space in each of the following cases and where should the working space be measured from (Face A or Face B)?

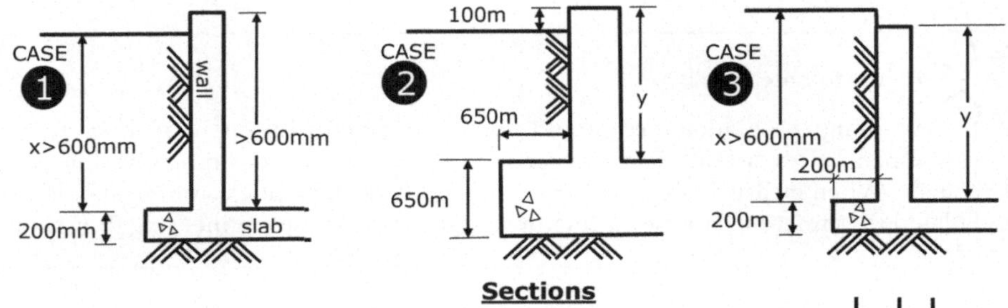

Case 1A: Only the slab requires formwork.
Case 1B: Only the wall requires formwork.
In cases 2 and 3, both the slab and the wall require formwork.
Case 2A: Height of y = 500mm
Case 2B: Height of y = 800mm
Case 3A: Height of y = 500mm
Case 3B: Height of y = 800mm

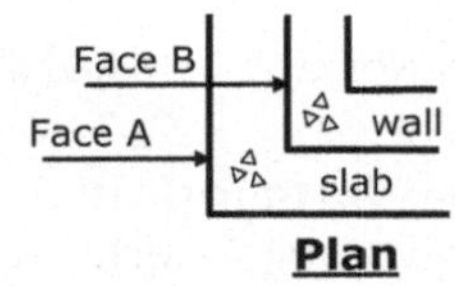

Other item coverage for excavation

According to VI(a).C.3, excavation items are deemed to include excavation in any soil type and sloping profile. The 2018 revision extended the coverage to include a wide range of operations in connection with excavation:

- excavating according to specified profiles and stages
- grubbing up old roots
- removing disused cables, drain pipes, manholes and the like
- sealing connections
- planking and strutting to sides of excavation
- temporary support to cables, drain pipes and the like during excavation
- excavating and backfilling extra working space that is necessary and beyond the working space rule as set out in VI(a).M.2

The above coverage rules apply to all excavation work such as surface excavation, oversite excavation, trench excavation, basement excavation and so forth.

Surface excavation and oversite excavation to reduced level

VI(a).3.1 Surface excavation ≤ 0.20m average depth

Topsoil, usually in a range of 150–300mm, has a high concentration of roots and other decaying materials that make it unsuitable for bearing structural loads. In most circumstances, topsoil is removed before construction. This is often measured under the surface excavation.

- Measured in m^2.
- As shown in **Figure 12.3**, the existing ground level may be uneven, and therefore the average depth is stated.

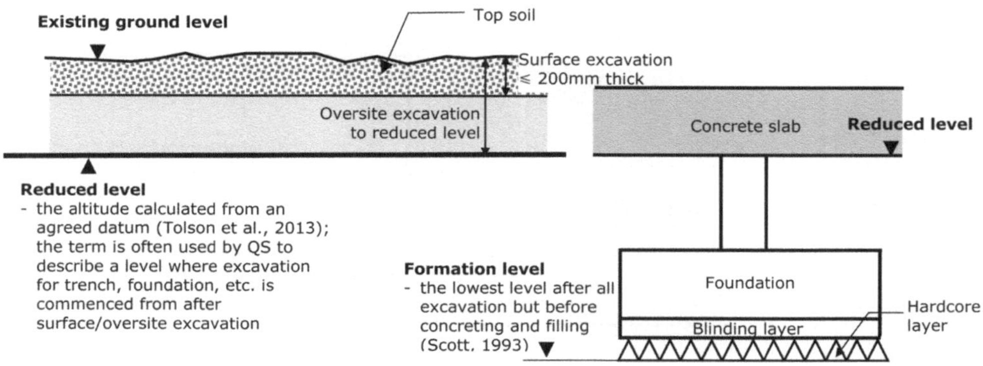

Figure 12.3 Surface excavation and oversite excavation to reduced level.

- Disposal of excavated soil has to be measured separately (VI(a).M.3) in volume (VI(a).6.1.1–6.1.3).

VI(a).3.2 Oversite excavation to reduced level

- Measured in m^3.
- Average depth has to be stated if thickness is ≤ 0.20m.
- As shown in **Figure 12.3**, an oversite excavation to reduced level may be desirable to have other excavation works such as column base excavation started from there. The depth of oversite excavation is measured from the existing ground level to the desired reduced level.
- Disposal of excavated soil has to be measured separately in volume (VI(a).M.3).

Excavation for basements and foundations

VI(a).3.4–3.7 Excavation for basements, trenches below basement level, foundations, column bases, ground beams and the like

- Measured in m^3.
- Excavations for the following are measured separately:
  - basements and similar
  - surface trenches
  - trenches below basement level (see **Figure 12.4**)
  - column bases
  - isolated piers
  - manholes

Points to note

The minimum dimensions for excavating surface trenches, column bases, isolated piers, manholes and the like have been removed from the HKSMM4 Rev 2018 edition. The revision simplifies the measurement of excavation works for these items and facilitates the grouping of similar items together in the BQ.

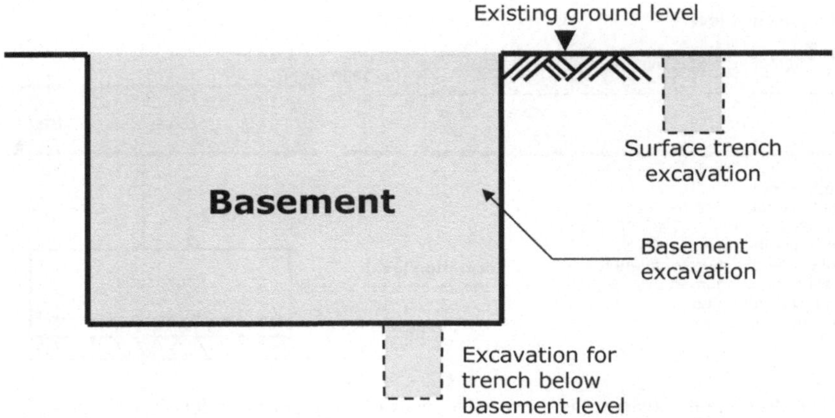

Figure 12.4 Examples of excavation work.

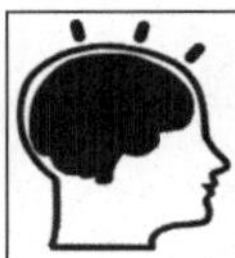

Question 12.4

i 'Basements and similar' – what does 'similar' refer to?
ii Is boring holes for piles regarded as excavation for 'basements and similar' or 'isolated pier'?

- State the commencing level.
- Measure in successive stages of 1.50m. In the example shown in **Figure 12.5**, the excavation work will be measured in three items:

  1 Excavate trench commencing at existing ground level not exceeding 1.50m deep.
  2 Excavate trench commencing at existing ground level 1.50–3.00m deep.
  3 Excavate trench commencing at existing ground level 3.00–4.50m deep.

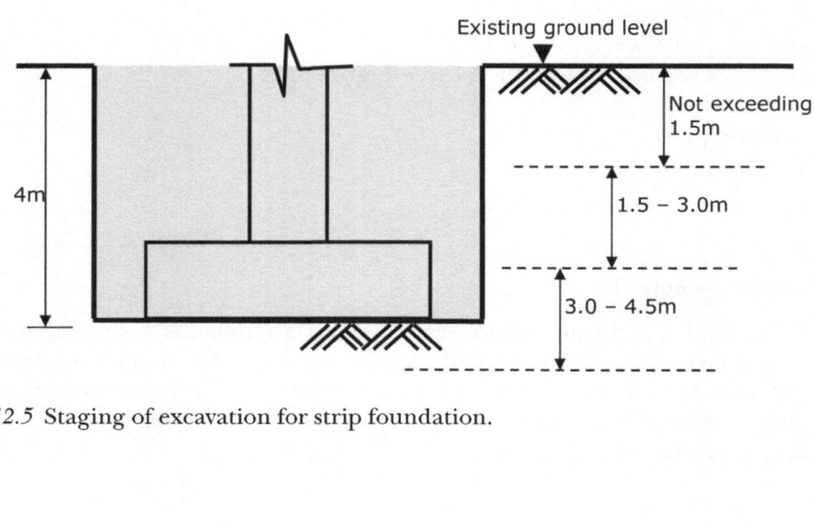

Figure 12.5 Staging of excavation for strip foundation.

Question 12.5

If the basement excavation will be done after 300mm oversite excavation, can we describe the commencing level of basement excavation as 'existing ground level'? If not, how to describe?

Question 12.6

An R.C. column base as in **Figure 12.5** has a size of 3.0mL×3.0mW×0.5mD.

i How should we allow the working space for the excavation of this column base?
ii Take-off the excavation quantities.

Excavation for post holes

VI(a).3.8 Excavation for post holes
If post holes ≤ $0.30m^3$

- Measured in number.
- Backfilling and disposal are deemed to be included in the excavation item (VI(a).C.4). Filling materials to be stated (VI(a).S.4).

If post holes > $0.30m^3$

- Measured in m^3, the rules for column base/isolated piers excavation (VI(a).3.7) will be applied instead.

Excavate trenches for curbs

VI(a).3.9 Excavate trenches for curbs and the like

- Measured in m.
- State the average depth in multiples of 0.25m.
- State the commencing level and size of curbs.
- Curbs without beds and those on beds are each measured separately.
- Backfilling and disposal are deemed to be included (VI(a).C.4). Filling materials to be stated (VI(a).S.4).

Excavate trenches for service pipes and cables

VI(A).3.10 – 3.11 Excavate trenches for service pipes, cables and the like

- Measured in m.
- Trenches for pipes or cables laid on beds and without beds are each measured separately. In case of trenches for drain pipes and pipe ducts, they should be measured according to IX.2–3 under Section IX Drainage.

- State the average depth in multiples or stages of 0.25m.
- State the commencing level.
- In case of trenches for a single pipe, state the size of pipes or cables according to the following groups (VI(a).3.10.2–10.4):
  - Pipes or cables $\leq$ 55mm diameter
  - Pipes or cables $>$ 55mm and $\leq$ 110mm diameter
  - Pipes or cables $>$ 110mm diameter with the sizes stated
- In case of trenches for multiple pipes, state the sizes, numbers and relative positions of the pipes or cables.
- Backfilling and disposal are deemed to be included (VI(a).C.4). Filling materials to be stated (VI(a).S.4).

Break up and remove concrete, blockwork and masonry work

VI(a).4 Extra over any types of excavation – breaking up and removing brickwork, masonry work and concrete met with in excavation

A contractor may need to excavate existing concrete pavings, brickwork and the like on or under the ground. Excavation of these existing structures are classified as below, and each measured separately as extra over the respective type of excavation. Note that as stipulated in the SMM, the extra over measurement is taken irrespective of the excavation depth (VI(a).4.*.0.1), and any additional cost of disposal is deemed to be included (VI(a).C.5).

- pavings, surface concrete and the like (measured in m^2 with the thickness stated if known).
- brickwork or blockwork met with in excavation (measured in m^3).
- stone masonry walls, slab foundations and the like met with in excavation (measured in m^3).
- unreinforced concrete and the like met with in excavation (measured in m^3).
- reinforced concrete met with in excavations (measured in m^3).

Points to note

Extra over items – SMM stipulates that some items have to be measured as extra over other items of work previously measured. The extra over item will not be priced at the full value of the labour and materials required. It is understood that the price for the extra over item is only the nett additional cost involved on top of the previous item that the extra over item has referred to. For instance, pipes are measured in metres whereas the fittings such as bends are measured in number as extra over items. The full price for a bend ($\$B$) is more expensive than the same length of pipe, which can be considered as the price for the supply and installation of the same length of pipe ($\$p$) plus the additional cost for the bend ($\$b$), i.e. $B = p + b$. When pricing the bend as an extra over item, we will enter $\$b$ (per number) as the unit rate in the BQ.

Question 12.7

If a plain concrete slab lies in a trench excavation (to its full length of 10m long) as the diagram shows, how should we take-off the excavation work?

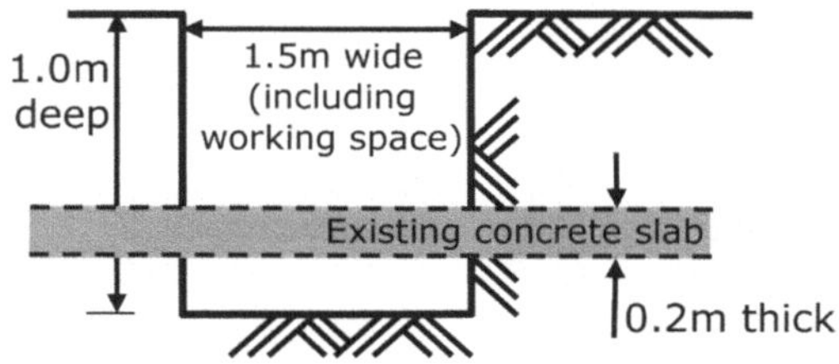

Earthwork support

VI(a).5.1 Temporary shoring and supports for sides of excavation into banks for retaining walls with face height > 1.50m (as shown in **Figure 12.6**)

- Measured as an item.
- State the minimum and maximum face heights above the reduced level.
- State the approximate overall length.
- Note that soil supports in the form of planking and strutting applied to an excavation with face height ≤ 1.5m is deemed to be included in the excavation item (VI.(a).D.1 and C.3(a)).

Disposal

VI(a).6.1–6.2 Disposal of excavated material and spoil heaps left by the previous contractor

- Measured in m^3.
- Multi-handling to and from temporary spoil heaps is deemed to be included.
- All charges in connection with disposal including the dumping charges as stipulated under the construction waste disposal charging scheme are included (VI(a).C.9(c)).

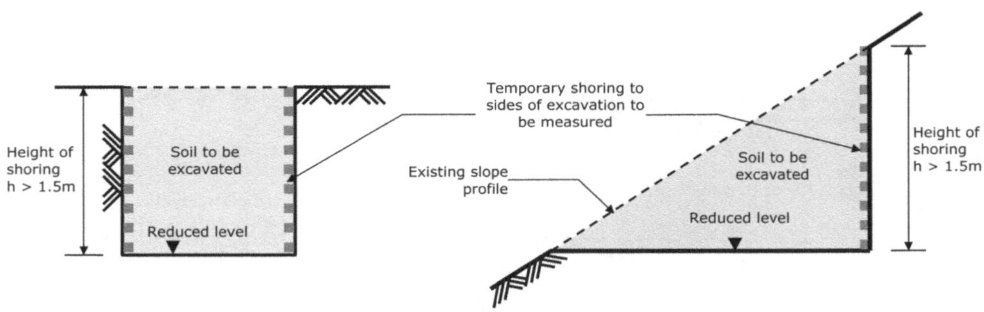

Figure 12.6 Temporary shoring and supports to sides of excavation.

- This item is only applicable to excavations where VI(a).C.4 rule does not apply. HKSMM4 VI(a).C.4 applies to small-scale excavations such as excavation for post holes ≤ 0.3m^3 and trench excavations for curbs, pipes and the like where disposal is deemed to be included in the excavation items.
- State the destination of disposal (VI(a).6.1.1–3) and each measured separately:
  - deposited on site (location specified)
  - to a dumping area away from the site provided by the employer (dumping area stated)
  - to a dumping area away from the site provided by the contractor

Backfilling and filling

VI(a).7–9 Backfilling to excavation and filling

Backfilling around walls and foundations is often required after excavation (as the shaded area in case 1, **Figure 12.7**). For filling to make up levels, it usually occurs when the existing ground level is lower than the formation level. Areas of fill will be placed with excavated material or imported filling material (cases 2 and 3 in **Figure 12.7**), in which the fill material is deposited and compacted as part of the permanent work.

- Measured in m^3.
- Backfilling to excavation, filling to make up levels and filling into planters are measured separately.
- State the source of fill materials, (VI(a).7.0.1–4), which can be:
  - Arisen from excavation
  - Obtained from materials on site
  - Obtained from specified off-site area
  - Provided by the contractor from its own source

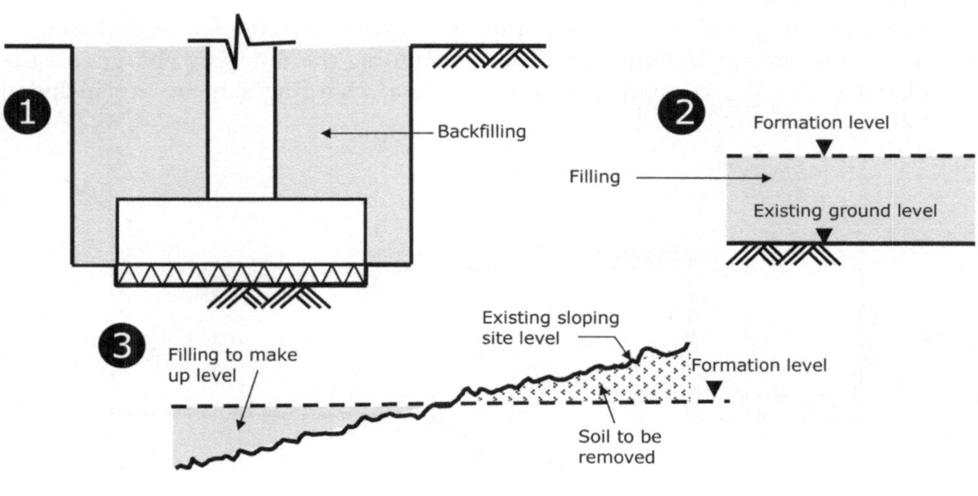

Figure 12.7 Backfilling and filling.

Points to note

In HKSMM4 Rev 2018, backfilling is no longer classified into two groups (depth $\leq$ 0.30m and >0.30m). All backfilling can be grouped together and measured in m^3 irrespective of the depths.

- Works in connection with backfilling or filling such as additional labour and material for filling in layers, spreading, compacting, levelling, watering, placing temporary boards to edges etc. are deemed to be included (VI(a).C.10).

Points to note

Soil dug out from an excavation is looser than before it was excavated. Therefore, the volume will be larger than when in the ground before. For instance, if a hole of $5m^3$ is excavated, the heap of soil may occupy $6m^3$. We refer this as bulking of soil. The volume of soil to be transported from site is therefore increased. In the case of backfilling, if a hole of $5m^3$ is to be filled with gravel, $7m^3$ loose gravel may be required because the gravel will be compacted. When measuring the quantities of disposal or backfilling of soil, the principle of measure nett quantities has to be followed. Therefore, the quantities taken should be the volume of the hole being excavated or backfilled, which is $5m^3$ in the current cases.

Question 12.8

Take-off the excavation, backfill and disposal work for the column base (plan size: 1.2m × 1.2m) as shown in the diagram.

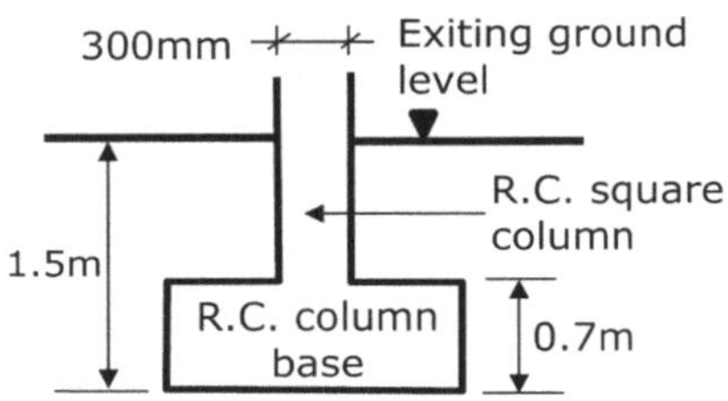

Hardcore or granular filling

VI(a).10 Hardcore or granular filling

Hardcore is usually made from well-graded gravels and crushed stones that are free from weeds, roots, vegetable soil or other unstable materials to give a dense, compact mass to:

- prevent the rising of underground water to the ground floor; or
- make up levels to provide a level base on which to lay a ground floor slab or to provide a firm base to carry construction traffic.
- Describe the kind of material used (VI(a).S.7).

- Measured separately according to the following groups:
  - beds
  - vertical beds
  - beds sloping > 15° from the horizontal
  - filling to make up levels
- If depth ≤ 0.30 m (VI(a).10.1):
  - Measured in m^2.
  - State the thickness.
- If depth > 0.30 m (VI(a).10.2):
  - Measured in m^3.
  - Where part exceeds 0.30m in depth, the whole is measured in m^3.

Blinding to hardcore surfaces

VI(a).12 Blinding to surfaces of hardcore or other filling materials

The function of the blinding layer is to prevent the wet concrete running down between the crushed stones of the hardcore layer and to provide a level base for further construction. Weak concrete, sand, or quarry dust can be used to blind, or seal, the top surface of the hardcore.

- Only the blinding layer that uses fine-graded materials such as sand is measured according to this rule, VI(a).12.
- Measured in m^2.
- State the materials used.

Reminder

If the blinding layer material is weak concrete, the item will be measured under the concrete trade in m^3.

Preparation of ground or slope surfaces

Surface treatments

VI(a).13.1–13.3 Formation of slopes in fillings, to cuttings or trimming rock to produce fair face

- Measured in m^2.
- The measured area should be the finished surface area of the slope.

VI(a).13.4 Preparation of existing ground or subsoil areas for application of top soil for turfing

- Measured in m^2.
- State the method of preparation.

VI(a).13.5 Consolidation of sub-grade for roads and pavings

- Measured in m^2.
- State the method of compaction, e.g. by a vibratory roller.

VI(a).13.6 Herbicide, selective weed killer or insecticide treatments

- Herbicides are chemicals applied over the area to stop the growth of vegetation.
- Measured in m^2.

Turfing and grass seeding

To plant a lawn, sowing grass seed over the area is the simplest way. There are different methods to plant grass seed such as hydroseeding and broadcast seeding. Another option for planting a lawn is turfing. Turfing is the direct application of grass with developed roots. Since the grass is already developed, it will grow easier. The grass roots will extend into the soil for strengthening.

VI(a).14 Turfing and grass seeding

- Measurement rules of this part must follow the Landscaping Section (XXII(b))

XXII(b).1 Soiling for turfing or grass seeding

- Measured in m^2.
- State the average thickness after compaction.
- State the type of soil, e.g. vegetable soil and method of preparation.

XXII(b).2 Turfing

- Measured in m^2.
- State the types of grass.
- State the details of pre-turfing and the like.

XXII(b).3 Broadcast seeding for grassed areas

- Broadcast seeding involves scattering seeds, by hand or mechanically, over a relatively large area.
- Measured in m^2.
- State the grass seed mix and the rate of application.
- State the details of pre-seeding fertilisers and the like.

XXII(b).4 Hydroseeding for grassed areas

- Hydroseeding involves spraying a slurry of seed, mulch (materials applied to the soil surface to retain moisture and improve fertility), fertiliser and water over prepared ground in a uniform layer. The grass seed will grow and the grass root will serve as reinforcing fibre to hold the surface soil tightly.
- Measured in m^2.
- State the details of seed mix, mulch and the like, as well as the rate and method of application.
- State the details of protective layer and the method of securing it in position.

Chunam surfacing

VI(a).15 Chunam surfacing to slopes

Chunam surfacing is a slope protection method used extensively in the past by applying a mixture of lime-cement plaster on the slope surface.

VI(a).15.1–15.2 Chunam surfacing

- Measured separately according to the following groups:
  - Sloping gradient ≤ 1 in 7
  - Sloping gradient > 1 in 7
- Measured in m^2.
- State the details of mix and thickness.
- Scaffolding and hoisting of materials are deemed to be included.

VI(a).15.3 Formation of weep holes

- Measured in number.
- Describe the details such as material, size and length.
- Scaffolding and hoisting of materials are deemed to be included.

Sprayed concrete surfacing

VI(a).16 Sprayed concrete to surfaces of slopes

Sprayed concrete (often regarded as shotcrete) is to spray cementitious mixture or concrete to the slope surface for stabilisation.

- Measured in m^2.
- State the mix and thickness.
- Steel mesh reinforcement, if used, can be included in the description or measure it separately in accordance with Section VII (c).5, Fabric reinforcement.

Bulk checking on excavation, backfilling/ filling and disposal quantities

Bulk checking is often recommended when measured items involve large quantities or major cost (Civil Engineering and Development Department, 2018). In the case of measuring the excavation trade, bulk checking on the quantities of excavation, backfilling, filling and disposal is very common as the quantities involved are usually large and the procedures are straightforward.

Diagrammatically, excavation, backfilling and disposal can be illustrated as **Figure 12.8** below:

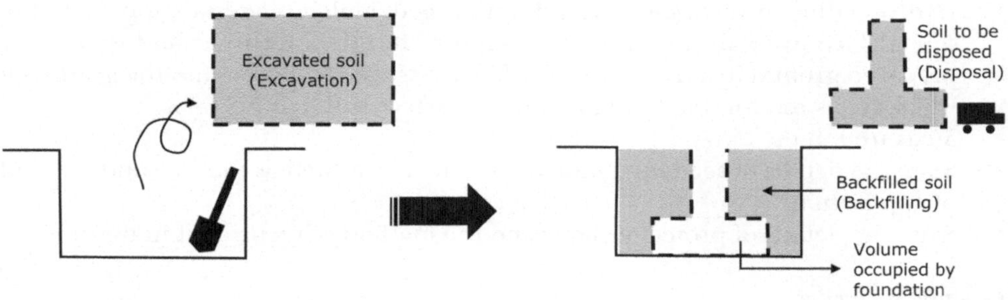

Figure 12.8 Relationship between excavation, backfilling and disposal quantities.

Whilst filling may be required to make up levels, the relationship between excavation, backfilling, filling and disposal can be expressed by the following equation:

Volume of excavation = volume of backfilling + volume of filling + volume of disposal

By abstracting the quantities from the taking-off, we can bulk check the quantities to see if there is any error. The following examples illustrate the bulk check in a simple spreadsheet format.

Example 1: If excavation volume is larger than backfilling/filling volume

Excavation (m^3)		*Backfilling / Filling (m^3)*		*Disposal (m^3)*	
Oversite excavation:	300	Filling to make up level:	160		140
Sub-total:	**300**	**Sub-total:**	**160**	**Sub-total:**	**140**
Foundation excavation:		Backfilling:			
Portion 1	700		550		150
Portion 2	350		200		150
Portion 3	200		150		50
Sub-total:	**1,250**	**Sub-total:**	**900**	**Sub-total:**	**350**
		Total backfilling / filling:	**1,060**	**Total disposal:**	**490**
Total (Excavation):	**1,550**	**Total (Backfilling / Filling + Disposal):**			**1,550**

Example 2: If backfilling/filling volume is greater than excavation volume

Excavation (m^3)		*Backfilling / Filling (m^3)*		*Disposal (m^3)*	
Oversite excavation:	300	Filling to make up level:	760		–460
Sub-total:	**300**	**Sub-total:**	**760**	**Sub-total:**	**–460**
Foundation excavation:					
Portion 1	700		550		150
Portion 2	350		200		150
Portion 3	200		150		50
Sub-total:	**1,250**	**Sub-total:**	**900**	**Sub-total:**	**350**
		Total backfilling / filling:	**1,660**	**Total disposal:**	**–110**
Total (Excavation):	**1,550**	**Total (Backfilling / Filling + Disposal):**			**1,550**

In example 2, the volume of oversite excavation (300m^3) is not sufficient to cover the volume required for filling. By using the 350m^3 surplus soil from the foundation excavation (which is 1,250m^3 – 900m^3) to make up the level, there is still a shortfall of 110m^3 soil. The contractor needs to import filling material of 110 m^3 to the site, which is denoted by a negative sign.

Reminder

'Filling to make up levels with excavated material' (300m^3 + 350m^3 = 650m^3) and 'filling to make up levels with imported materials obtained from specified off site area' (in 110m^3) will appear as two separate items in the BQ.

Suggested answers

Question 12.1

No, we do not need to allow working space if the foundation does not involve any formwork for the concrete construction.

Question 12.2

600mm working space (the white area) should be allowed on both sides of the footing.

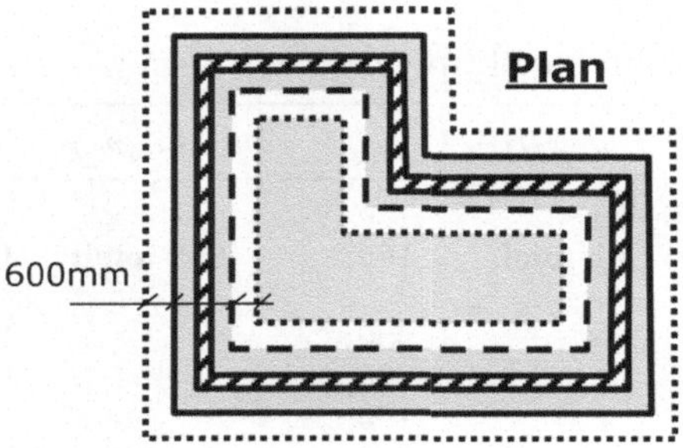

Question 12.3

When considering excavation and working space allowances, only profiles A and B are possible (sloping profile is ignored as it is irrelevant here).

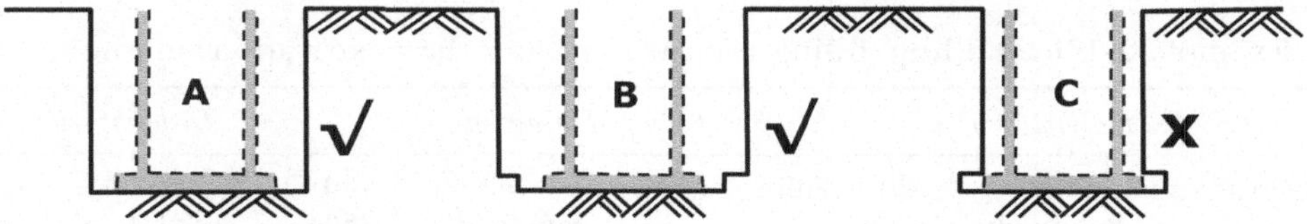

Case 1A: 250mm working space measured from Face A for the full height of excavation ('x' + 200mm).

Case 1B: 600mm working space measured from Face B for height 'x'. No working space for the slab (assuming projection of the slab is < 600mm).

Case 2A: 600mm working space measured from Face A for the full height of excavation.

Case 2B: 600mm working space measured from Face A for the full height of excavation.

Case 3A: 250mm working space measured from Face A for the full height of excavation.

Case 3B: 600mm working space measured from Face B for height 'x' and then 250mm working space measured from Face A for the slab (200mm thick).

Question 12.4

i Excavations for swimming pools, cellars etc. are under the category of 'basement and similar excavations'.

ii No, boring holes for piles is neither excavation for basements and similar nor excavation for isolated piers. It is measured under Section V, Piling and Caissons.

Question 12.5

No, after oversite excavation, the commencing level of basement excavation should be described as 'reduced level'.

Question 12.6

i 0.25m working space is allowed on each side of the column base and therefore the plan area of excavation becomes 3.50m × 3.50m.

ii Taking-off of the excavation is:

	3.50 3.50 1.50	Excav for col base comm. at existing grd lev. n.e. 1.5m dp	*Whilst the total depth of excavation is 4.0m and the excavation work is measured in successive stages of 1.5m deep, we need three items to measure. Note that the description of each item is kept at 1.50m successive stages (so the last stage is not '3.0–4.0m deep'). Also note that the dimension for the depth of the last item should be the remaining depth after the first two stages, which is 1.0m (not 1.50m). This gives the total quantity of excavation exactly as 3.50 × 3.50 × 4.00m³.*
	3.50 3.50 1.50	Ditto 1.5 – 3.0m dp.	
	3.50 3.50 1.00	Ditto 3.0 – 4.5m dp.	

Question 12.7

	10.00 1.50 1.00	Exc. surf. tr. comm. at existing grd lev. n.e. 1.5m dp	*When measuring the extra over item here, no deduction of concrete volume from the excavation quantity should be made. The volume measured for the trench excavation remains as 10m × 1.5m × 1m.*
	10.00 1.50 0.20	E.o. excavn and disposal at any depth for breaking up and removing unreinf. conc. slab met with in excavn	

Question 12.8

	2.40 2.40 1.50	Exc. col. base comm. at existing grd lev. n.e. 1.5m dp & Backfill sel. exc. matl to excavn	*600mm working space is allowed as the excavation depth and the height of formwork both exceed 600mm.* *It is a common practice to deal with backfill and disposal in this manner. First take-off the backfill in the same volume as the excavation. Then adjust the volume of backfill when measuring the underground features such as column base and column. At the same time, the volume of disposal is added based on the volume of underground features.* *With this approach, we can avoid calculating the irregular-shaped volume of backfill or disposal.* *However, there are also situations where it is more straightforward to measure the full volume of disposal first, and then adjust it when measuring the underground features such as foundation. An example can be found in the* ***Worked Example 19.4*** *of* ***Chapter 19*** *when dealing with surface excavation.*
	1.20 1.20 0.70	R.C. grade */* to col. base & <u>Ddt</u> Backfill a.b. & <u>Add</u> Disposal of exc. matl away from site, depositing to dumping area provided by the Employer	
	0.30 0.30 0.80	R.C. grade */* to col. & <u>Ddt</u> Backfill a.b. & <u>Add</u> Disposal of exc. matl a.b.	

13 Measurement of concrete works

Introduction

Concrete works is one of the most important trades in building measurement as reinforced concrete is the dominant structural material used in Hong Kong. Major work items covered in Concrete Works, Section VII, include:

- Concrete
- Gun applied reinforced concrete
- Reinforcement
- Formwork
- Precast concrete works
- Prestressed concrete works
- Precast prestressed concrete works

In this chapter, measurement rules for concrete, reinforcement, formwork and precast concrete will be elaborated in detail. Since prestressed concrete is less popular in building works, prestressed concrete and precast prestressed concrete will not be discussed here.

Concrete

Concrete in general

VII(a).1 Concrete

Measurement of concrete is quite straightforward, except that concrete works has to be classified and measured separately according to their mixes, types and components.

- Measured in m^3 unless otherwise stated.
- Different concrete mixes should be measured separately.
- The following concrete types and components should be measured separately:

Concrete Types	*Concrete Components*
• Reinforced concrete • Plain concrete • Watertight reinforced concrete • Concrete in casings to structural steelwork • Precast concrete • Prestressed in-situ concrete • Precast prestressed concrete • Lightweight concrete • Gun applied reinforced concrete (measured in m^2)	• Blinding layer laid under foundations and beams with thickness stated • Small bases for fencing posts, etc. • Beds (including on-grade slabs) • Water channels • Foundations • Column bases • Pile caps • Suspended slabs and roofs • Horizontal • Coffered and troughed slabs as shown in **Figure 13.1** • Transfer plates • Beams • Ground beams • Transfer beams • Walls • Columns, piers etc. • Steps and staircases • Curbs • Copings • Projecting cills • Strings

VII(a).C.1 Concrete items are deemed to include:

- Concrete hoisting, placing in position, vibration, curing, protection and complying with temperature control requirements
- Forming holes for pipes and the like ≤ 150mm diameter, and openings ≤ $0.50m^2$ sectional area
- Forming construction joints and provision of approved waterstops (except those joints that are designed by the engineer)
- Provision of test cubes and test certificates

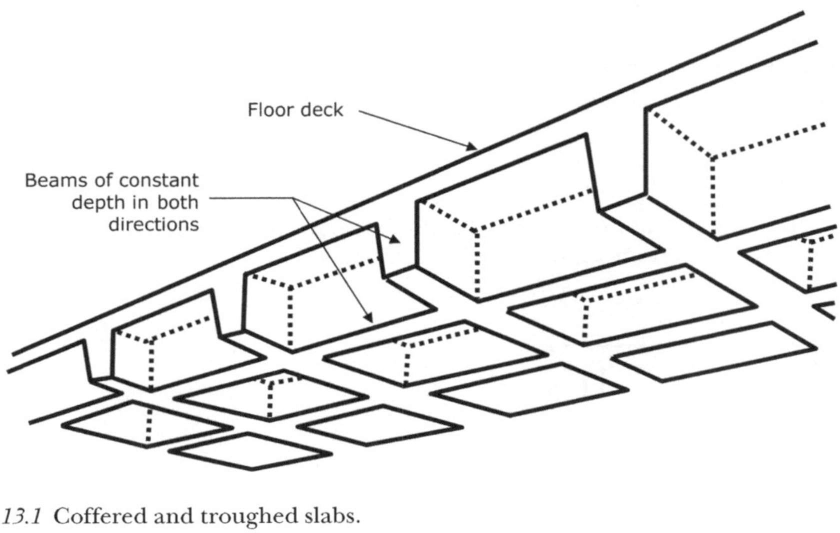

Figure 13.1 Coffered and troughed slabs.

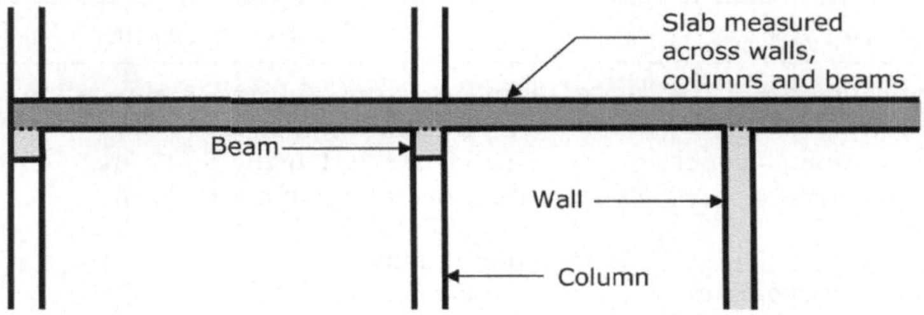

Figure 13.2 Measurement of concrete in slabs and beams.

VII(a).M.10 No deductions of concrete volume for the following:

- reinforcement or structural steel sections which are put inside concrete members
- voids ≤ 0.10m³ when concrete work is measured in volume
- openings in walls, floors, roof slabs and the like ≤ 0.50m²

Points to note

The deduction rules for voids and openings in concrete works has been revised to larger sizes in the HKSMM4 Rev 2018.

Concrete in beams and slabs

VII(a).17 Suspended slabs

Slabs are taken across walls, columns and beams (as shown in **Figure 13.2**), unless the concrete mix for walls, columns and beams is different from that of the slab (VII(a).M.12).

- Measured in m³.
- State the thickness of slab. In case of coffered and troughed slabs, state the overall thickness.
- Horizontal slabs, sloping ≤ 15° and sloping > 15° (with respect to the top surface), should be measured separately.

Points to note

For coffered and troughed slabs, if the margin is > 500mm wide, the shaded portion will be measured as ordinary slabs (VII(a).D.1).

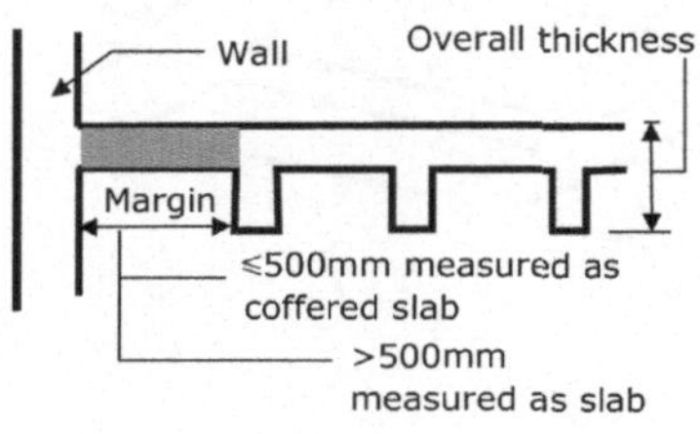

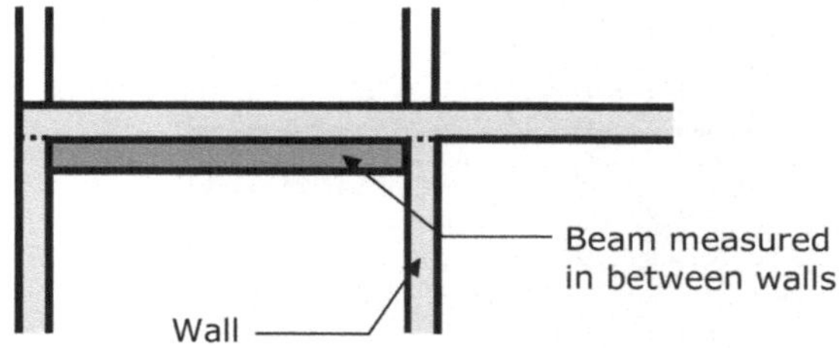

Figure 13.3 Measurement of concrete in beams between walls.

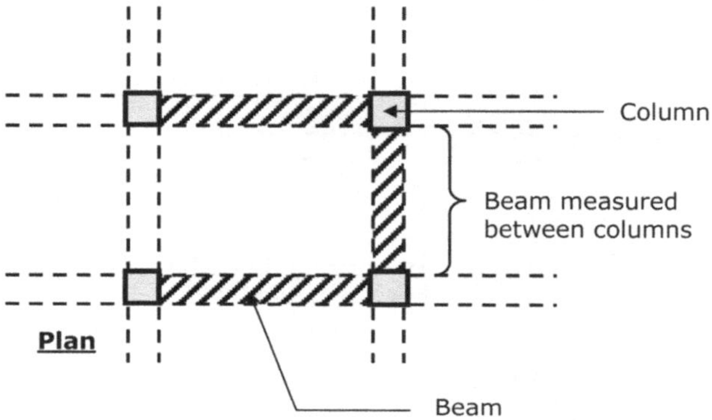

Figure 13.4 Measurement of concrete in beams between columns.

VII(a).M.14 Beams

Beams are measured below slabs and between walls (as shown in **Figure 13.3**) and columns (**Figure 13.4**).

- Measured in m^3.
- Horizontal beams, sloping $\leq 15°$ and sloping $> 15°$ should be measured separately.

VII(a).M.13 Shoulders

If the concrete grade for walls or columns is richer than that for slabs and beams, shoulders of a richer mix will be required as shown in **Figure 13.5**. The shoulders in these concrete items should be so described and any additional formwork or mesh required to form the shoulders is deemed to be included.

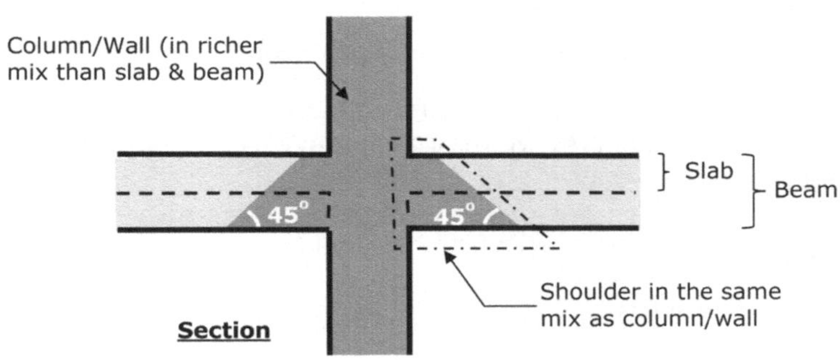

Figure 13.5 Typical shoulder at the junction of column/wall and beam/slab.

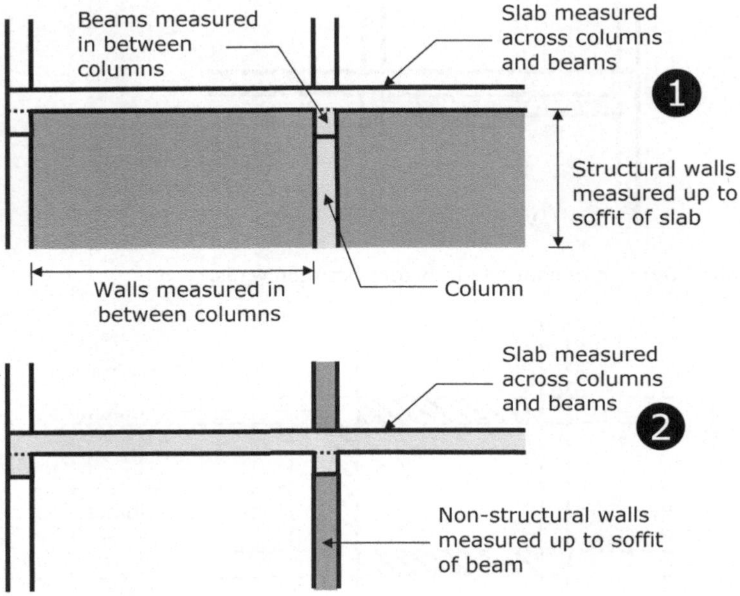

Figure 13.6 Measurement of concrete in walls.

> **Reminder**
>
> If the slabs/beams are having a different (weaker) mix (as in the case of **Figure 13.5**, concrete to the columns/walls will be taken through the slabs.

Concrete in walls and columns

VII(a).22–23 Walls and columns

Concrete walls are measured between columns or projections as illustrated in **Figure 13.6** case 1. Following the rules for concrete slabs and beams, concrete walls and columns are measured up to the soffit of slabs or beams (**Figure 13.6),** except where the beams or slabs are of a different mix (VII(a).M.16).

- Measured in m^3.
- State the thickness of concrete walls.
- Walls and retaining walls are measured separately.

VII(a).M.17 Distinction between walls and columns

As shown in **Figure 13.7**, if the width of a column exceeds four times its thickness, it is classified as a wall.

Concrete in stairs

VII(a).24 Concrete in stairs

- All concrete in stairs, including steps, support beams, landings and the like are measured together in the item of stairs (VII(a).M.18). However, landings that form part of the main floor slabs are regarded as suspended floors (as shown in **Figure 13.8**).

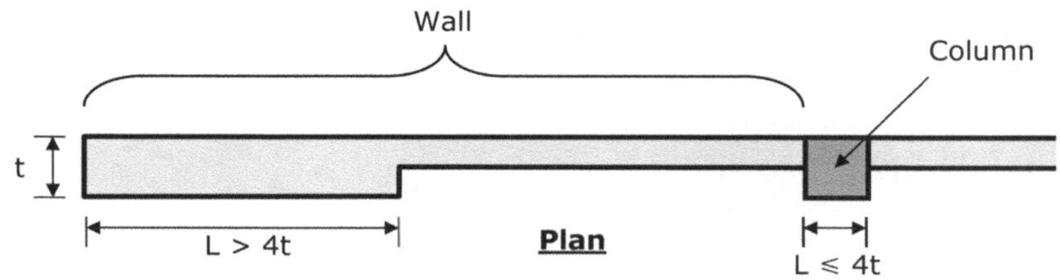

Figure 13.7 Distinction between walls and columns in measurement of concrete.

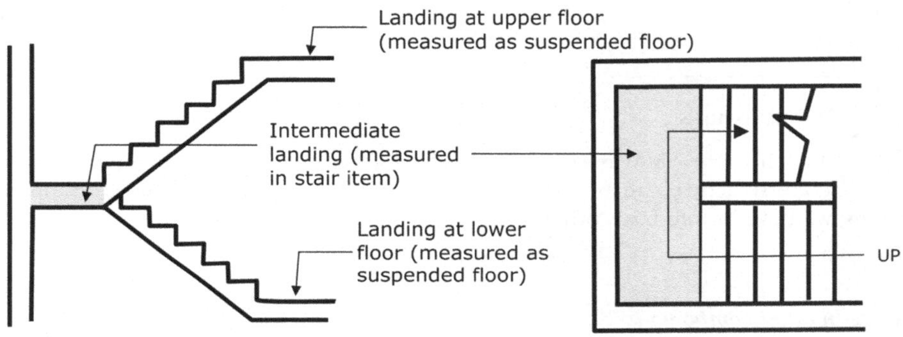

Figure 13.8 Concrete staircase landings.

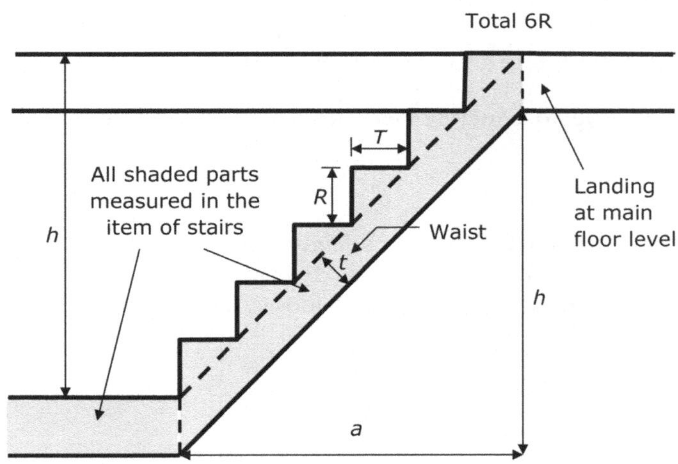

Figure 13.9 Concrete staircase.

- Referring to **Figure 13.9**, concrete to stair consists of three main parts: the waist, the steps and the landing. Measurement of concrete to landing is similar to the measurement of concrete to suspended slab. For the measurement of concrete volume to waist and steps, the following formulae can be used:

$$\text{Waist} = \sqrt{a^2 + h^2}(w)\,(t)$$

Where

w = width of stair on plan

$$Steps = \frac{1}{2}(T)(R)(w)(n)$$

Where

w = width of stair on plan
n = number of risers
$T = a/n$
$R = h/n$

Points to note

To follow the SMM measurement rules for various concrete members, a concrete structure can be measured in the following sequence: slabs, columns, beams, walls, stairs and lastly other items.

Concrete in other components

VII(a).8–10 Foundations, isolated column bases and pile caps

- Each type of component is measured separately in m^3.
- Note that in case of small bases for fencing posts and the like, they should be measured in number with the sizes stated (VII(a).2)

VII(a).4 Thickening to concrete beds under walls, partitions and the like and under channels

- Measured in m.
- State width and thickness. Referred to **Figure 13.10**, the average width $(w1 + w2)/2$ and thickness t should be stated in the description.
- Deemed to include extra excavation, disposal, hardcore and formwork.

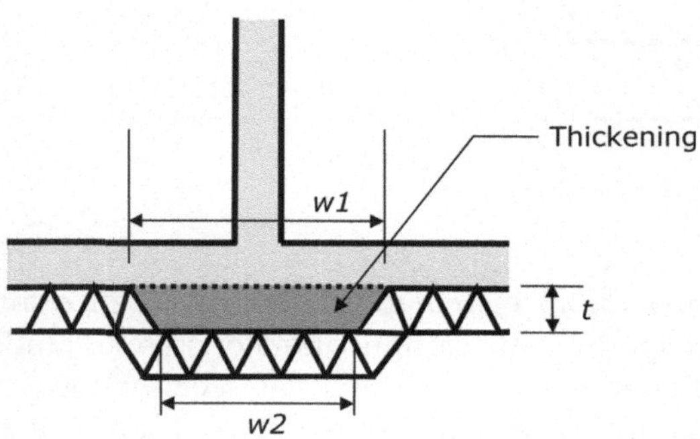

Figure 13.10 Concrete thickening under walls.

VII(a).5–7 Surface water channels, stepped channels to slopes and channels formed in concrete beds

Building steps into the drain or channel (**Figure 13.11**) can help to reduce the force of the falling water. Stepped channels are quite expensive and therefore, they are mainly used on steep slopes.

- All channels are measured in m.
- Surface water channels, stepped channels to slopes, channels formed in concrete beds (as shown in **Figure 13.11**), straight channels and curved channels shall each be measured separately.
- For surface and stepped channels, state the shape, width, thickness of sides and base and average depth; describe excavation, disposal, formwork and rendering.
- For stepped channels to slopes, state also the height of steps and gradient of channel between steps; describe also weep holes in steps.
- For channels formed in concrete beds, state the shape, width and average depth; describe the formwork.

Miscellaneous

VII(a).M.4 Treatment of the face of concrete beyond ordinary depositing, spreading and levelling

This is usually found in treating surfaces of concrete floor by trowelling or power floating. Trowelling by brush or bamboo broom can produce a slip-resistant surface which is often found in pavement surfaces. Power floating finishes the concrete surface using a power float, which is a concrete finishing machine to smooth and level the surface of the concrete.

- Measured in m^2.

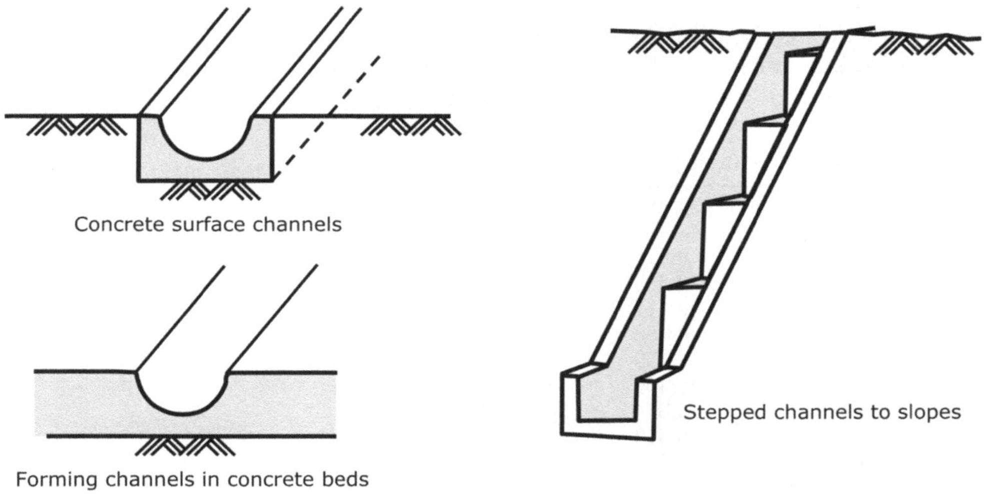

Figure 13.11 Different types of channels.

Reinforcement

VII(c) Reinforcement

- Reinforcing bars shall be measured separately for:
  - Different types of steel, e.g. high yield steel (often denoted as 'Y' (410 N/mm^2) or 'T' (460 N/mm^2) on drawings) and mild steel (denoted as 'R' (250 N/mm^2)) (VII(c).M.1)
  - Different bar size
  - Different nature of reinforcing bars, e.g. general reinforcing bars, links, stirrups, helical reinforcement and curved bars (VII(c).2.1.1–1.6)
  - Bars exceeding 12.00m long, measured in stages of 3.0m (VII(c).2.1.2)
  - Fabric reinforcement (VII(c).5)

Reminder

Unlike measurement of concrete work, reinforcing bars in different structural members such as foundations, walls, columns, beams etc. are not measured separately.

Reinforcement in general

Conversion of bar length to weight

All reinforcement shall be measured in kg (except for fabric reinforcement which is measured in m^2). However, we cannot measure the weight directly from drawings. We can only measure the lengths of bars and then convert the figures to weights by multiplying the length of each bar by the unit weight per meter for that bar size (as listed in **Table 13.1** below).

Allowance for concrete cover

Concrete cover is the distance measured from the exposed concrete surface to the nearest surface of the steel bar. It prevents steel bars from corrosion when exposed to air. Also, the concrete cover provides resistance against fire.

Table 13.1 Conversion Table for Reinforcing Bars.

*Nominal Diameter**	*Kg/m*	*Metre per Tonne*	*Approx. No. of 12m Bars per Tonne*
6mm	0.222	4,505	376
8mm	0.395	2,532	211
10mm	0.617	1,623	136
12mm	0.888	1,126	94
16mm	1.579	633	53
20mm	2.466	406	34
25mm	3.854	259	22
32mm	6.313	158	14
40mm	9.864	101	9

*Note: Nominal diameter refers to the bar diameter excluding deformations. Nominal diameter is used for design and weight calculations.

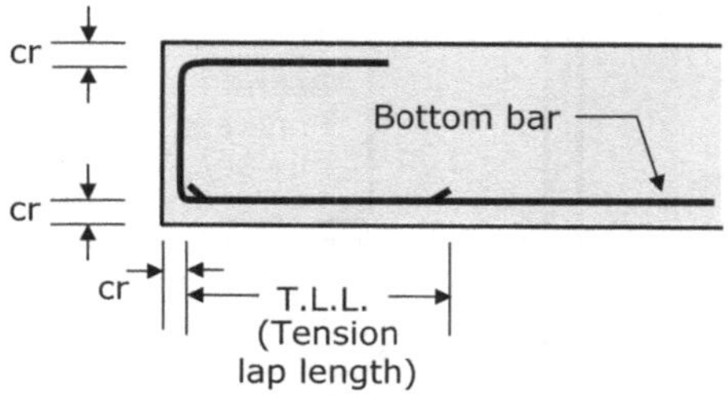

Figure 13.12 Slab without top reinforcement.

Minimum concrete cover will be applied by engineers to all rebar design, resulting in shorter/fewer rebars. Therefore, when calculating the bar length, concrete cover has to be deducted. For instance, the length of the bottom bar in **Figure 13.12** should be:

$$\text{Length of bottom bar} = L - 2cr$$

where

L = length of slab
cr = concrete cover

Allowance for lap lengths and anchorage lengths

Lap lengths and anchorage lengths are frequently applied in the calculation of bar lengths. A lap length is the length required to transfer the force from one bar to another, i.e. the length of bar overlap when two bars are tied together to extend the length. An anchorage length is the length of bar required to transfer the force in the bar to the concrete, required when one structural member (e.g. a beam) intersects another member (e.g. a column). Example of lap length and anchorage length can be found in **Fig. 13.13**.

When measuring reinforcement, the taker-off should ensure that a complete set of framing plans, typical details and general notes are available. The framing plans provide information about the clear span of slabs, beams and the like. From the typical details and general notes, minimum concrete cover, lap lengths and anchorage lengths can be identified. A thorough review and understanding of all the structural drawings is crucial to the correct measurement of reinforcement.

From **Figure 13.13**, the length of the top bar is not L, but $L + 2\ T.A.L.$ Exact lengths of the lap length and the anchorage length are dependent on the concrete grade and bar size as specified in the general notes.

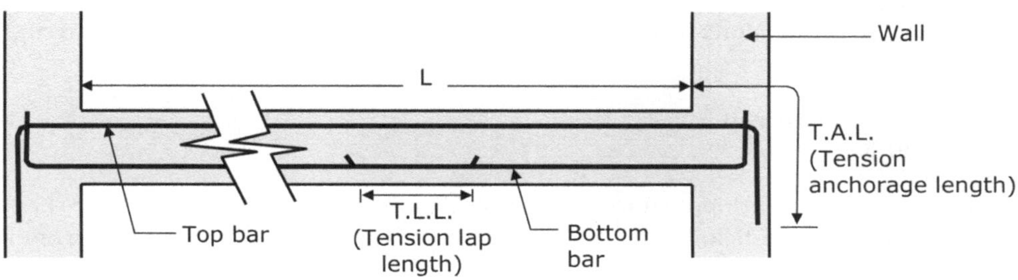

Figure 13.13 Typical sectional detail of a slab.

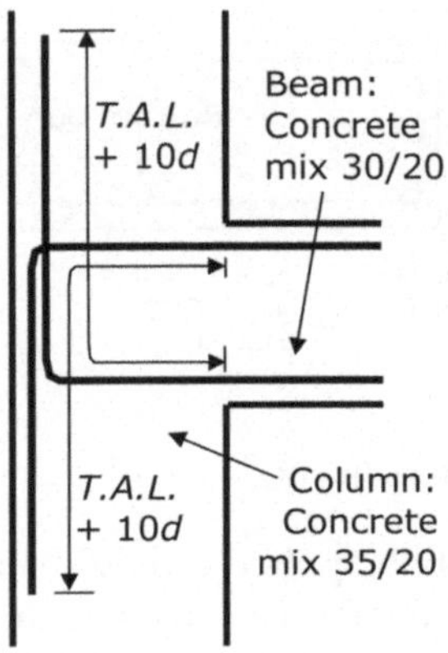

Figure 13.14 Section at column and beam.

If the concrete mix for one member is higher than that of the other member, T.A.L. should be calculated based on the stronger mix. For instance, in **Figure 13.14**, the length of T.A.L. for the beam rebar should be based on the concrete mix for the column (i.e. 35/20).

Points to note

In **Figure 13.14**, 35/20 Concrete mix represents 35MPa (grade strength) with 20mm nominal maximum aggregate size.

Calculation of bar length for different bending shapes

Reinforcing bars in concrete are bent into different shapes to increase the anchorage and strength. To calculate the exact bar lengths, BS4466:1989 or BS8666:2005 provide clear formulae for different bending shapes. Note that all the formulae in the British Standards are based on the external lengths or external perimeters of the rebar shapes. To illustrate the application of British Standards in the calculation of actual bar length, a single stirrup (**Figure 13.15**) is used as an example:

Based on BS4466 Shape Code 61,

$L = 2(A + B) + 12d$ (Note: A and B contribute to the external perimeter of the stirrup)

where

L = length of stirrup
d = bar size

Although A and B are not supplied in drawings normally, we can substitute them with H (the depth of beam) and W (the width of beam) respectively. The formula becomes:

$$L = 2(H + W - 4cr) + 12d$$

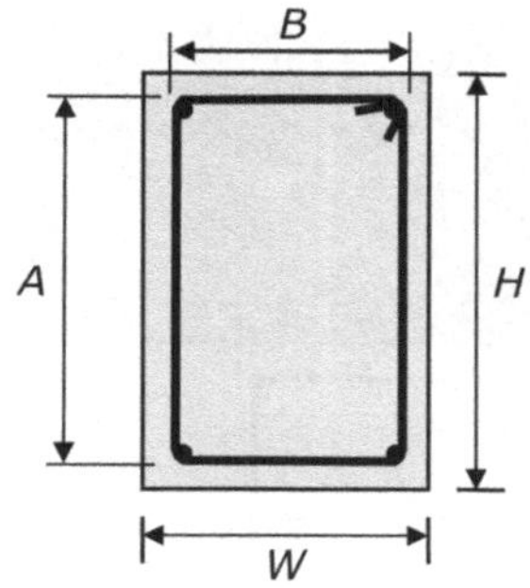

Figure 13.15 Stirrup (BS4466 Shape Code 61).

where

L = length of stirrup
H = depth of beam
W = width of beam
cr = minimum concrete cover

Number of bars

When calculating the number of bars, stirrups etc., the number is rounded up to the nearest whole unit as an industrial practice (e.g. 10.1 rounded up to 11).

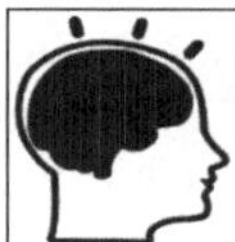

Question 13.1

How many horizontal steel bars (top bars only) should be placed in the floor slab as shown on the plan? (Given: minimum concrete cover is 50mm)

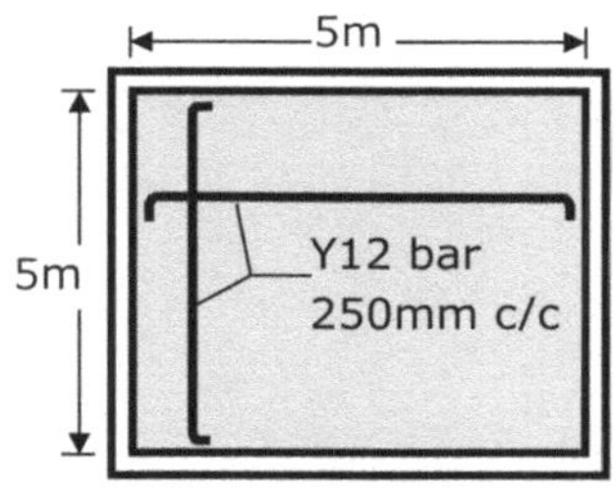

Reinforcement in slabs

On the whole, reinforcement in slabs is the easiest to measure. Except for slab-on-grade that may use fabric reinforcement, rebars in slabs normally involve top and bottom bars running in x and y directions. **Figure 13.16** shows the reinforcement details in a simple slab. The drawing shows the shape of bar and the other details such as diameter of bar, spacing etc.

The general principles for the calculation of bar lengths in slabs have been illustrated in the previous sections when discussing the basic rules for reinforcement

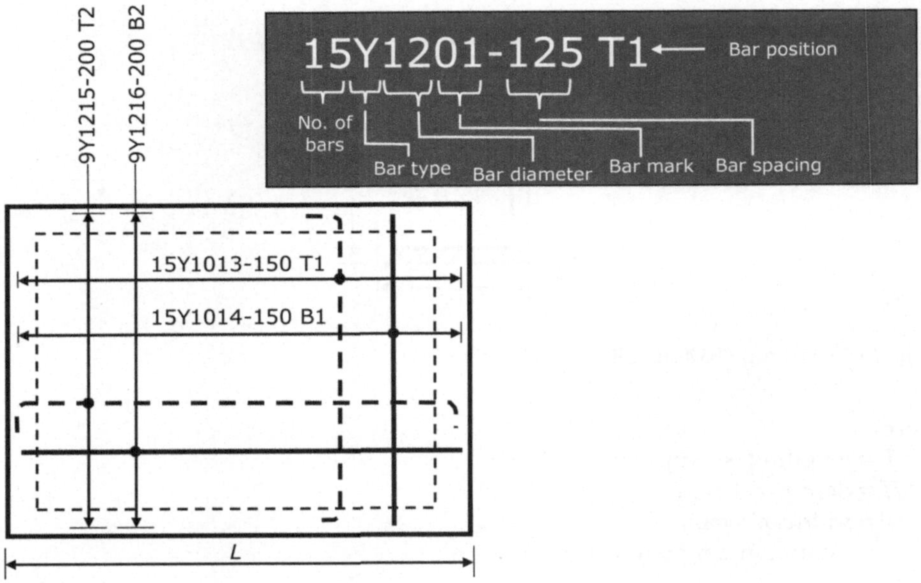

Figure 13.16 Typical reinforcement details in slab.

measurement and are therefore not repeated. Engineers will normally calculate the number of bars required and put the information against the bar. If the number of bars required is not provided, we can calculate the figure according to the drawing.

For example, assuming that the number of bars is not indicated on the drawing, the number of bars (for bar mark 13) **in Figure 13.16** should be:

$$N = \frac{L - 2cr}{s} + 1$$

where

N = number of bars in the slab
L = length of slab
cr = minimum concrete cover
s = centre-to-centre spacing of bars

Reinforcement in beams

Stirrups

Stirrups in beams can take different forms (open or closed forms), shapes and number. Two-legged stirrups (i.e. single stirrups as in **Figure 13.15**) or multiple-legged stirrups (e.g. four-legged as in **Figure 13.17**) can be found in beams to provide the desired level of shear resistance.

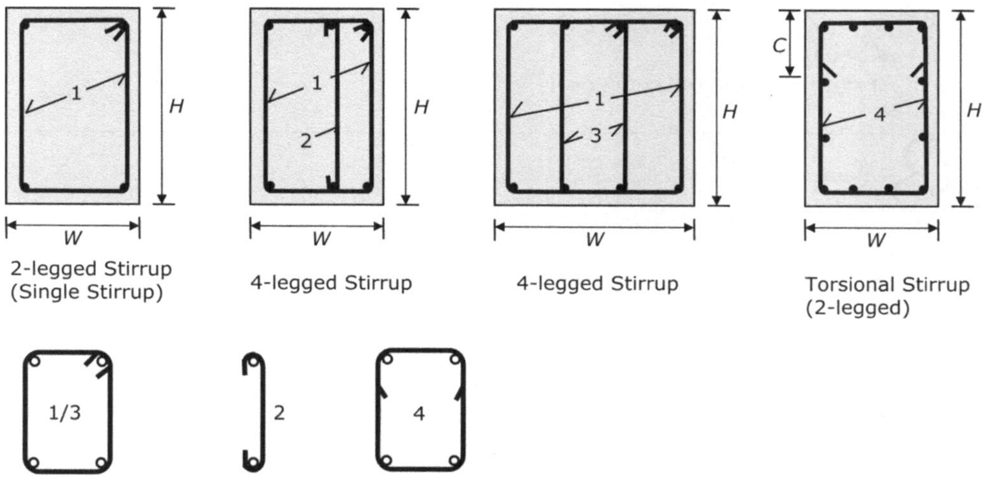

Figure 13.17 Examples of some common stirrups.

Referring to **Figure 13.17**,

- Type 1 (based on BS4466 shape code 61):

$$= 2\ (H + W - 4cr) + 12d$$

Where
d = bar size
cr = minimum concrete cover

- Type 2 (based on BS4466 shape code 33):

$$= H - 2cr + 2h$$

Where
cr = minimum concrete cover
h = constant value h based on BS4466 Table 3

Points to note

Table 3 in BS4466 lists out the lengths for various constants (r, n and h) used in the standard. The grade R steel in the table refers to grade 250 plain steel bars, i.e. mild steel bars. The grade T steel refers to grade 460 high yield steel bars.

- Type 3 (based on BS4466 shape code 61):

$$= 2\left(H - 2cr + \frac{W - 2cr - d}{3} + d\right) + 12d$$

Where

d = bar size

cr = minimum concrete cover

Reminder

The length of 'x' in Type 3 stirrup is $[(w-2cr-d)/3]+d$ as illustrated below.

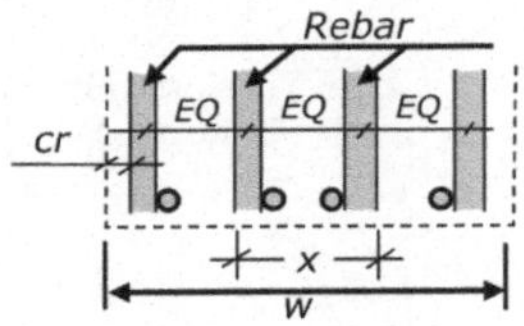

- Type 4 (based on B8666 shape code 63):

$$= 2H + 3W + 2C - 12cr - 3r - 6d$$

Where

d = bar size

r = minimum radius for scheduling (Refer to BS8666 Table 2)

Question 13.2

If all the stirrups are the same diameter and steel type, is it necessary to separately measure the various types of stirrups in **Figure 13.17** as 'single stirrups', 'double stirrups' and 'torsional stirrups'?

When calculating the number of stirrups required, spacings at ends (denoted as 'a' in **Figure 13.18**) have to be deducted appropriately. For instance, in **Figure 13.18**, the number of stirrups required for the secondary beam (the shaded part) is:

$$N = \frac{L - 2a}{s} + 1$$

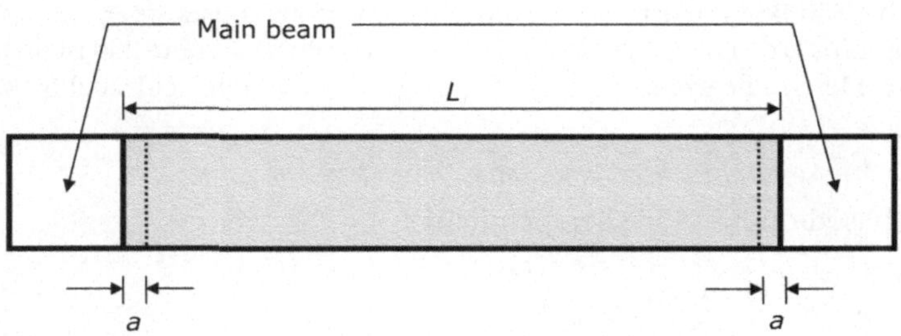

Figure 13.18 Section of a secondary beam with main beams at ends.

where

N = no. of stirrups
L = length of beam
a = spacing as specified in the detailed drawings or the same as concrete cover
s = centre-to-centre spacing between each stirrup

Reinforcement in columns

Vertical bars

When calculating the length of vertical bars for columns, it is recommended to measure it floor by floor rather than from pile cap to roof in one calculation. **Figure 13.19** shows an example of rebar design in interior columns and perimeter columns.

Points to note

Referring to **Figure 13.19**, the cranks in vertical rebars (the slight bending at the laps) are usually designed with a slope of 1:10 (or 1:12 in ASD standard design) and a minimum length of 300 mm. Since the bending angle is small, the crank can be considered as straight in the length calculation. However, if dealing with a larger bending angle such as the cranked bars in slabs or beams, shape code 41 of BS4466 should be referred to calculate the crank length.

Case 1 (interior columns):
Length of vertical bar:

$$L = H + T.L.L.$$

where

L = length of vertical bar
H = floor-to-floor height
$T.L.L.$ = tension lap length as specified on the drawings

Length of bent vertical bar:

$$L = H + T.A.L.$$

where

$T.A.L.$ = tension anchorage length

Length of dowel bar:

$$L' = T.A.L. + t + T.L.L.$$

where

L' = length of dowel bar
t = thickness of slab

Case 2 (perimeter columns):
Length of vertical bar:

$$L = H + T.L.L.$$

where

L = length of vertical bar
H = floor-to-floor height
$T.L.L.$ = tension lap length as specified on the drawings

Binders

The length and the number of binders can be calculated as in the case of stirrups in a beam. However, if facing the circled situation in **Figure 13.20** where the

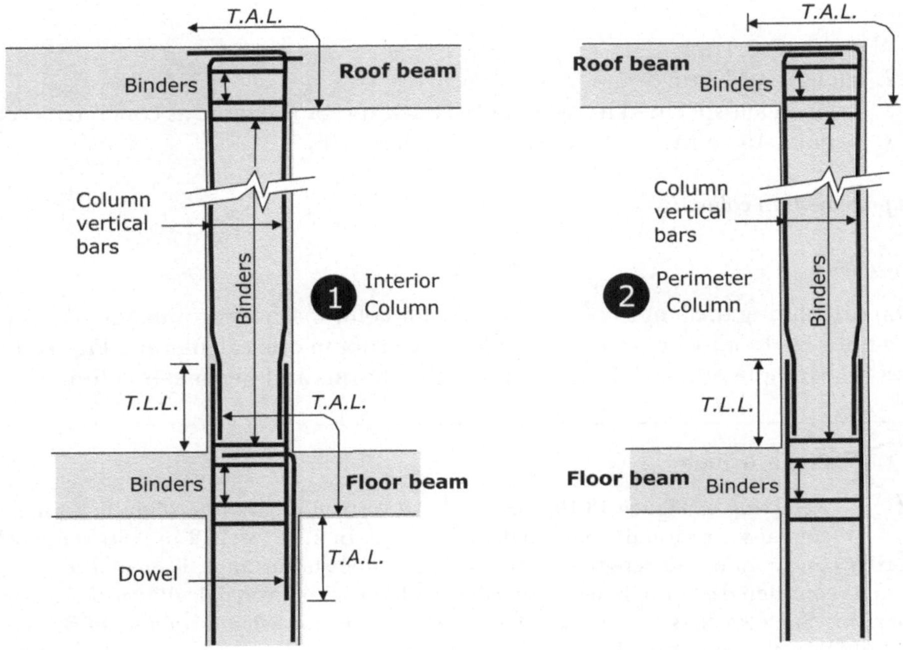

Figure 13.19 Vertical bars in columns (section).

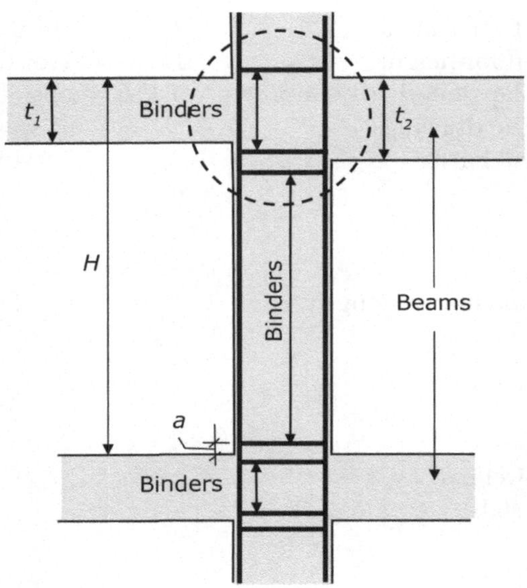

Figure 13.20 Binders in columns (section).

adjoining beams have different depths, the available height for the binders should take into account the deeper beam. As a result, the number of binders in a column should be:

$$N = \frac{H - t_2 - 2a}{s} + 1$$

where

N = no. of binders
H = floor-to-floor height
t_2 = depth of deeper beam
a = distance as specified in detailed drawings (or minimum concrete cover)
s = centre-to-centre spacing between each binder

Reinforcement for all floors can be obtained by multiplying the number of typical floors. Adjustment should be made for the different rebar layout in the roof slab.

Reinforcement in walls

Vertical bars

Reinforced concrete walls can be either structural walls or non-structural walls. The rebar details for these two types are slightly different, which can be illustrated in **Figure 13.21**.

Many reinforced concrete structural walls such as external walls and core walls extend from one floor to the next. In practice, the vertical bars of structural

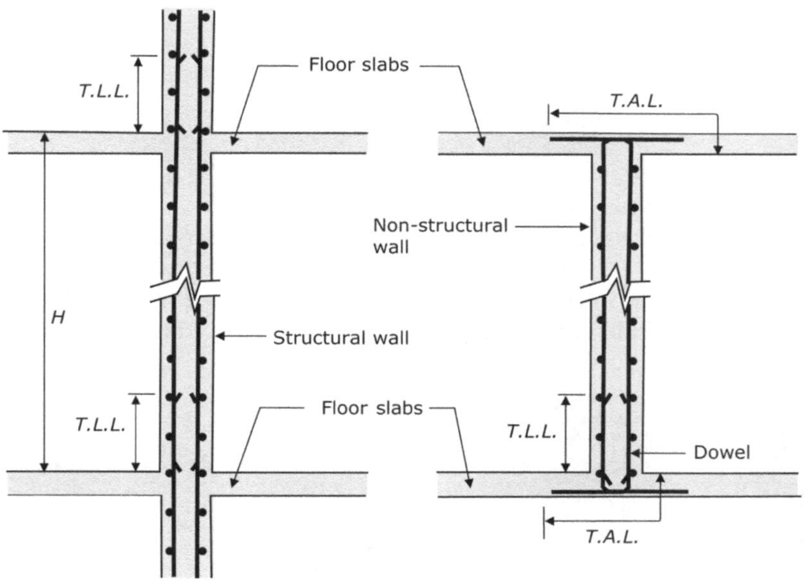

Figure 13.21 Vertical bars in concrete walls (section).

walls are fixed in a floor-by-floor manner. Non-structural walls are designed with proper anchorage to adjacent structural members such as the floor (as shown in **Figure 13.21**). Therefore, the length of vertical bars in both structural and non-structural walls should be calculated on a single floor basis.

Structural walls:

$$L = H + T.L.L.$$

where
L = length of vertical bar in structural wall
H = floor-to-floor height

Non-structural walls:

$$L = H + T.A.L. - t$$

where
L = length of vertical bar in non-structural wall
H = floor-to-floor height
t = thickness of upper floor slab

Length of dowel bar (L') in non-structural walls:

$$L' = T.L.L. + T.A.L.$$

Horizontal bars

The horizontal bars in walls are generally straight bars. At wall intersections and corners, separate hooked bars or u-bars will be used, as shown in **Figure 13.22**. The length of these bars can be calculated with reference to BS4466, which is quite straightforward. The **Worked Example 19.4** of **Chapter 19** contains an illustration of the calculation of various horizontal bar lengths and is not repeated here.

When the wall continues from one floor to the next, the number of horizontal bars in structural walls should be:

$$N = \frac{H}{s}$$

where
N = no. of horizontal bars in wall (one layer only)
H = floor-to-floor height
s = centre-to-centre spacing between each horizontal bar

Links are often used in the walls (as shown in **Figure 13.23**) to restrain the vertical bars. The length calculation for the links can be referred to the example in the

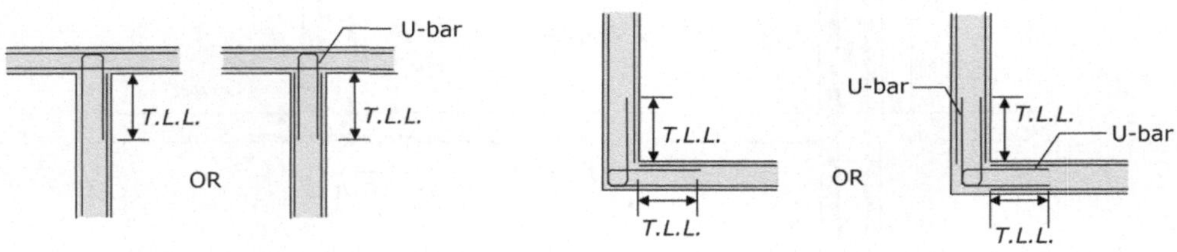

Figure 13.22 Horizontal bars at wall intersections and corners (plan).

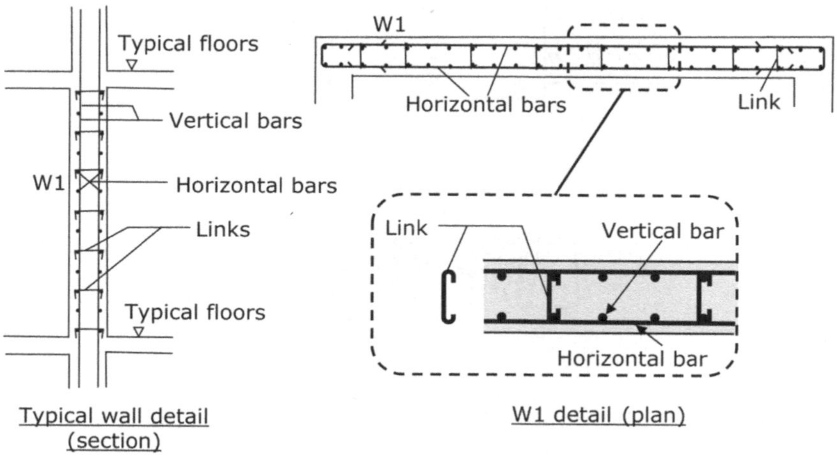

Figure 13.23 Typical detailing of links in wall.

Reinforcement in Beams section. The number of links required and their respective length can be obtained by referring to the structural drawings.

Question 13.3

How many links are required for the wall (W1) as shown in **Figure 13.23**?

Trimming bars for openings

Openings have to be formed on the walls for windows, doors or for passage of building services. If the size of the opening is large, trimming bars must be provided for strength. The maximum allowable size of an opening without trimming bars and the layout of trimming bars are usually provided in the detailed drawings. Example of trimming bars measurement can be found in the **Worked Example 19.4** of **Chapter 19**. Since the measurement of trimming bars is simple, the calculations are not elaborated here. However, it is worth noting that the calculation of trimming bars is often overlooked by takers-off.

Miscellaneous

VII(c).5 Fabric reinforcement

Fabric reinforcement is mainly used for on-grade slabs, i.e. ground slabs built on top of the soil.

- Measure the nett area in m^2 and state the weight.
- Description shall include the extra material at laps that is not measured in the quantities.
- No deduction for openings $\leq 1m^2$.

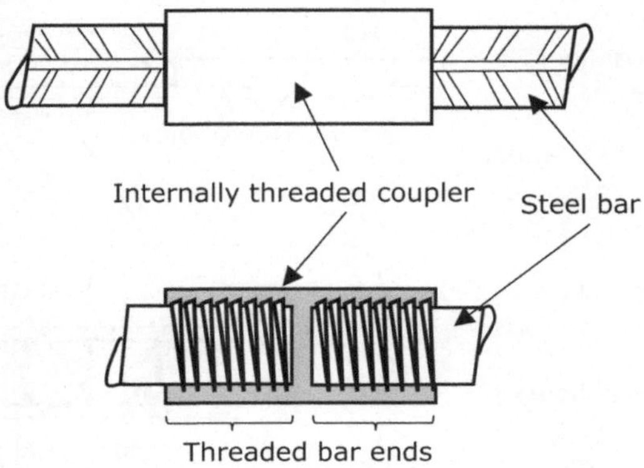

Figure 13.24 Coupler for connecting reinforcing bars.

VII(c).4 Couplers

Couplers provide a cost-effective method for connecting two reinforcing bars together, as shown in **Figure 13.24**.

- Only couplers included in engineer's design are measured.
- Measured in number with full description.
- The threading ends of the bars are deemed to be included (VII(c).C.5). However, if the couplers are to be provided by others, the threading ends of bars will be measured in number (VII(c).3).
- If couplers are used, no lap length shall be measured at the connecting ends of reinforcing bars.

Formwork

Formwork in general

VII(d).1 Formwork

- Measured in m^2 unless otherwise stated.
- Formwork is measured to the concrete surfaces of finished structure which require support during casting (VII(d).M.1).

The following types of formwork shall be measured separately:

- Standard formwork (i.e. sawn formwork)
- Formwork to produce a fair faced finish or other formed finish (VII(d).M.5)
- Left-in formwork (formwork not designed to remain in position but impossible to remove) (VII(d).M.4 & D.1)
- Permanent formwork (formwork designed to remain in position, usually performing a retaining or supporting function) (VII(d).M.4 & D.2)
- Permanent formwork to one face only (for instance, permanent formwork to one face of basement wall) (VII(d).5)

- Formwork to different members of the building:
  - Slabs: slab soffits, top formwork to slabs and soffits to coffered or troughed slabs each measured separately
  - Stairs: landing soffits, soffits of stairs, risers and strings each measured separately
  - Wall faces
  - Sides of piers, columns and stanchions
  - Sides and soffits of beams
  - Forming openings: edges and breaks in slabs, edges and breaks in walls, soffits of openings in walls and boxing for openings each measured separately
  - Sides of curbs
  - Edges and soffits of copings, cills, cornice, eaves, overhangs and the like
  - Small surfaces: cantilever ends, brackets, ends of steps and the like
  - Complex shapes (for work curved in more than one direction)
- No deductions made for openings ≤ 1.00m^2 (VII(d).M.3).
- All formwork is deemed to include forming holes for pipes and the like ≤ 150mm diameter and boxing for openings ≤ 0.50m^2 sectional area (VII(d).C.4(c)).
- All formwork is deemed to include all necessary materials and work for erection, supporting, easing, striking and removal, repropping and coating with release agents (VII(d).C.1 and C.2).

Formwork to slabs

VII(d).2–3 Soffits of slabs and landings

- Measured in m^2.
- Measured separately for slab/landing thickness ≤ 200mm and > 200mm.
- Formwork to soffits of slabs and landings that are horizontal, sloping, curved and curved in more than one direction have to be measured separately.

Normally, any beams attached to the soffit of slab are poured together with the slab. Although the formwork to slab and beam is fixed at the same time, formwork to the two members has to be measured separately. As shown in **Figure 13.25**, the formwork to soffit of slab should not include the area covered by the beams.

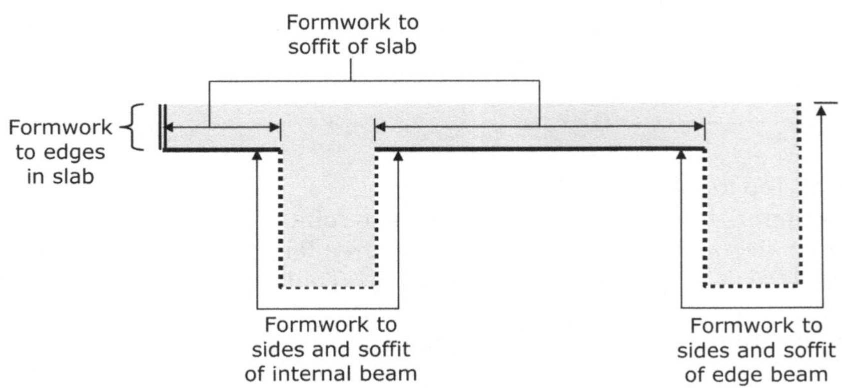

Figure 13.25 Typical section of formwork to slabs and beams.

Points to note

When taking-off the quantities for formwork to soffit of slab, we often measure the entire area of slab from the floor plan. The following steps are suggested for the formwork adjustment:

1 Measure the overall slab area.
2 Deduct area of beam soffits.
3 Deduct wall and column areas (on plan).
4 Adjust for other items (e.g. deductions for openings).

Question 13.4

How should the area of formwork to the soffit of coffered slab be measured?

VII(d).2.1.1.1–2 Formwork to soffits at different heights

- Strutting or propping is required to support the formwork to soffits, and this is deemed to be included in the formwork item.
- Formwork to soffits with strutting height > 3.5m shall be measured separately in stages of 1.5m. **Figure 13.26** illustrates the situations in horizontal slab and sloping slab.

Question 13.5

If measuring the sawn formwork to a horizontal slab (10m long × 10m wide × 200mm thick) and the height to soffit is 5.00m, is the taking-off on the right-hand side correct?

	10.00 10.00		Sawn fwk to horiz. soff. of suspended slabs ≤ 200mm thk; strutting n.e. 3.5m high
	10.00 10.00		Ditto; strutting 3.5 – 5.0m high

VII(d).8 Top formwork

Top formwork is not the same as formwork to soffit of sloping slab. If the slope of a slab is gentle, only formwork to soffit is required for concrete casting. However, if the slope is steep (> 15° from the horizontal as required in SMM VII(d).M.10), top formwork has to be measured (as in **Figure 13.27**).

VII(d).9 Formwork to edges and breaks in slabs and walls

- These are measured separately (**Figure 13.25** and **Figure 13.30**), neither with soffit of slabs nor with sides of walls (which is different from the case of beams).

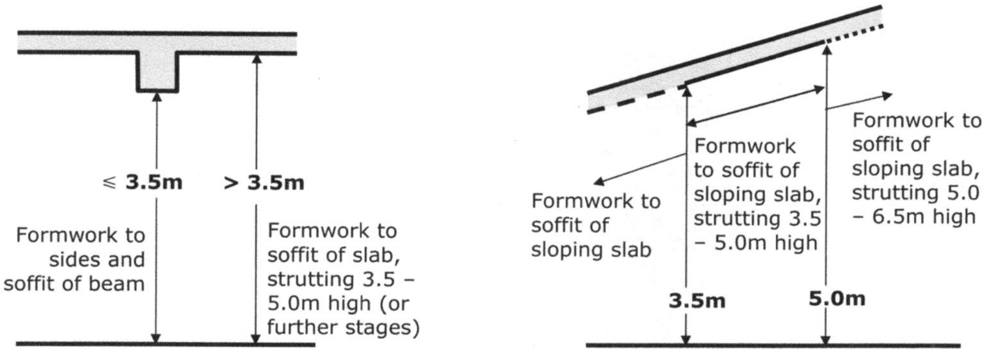

Figure 13.26 Formwork to soffit of horizontal slab and sloping slab.

- Measured separately for:
  - width (i.e. thickness of slabs/walls) > 300mm; in m^2
  - width ≤ 300mm; in m

> **Reminder**
>
> The classification for edges and breaks has been simplified in HKSMM4 Rev 2018.

Formwork to beams

VII(d).7 Formwork to sides and soffits of beams

As mentioned above, formwork to soffits and formwork to sides of slabs are measured separately. However, for sides and soffits of beams, they are measured together in m^2. Referring to **Figure 13.25**, formwork to the internal beam is measured from the soffit of slab, whereas formwork to the edge beam includes the edge of slab.

VII(d).M.8 and M.9 deductions

- Formwork to secondary beams is measured up to the sides of main beams, but no deduction is made from the formwork of main beams where secondary beams intersect them (**Figure 13.28**).

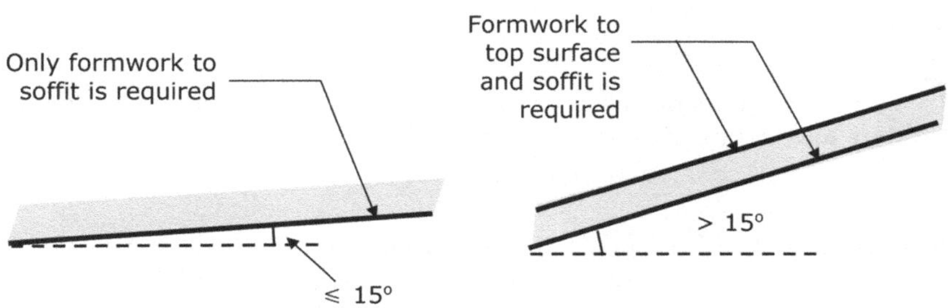

Figure 13.27 Formwork to sloping slabs.

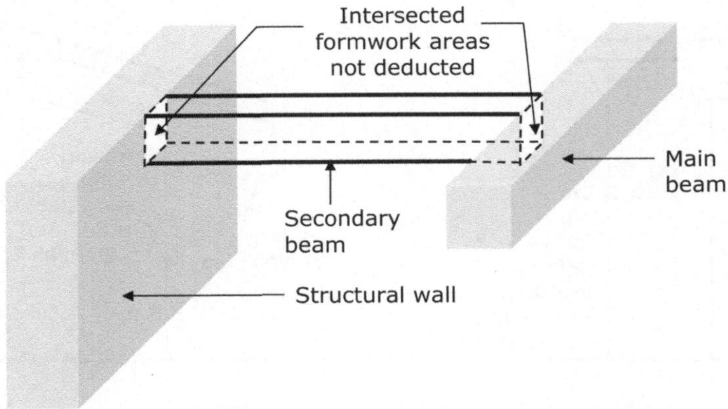

Figure 13.28 Secondary beam intersects wall and main beam.

- Similarly, when beams intersect columns, walls and the like, the formwork to sides and soffit of beams is measured up to the columns and walls and no deduction is made from the formwork to columns, walls and the like at these intersections (**Figure 13.28**).

> **Reminder**
>
> Referring to the case shown in **Figure 13.28**, we will not place any formwork over the white areas in practice. However, when measuring the formwork to walls and main beam, formwork to those areas is measured. The SMM rules, VII(d).M.8 and M.9, allow us to neglect the deduction of these formwork areas, which are relatively small.

Formwork to walls and columns

VII(d).5–6 Formwork to wall faces, columns and the like

Formwork is measured to each face of walls requiring support (VII(d).M.7). In general, formwork to sides of columns and walls is measured up to soffits of slabs. In the case of the external walls, formwork to interior sides is measured up to the soffits of slabs, whereas formwork to the exterior sides should be measured across the slabs (see **Figure 13.29**).

> **Reminder**
>
> Formwork to both the interior sides and the exterior sides of the walls are 'formwork to sides of walls' and no need to measure them separately.

As shown in **Figure 13.30**, no formwork should be measured to the sides of columns or edges of walls where the column and wall W1 intersect. Similarly, no formwork should be measured to the sides or edges of walls where W1 and W2 intersect. Here, we simply follow the measurement rule that formwork is measured to the concrete surfaces which require support during casting (VII(d).M.1).

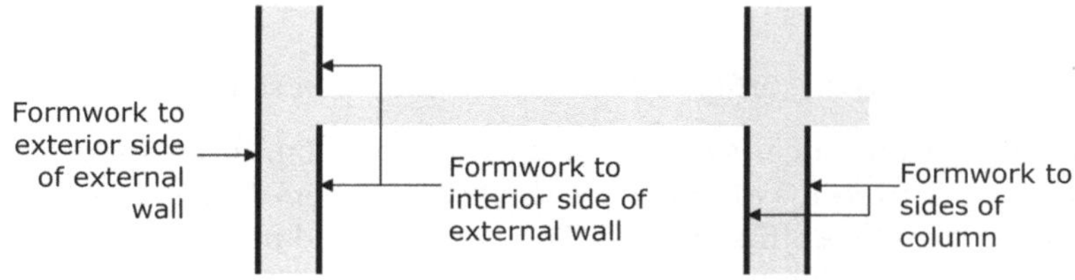

Figure 13.29 Typical section showing formwork to sides of external walls and columns.

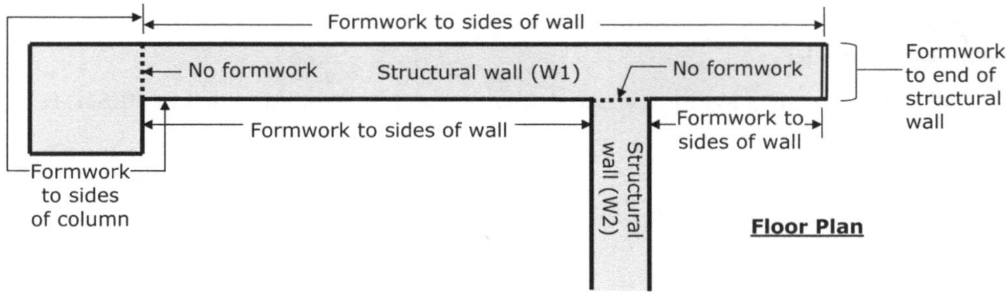

Figure 13.30 No formwork to sides of walls/columns at the intersections of structural wall and column.

Points to note

The following steps are suggested for measuring the formwork to walls:

1 Measure the total wall area by multiplying the centreline of walls by the height.
2 Multiply the total wall area by two to obtain the formwork area on both sides of walls.
3 Deduct the intersections with walls, slabs and beams where appropriate.
4 Adjust other items (e.g. deductions for openings) where appropriate.

Question 13.6

Do we need to measure the formwork to:

i soffit of slab at 'a'?
ii soffit of beam at 'b'?
iii side of wall at 'c'?
iv side of pilaster at 'c'?

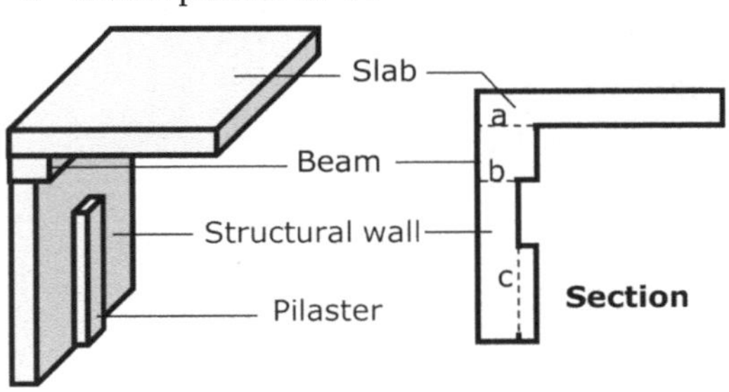

Formwork to stairs

VII(d).3a, 10 Formwork to soffits of stairs, risers and open strings of stairs

- Formwork to stairs includes formwork to soffits of landings (explained earlier with formwork to soffits of slabs), soffits of stairs, risers and open strings.
- Sloping surfaces for soffits of stairs shall be measured in m^2.
- No top formwork is required for treads.
- Formwork to risers and open strings shall be measured in m if width ≤ 300mm and measured in m^2 if width > 300mm.

> **Reminder**
>
> The classification for risers and open strings has been simplified in HKSMM4 Rev 2018.

- The length of open string can be calculated by the following formula (referred to **Figure 13.31**).

$$L = \sqrt{a^2 + h^2}$$

where
L = length of open string

- The width of open string can be obtained as follows:

$$W = t_1 + t_2$$

where
W = width of open string
t_2 = thickness of waist which is usually given in drawings

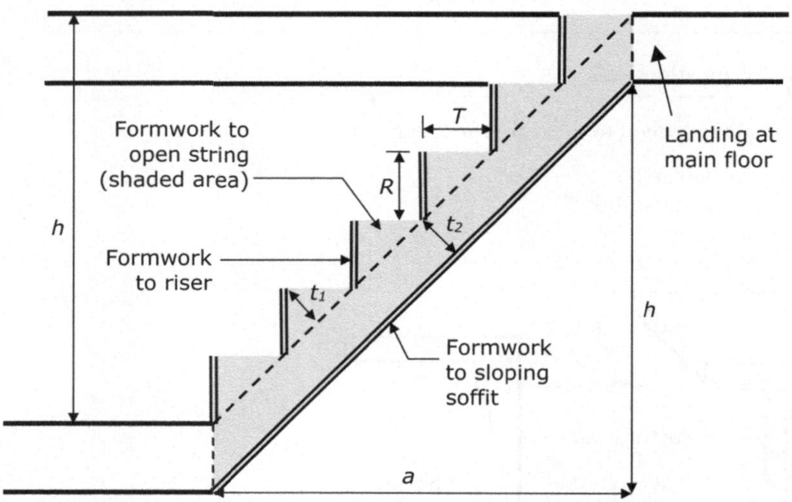

Figure 13.31 Staircase formwork.

t_1 is normally unavailable and has to be calculated by the following formula:

$$\frac{1}{2} t_1 \sqrt{T^2 + R^2} = \frac{1}{2} TR$$

$$t_1 = \frac{TR}{\sqrt{T^2 + R^2}}$$

Question 13.7

Referring to **Figure 13.31**, should the formwork to the 'landing at main floor' be regarded as formwork to soffits of landings or formwork to soffits of slabs?

Question 13.8

Referring to **Figure 13.31**, if the width of open string (W) is > 300mm, what is the area of the open string formwork? (Present the answer by using the variables on the drawing.)

Forming openings

In a concrete structure, openings are formed for different purposes. Where windows and doors are installed in the walls, there is a need to form openings in the walls for these elements. Stairwells, lift shafts, refuse chutes etc. also require openings to be formed in the slabs. Furthermore, openings must be allowed for pipework, cables and the like that run through the structure.

Several SMM rules are related to forming openings in the walls and slabs.

- No deduction of (slab or wall) formwork is made for openings ≤ 1.00m^2 (VII(d).M.3).
- All formwork is deemed to include boxing for openings ≤ 0.50m^2 sectional area (VII(d).C.4(c)).

VII(d).15 Forming holes for pipes, cables and the like, > 150mm diameter

- Measured as an item.
- No deduction of formwork to slabs/walls for opening area ≤ 1.00m^2.

Reminder

Forming holes ≤ 150mm diameter are deemed to be included in the formwork items, and no deduction of the formwork is required.

VII(d).9 & VII(d).16 Forming openings > 0.5m^2

To form a large opening in a wall or slab, extra formwork is required for the boxing out. For openings > 0.5m^2 and ≤ 1.00m^2, boxing for openings is applied. For

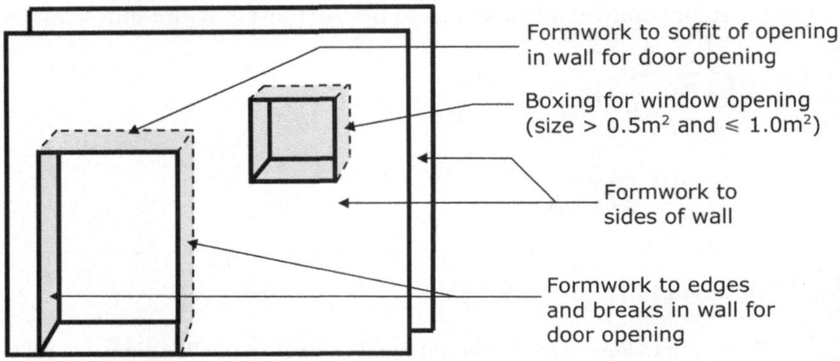

Figure 13.32 Formwork to walls with openings.

openings larger than $1.00m^2$, formwork to edges and breaks of walls/slabs and soffits of openings will be applied (as shown in **Figure 13.32**).

Forming openings $> 0.5m^2$ and $\leq 1.00m^2$ (VII(d).16)

- Measure as boxing for openings in number.
- State thickness of concrete.
- State the size of opening in stages of $0.10m^2$.

Forming openings $> 1.00m^2$ (VII(d).9)

- Measure as edges and breaks in slabs (or walls) for the vertical surfaces, and soffits of openings in walls for the flat surfaces.
- If width > 300mm, measure in m^2.
- If width ≤ 300mm, measure in m.

Reminder

When the sectional area of an opening is $\leq 0.5m^2$, neither boxing for opening nor edges and breaks in slabs/walls are measured. The forming of such small opening is held to be included in the formwork measurement and no deduction is made from the formwork to walls/slabs (VII(d).C.4(c)).

The classification ranges for the opening deductions, boxing for openings, and edges and breaks have been widened in HKSMM4 Rev 2018.

Question 13.9

When measuring the edges and breaks in walls and soffits of openings in walls (SMM item VII(d).9) for a door opening that is $>1.00m^2$, should we group them as one item or measure them separately?

Application of the HKSMM clauses in relation to forming openings of different sizes (VII(d).M.3, VII(d).16 and VII(d).9) is summarised in **Figure 13.33**.

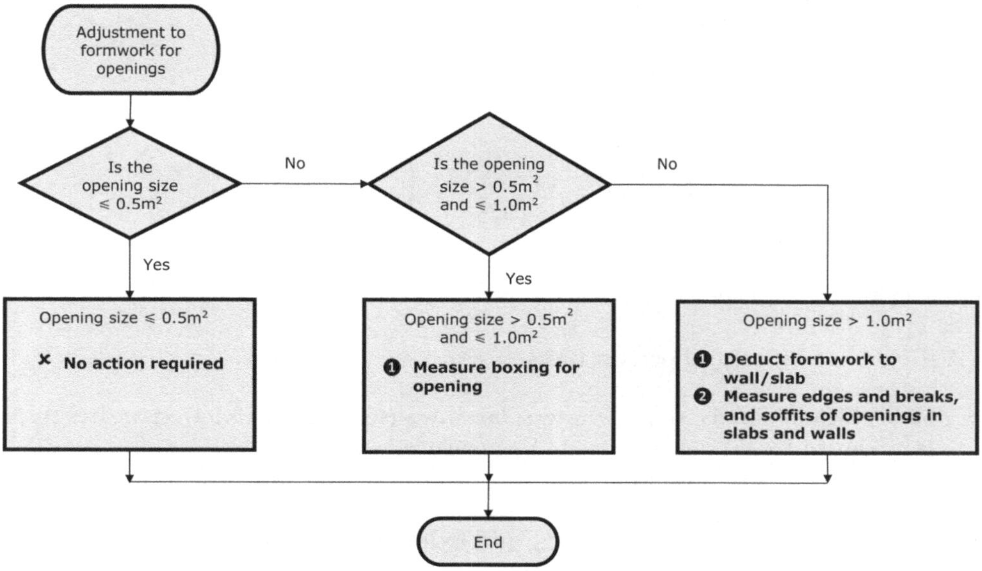

Figure 13.33 Flowchart of making formwork adjustment for openings.

Precast concrete

Precast concrete units such as facades, slabs, spandrel steps, curbs and lintels are widely adopted in public housing and private developments. Considering the boost to Modular Integrated Construction (MiC) techniques by the local government, precast concrete works will be more popular.

> **Points to note**
>
> Modular Integrated Construction (MiC) refers to
>
> a construction whereby free-standing integrated modules (completed with finishes, fixtures and fittings) are manufactured in a prefabrication factory and then transported to site for installation in a building
>
> (Buildings Department, 2019).
>
> The Disciplined Services Quarters for the Fire Services Department at Pak Shing Kok, Tseung Kwan O, is the first government project to use the MiC technique. The project comprises five blocks with eight MiC units per floor. Each MiC unit is about 50m^2, providing three bedrooms, a living room, a dining room, a kitchen and a bathroom.

VII(e).1 General

- All units if >2.00m long are given separately and the number stated.
- The following are deemed to be included:
  - reinforcement in the unit
  - specified surface finishes
  - additional reinforcement not detailed but necessary for handling purpose
  - necessary bedding and fixing
- Cast-in accessories, if any, should be stated.
- Advisable to state the concrete composition.

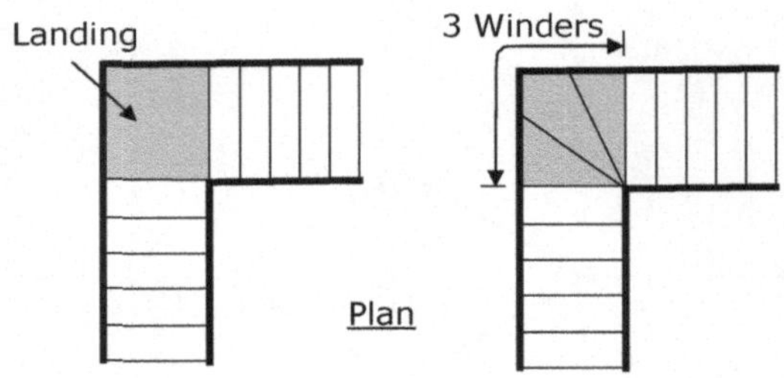

Figure 13.34 Landings and winders.

VII(e).2, 5, 6 Structural precast units

- Slabs, façade panels, steps, winders, landings (see **Figure 13.34**), spandrel steps (see **Figure 13.35**) etc. all have to be measured separately.
- All items measured in number. Steps and landings can be measured in number or m (VII(e).5).
- Dimensioned description, bedding and fixing should be stated.
- Extreme sizes of spandrel steps, circular steps, winders and landings should be stated.

VII(e).3, 7, 8

- Cills, lintels, caps, channel covers, path edgings, road curbs, dropper curbs (**Figure 13.36**) and the like each have to be measured separately.
- Dimensioned description, bedding and fixing should be stated.
- Cills, cornices, lintels, caps etc. are measured in m; straight units and curved units have to be measured separately.

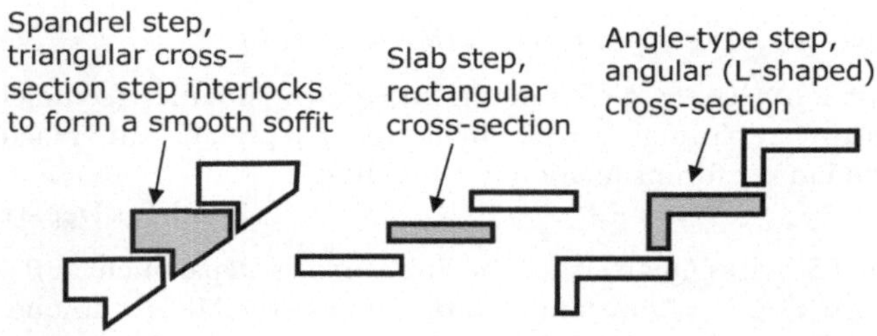

Figure 13.35 Different types of precast steps.

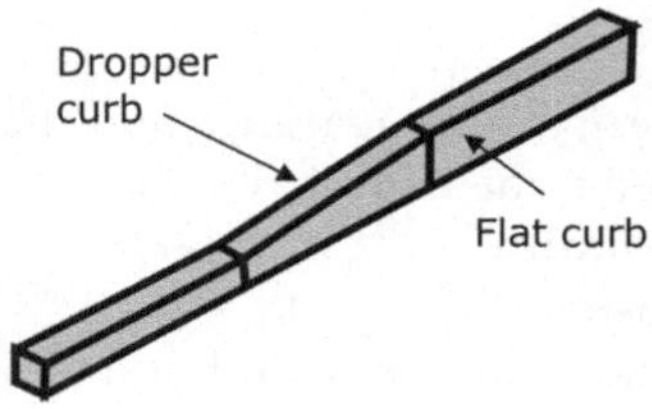

Figure 13.36 Precast Concrete Curbs.

- Channel covers, path edgings and road curbs are measured in m; straight units, curved units and straight units laid to curve have to be measured separately.
- Dropper curbs are measured in number.

Points to note

Since dimensioned description, fixing method, surface finishes and cast-in accessories have to be stated, the need for separate measurements can be far more extensive than the classification table prescribed in the SMM. For example, having a different overall dimension or a different rebar size inside a precast unit will require separate measurements, as illustrated in the lintel example below:

<u>Precast concrete lintels; grade 20/20; bedding in cement mortar (1:3)</u>

A 100 × 150; reinforcing with one 12mm diameter plain steel bar (In 10 nr.)
B 100 × 150; reinforcing with one 16mm diameter plain steel bar; units exceeding 2.00m long (In 10 nr.)
C 150 × 150; reinforcing with one 20mm diameter plain steel bar; units exceeding 2.00m long (In 10 nr.)

Other units such as cills, facades etc. may require fair face treatment or additional finishes to be affixed in the factory. In that case, the description will include such detail, which gives more variations among the units. As a result, more items will need to be measured separately.

Question 13.10

If a precast toilet unit is fabricated with finishes, fittings and building services, how should all the details be described clearly to meet the SMM requirements?

Suggested answers

Question 13.1

A total of 42 steel bars are required for the top bars.

Dividing the width of the slab by the spacing of bars:
= (5m – 0.05m × 2)/0.25m
= 19.6 no.
= 20 no. (rounded up to the nearest whole number)
Total no. of bars required:
= 2 × (20 + 1)
= 42 no.

When distributing objects at a specified distance, you can have three conditions: placing at both ends, at one end and not placing at the two ends, as illustrated in the three case diagrams respectively. When fixing steel bars in slabs, we apply condition 1 (placing at both ends). Therefore, one more steel bar is added to each layer of rebar that runs in one direction.

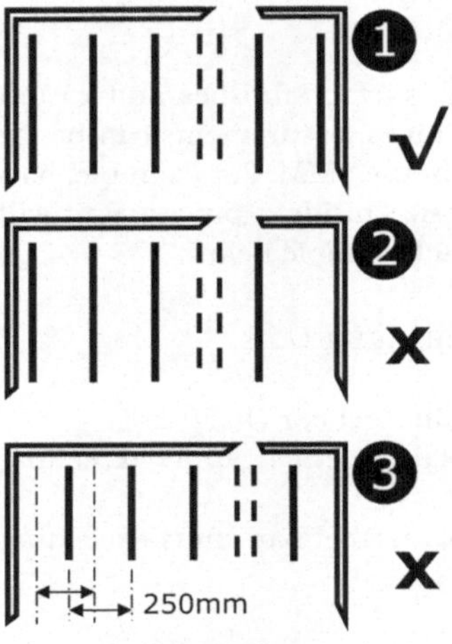

Placing of steel bars on plan

Question 13.2
No, we only need to describe the rebar as 'stirrups' in the description. Noted that rebar in different structural members such as columns, beams and walls are not measured separately. The description can be written as, for example, '32mm diameter high yield steel bars in links, stirrups and binders'.

Question 13.3
The total number of links required should be 56.

$$8(\text{horizontally}) \times 7(\text{vertically}) = 56 \text{ no.}$$

Question 13.4
While there is no definition/coverage rule for soffits of coffered slab and the measurement principle for this item is the same in both UKSMM7 and HKSMM, reference can be made from the UKSMM. According to the E20 Rule M5 of UKSMM7, soffits of coffered slabs are measured 'as if to a plain surface' (as shown below), which is reasonable and consistent with the measurement principle in HKSMM. QS can include a relevant preamble clause to clarify the measurement principle applied.

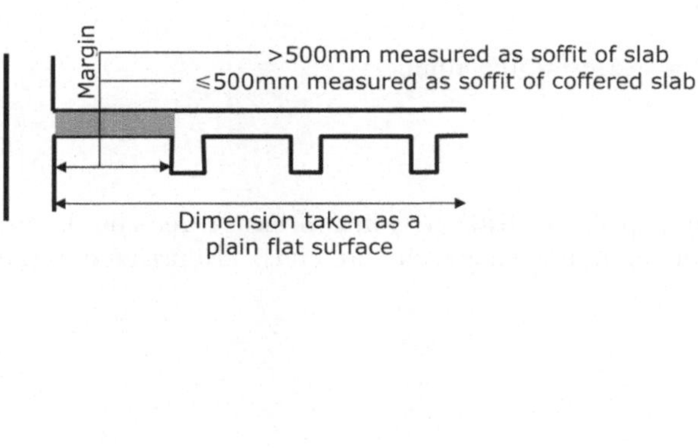

Question 13.5

The taking-off as shown is wrong. Only the second item is required.

The SMM rule, VII(d).2.1.1.2, requires formwork to soffits to be measured in stages of 1.5m high if the soffit height is > 3.5m. This rule should not be mixed up with the measurement rule for excavation that requires excavation to be measured in successive stages of 1.5m deep. When measure excavation, we slice the excavation volume into 1.5m thick layers and measure them one by one. Here, we simply choose the correct stage that represents the strutting height (3.5–5.0m) and put it in the item description.

Note that the requirement of describing the work in a specific stage appears frequently in many trades.

Question 13.6

No formwork is required to all areas from (i) to (iv). The deduction rule VII(d).M.9 only applies to the case when beams intersect columns, walls and the like. In the case of Question 13.6, the beam is running along the wall, not intersecting. Therefore, the deduction rule does not apply and the formwork should only be measured where concrete surfaces require support for casting.

Question 13.7

Formwork to the landing of main floor should be counted as the formwork to soffits of slabs.

Question 13.8

The area of open string formwork is:

$$= \left(\sqrt{a^2 + h^2}\right) W$$

Here, we are not calculating the nett size of the string board but the extreme size of formwork required to prepare the string board.

Question 13.9

The intention of the formwork classification in the HKSMM is to separately measure the formwork to vertical surfaces and flat surfaces (and curved surfaces as well). This is because the formwork design, temporary support and level of construction difficulty are different in each category. Although 'edges and breaks in walls' and 'soffits of openings in walls' both appear in item HKSMM VII(d).9, it is a common practice to measure them as separate items.

Question 13.10

To list out the details of the precast unit clearly, it is advisable to include the drawing reference in the item description as well.

14 Measurement of brickwork and blockwork

Introduction

Unlike concrete, the applications of brickwork and blockwork are more limited in Hong Kong, mainly in walls, columns and pavings. Major work items that are covered in the Brickwork and Blockwork Section VIII of the HKSMM include:

- Brick walls
- Arches
- Isolated piers
- Block walls
- Glass block walls
- Brick or block pavings
- Brick steps
- Brick cills, copings etc.
- Sundries

Certain classes of brick/block walls are measured separately as indicated below:

- Brick or block walls of different thickness, kind, quality, bonds (e.g. English/Flemish/Stretcher), mix of mortar (e.g. cement and sand or cement, lime and sand) and types of pointing (e.g. flush/recessed/weathered)
- Vertical walls
- Tapering walls with one face battering (see **Figure 14.1**)
- Tapering walls with both faces battering (see **Figure 14.1**)
- Walls built battering (sloping walls with parallel sides, **Figure 14.1**)

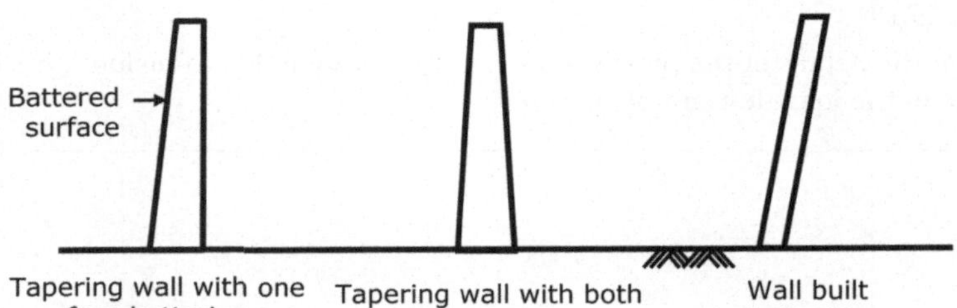

Figure 14.1 Elevations of different walls.

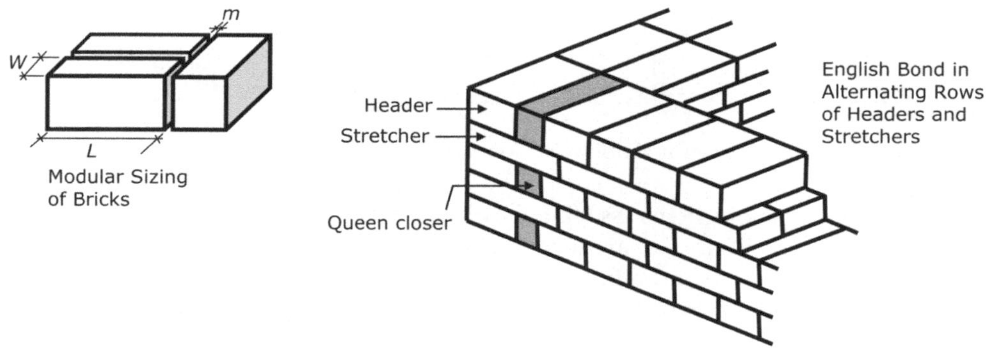

Figure 14.2 Design of standard bricks.

- Dwarf walls
- Hollow walls (refers to cavity walls)
- Brickwork in backing to masonry
- Brickwork in filling in openings
- Curved works
- Glass block walls

Brick walls in general

With reference to the HKSMM, thicknesses of brick walls are classified according to the multiple of half a brick. This is related to the common practice of standardisation in brick dimensions. Although different countries have different standard brick sizes, most bricks are made in an approximately 2:1 ratio for the length and width – two widths (W) and one mortar joint (m) is the same as the length (L) of one brick as in **Figure 14.2**. When the common red bricks are laid crosswise, two 102.5mm widths and one 10mm mortar joint will give 215mm, the same length as that of a single brick. Different bonds such as English bond (**Figure 14.2**), Flemish bond etc. can be laid easily within the standard co-ordinating size.

When using common bricks of 215mm × 102.5mm × 65mm, a half brick wall is 102.5 mm thick laid in stretcher bond. Other modular thicknesses are shown in **Table 14.1**.

If other types and sizes of bricks are used, the 'half a brick' module will give different dimensions. Therefore, when measuring brickwork, the type, quality and size of brick must be described.

Table 14.1 Nominal Size of Brick Wall Using Common Red Bricks.

Modular Description	*Overall Wall Thickness (including mortar thickness) mm*
Half brick thick	102.5
One brick thick	102.5 × 2 + 10 = 215
One and half brick thick	102.5 × 3 + 10 × 2 = 327.5
Two brick thick	102.5 × 4 + 10 × 3 = 440

Figure 14.3 Forming reveals for windows or doors.

- Most sundry labour items associated with brickwork are deemed to be included. Some of the popular ones are listed below:
    - Forming square and rebated reveals (**Figure 14.3**)
    - Cutting chases, grooves, holes and similar including any grouting and making good
    - Cutting splayed, chamfers and similar
    - Leaving pockets in the work to receive built-in items including subsequently blocking up

VIII(a).M.2 and M.3 deductions

- Nett sizes of the openings are measured in deductions.
- No deductions are made for openings ≤ 0.50m^2.
- No deductions are made for lintels, cills and the like.

Points to note

The deduction rule for openings in brickwork has been revised to larger sizes in the HKSMM4 Rev 2018.

VIII(a).M.4 Brickwork and block walling circular on plan

- Measured separately in m^2.
- Measure the mean length for curved walls (**Figure 14.4**).

Brick walls and block walls

VIII(a).2–5 Brick walls, isolated piers and dwarf walls

- Measured in m^2.
- State the kind and size of bricks, bond, mortar mix and type of pointing (VIII(a).S.1–S.4).
- If the brick wall ≤ one and a half bricks thick and is a multiple of half a brick thick: state the thickness.
- If brick wall > one and a half bricks thick and is a multiple of half a brick thick:
    - Various thicknesses grouped together.
    - Reduced to one brick thick and measured as reduced brickwork. For instance, a two brick thick wall of 3m^2 will be measured as 6m^2 reduced one brick wall.

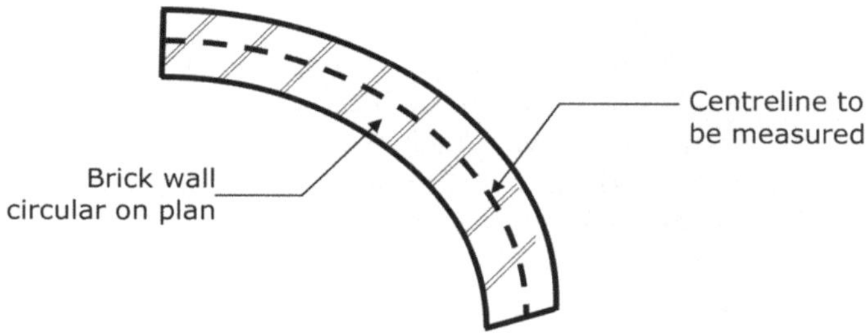

Figure 14.4 Measurement of brickwork circular on plan.

- If brick wall is not a multiple of half a brick thick:
  - State the thickness and different thicknesses shall be measured separately.
  - Description includes for all cutting.

Question 14.1

Take-off the following brick walls if all of them are 10m L × 3m H, using common bricks (215mm × 102.5mm × 65mm) laid in English bond in c/s mortar 1:3.

i 150mm thick
ii 215mm thick
iii 440mm thick
iv 552.5mm thick

VIII(a).12 Block walls

- Measured in m^2.
- State the kind and size of blocks, bond, mortar mix and type of pointing (VIII(a).S.8–S.11).
- Measured separately for:
  - general block walling and partitions
  - block wall in isolated positions in pipe ducts and the like
  - concrete block wall
  - hollow block wall
  - patent block wall
- Block walls are deemed to include for filling in hollow blocks at ends and intersections (VIII(a).C.3).

VIII(a).13 Glass block walls and panels

- Measured in m^2.
- State the kind and size of blocks, bond, mortar mix or jointing compound and type of pointing (VIII(a).S.12–S.15).

VIII(a).6 Fair face brickwork

- Fair face brickwork, which is quite popular for exteriors, refers to the brick wall where the material is left on view with no further finishing. To achieve this, more work is involved during bricklaying including choosing better bricks and laying them in a neat manner with specified pointing.
- Measured as extra over brickwork in m^2.

Reminder

When measuring the extra over brick wall for fair face work, no deduction should be made to the common brickwork.

- State the bond, mortar mix and type of pointing of the fair face work (VIII(a).S.5–S.7).
- Fair face work applied to the following has to be separately measured:
  - general faces;
  - built overhand (As shown in **Figure 14.5,** the work needs to be carried out over another wall (or obstruction); and therefore, requiring the bricklayer to lean over the wall (or obstruction) to complete the work on wall B. Usually, this happens when the bricklayer works from inside the building without the use of external scaffolding);
  - battering faces; and
  - curved faces.

Question 14.2

The diagram here shows a layout plan of a brick wall. What is the length for the measurements of (i) to (iii)?

(Given: G_e = external girth of the wall
G_m = mean girth of the wall
G_i = internal girth of the wall)

i the brick wall
ii extra over the brick wall for fair face work on the external side
iii extra over the brick wall for fair face work on both sides

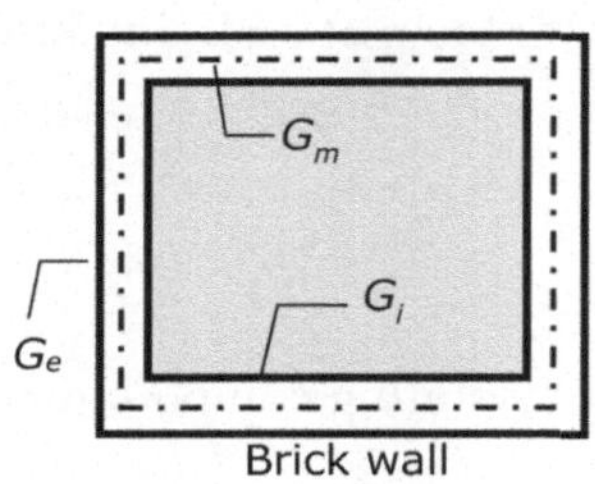

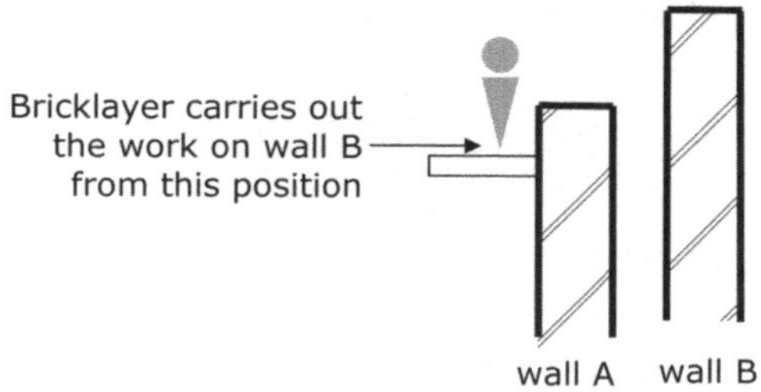

Figure 14.5 Overhand work on brick wall.

VIII(a).15 Bedding and pointing in sealant compound other than cement or cement lime mortar

- Pointing in sealant is to leave the mortar joints recessed to receive some sealant in which the joint profile (such as weathered joint) is formed.
- Typical pointing in cement/cement lime mortar is deemed to be included in the brickwork item. However, if non-cementitious material is used for pointing, it has to be measured separately.
- Measured as extra over brickwork/blockwork in m^2.

VIII(a).17 Cutting and bonding walls of different construction

- Measured in m.
- Describe both types of walls.

VIII(a).19 Reinforcement in brickwork or blockwork

- Measured in m.
- Measured nett.
- State the material and width.

Miscellaneous

VIII(a).10 Cills, copings etc.

- Measured in m.
- State the number of courses and projection (if any).
- Angles, returned ends and the like (as **Figure 14.6** below) are deemed to be included.

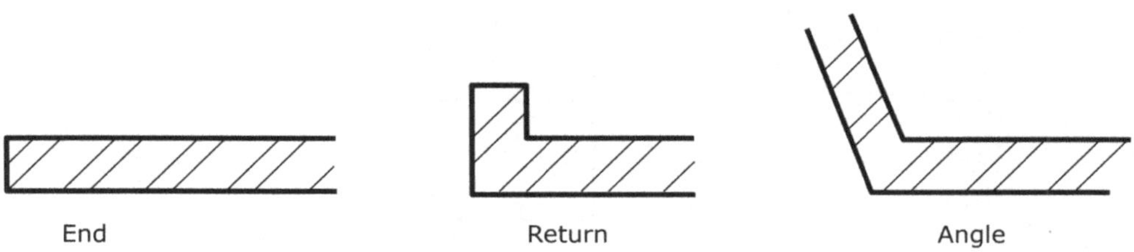

Figure 14.6 Terminologies for different wall layouts.

- The following should be measured separately:
    - Straight work
    - Curved work
    - Cills
    - Cornices
    - Plinth cappings
    - Copings
    - Others, described

Points to note

Copings and cappings are placed at the head of a wall or a column to protect the top of these members from exposure to changing climate. The main difference between copings and cappings, as shown in **Figure 14.7**, is that copings are designed to shed rainwater but cappings are not (PD 6697:2010). To perform such a function, it is critical that the throat at the underside of the overhang must be continuous over the mortar joints to keep the rainwater away effectively.

VIII(a).20 Damp-proof courses (except asphalt)

- Describe the materials and thickness of damp-proof course (dpc).
- Measured nett.

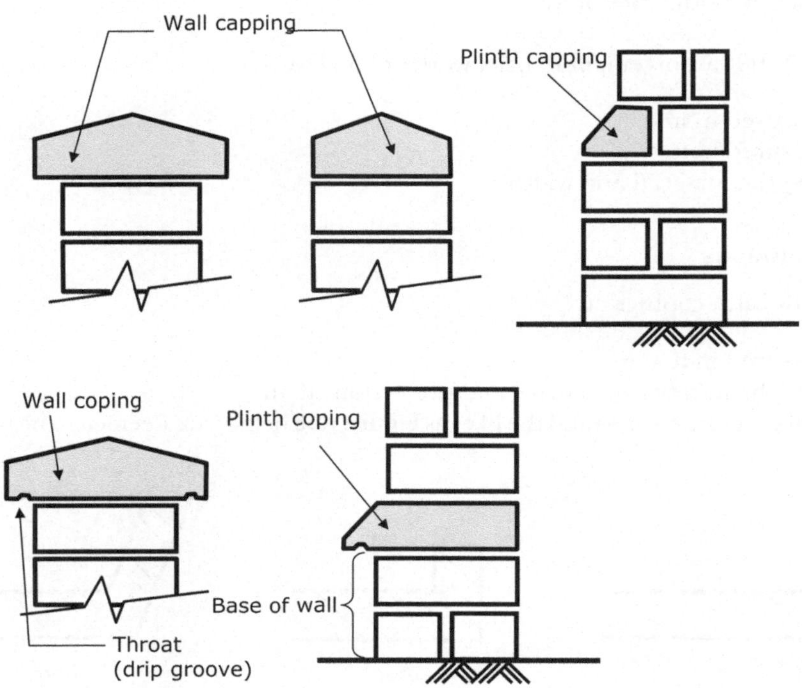

Figure 14.7 Cappings and copings to walls (or columns).

- If dpc ≤ 225mm wide:
  - Measured in m.
  - State the width.
- If dpc > 225mm wide:
  - Measured in m^2.
  - Separately measure the following:
    - Horizontal dpc
    - Vertical dpc
    - Dpc on curved walls

Brick and block pavings and steps

- No deductions are made for openings ≤ $0.50m^2$, measured nett (VIII(a).M.2).

VIII(b).2 Brick and block pavings

- Measured in m^2.
- State the quality of materials, specifications of bricks/blocks, nature of base, preparatory work, bedding, joints and pattern (if required).
- The following works should be measured separately (see **Figure 14.8**):
  - Level or to falls only ≤ 15° from horizontal.
  - To falls and crossfalls and to slopes ≤ 15° from horizontal.
  - To slopes > 15° from horizontal.

VIII(b).3 Brick steps

- Measured in m.
- State the width and height of steps.
- The following works to be measured separately:
  - Straight
  - Curved

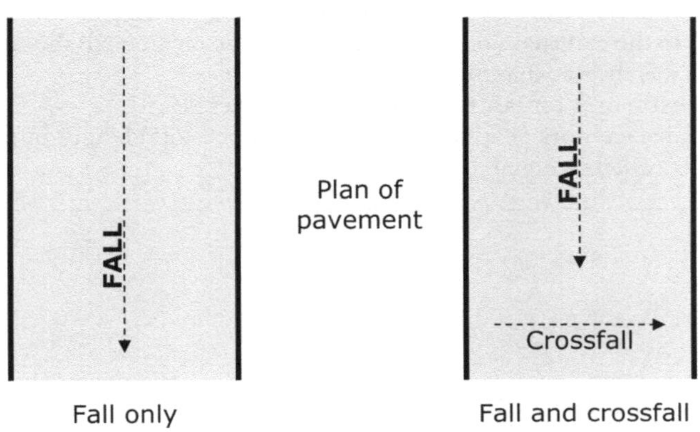

Figure 14.8 Falls of pavement.

Suggested answers

Question 14.1

	10.00 3.00		150mm thk vert. bk wl in comms. in English bond; c.m (1:3) incl. all cutting (i
	10.00 3.00		1B thk vert. bk wl in comms. in English bond; c.m (1:3) (ii
2/	10.00 3.00		Red. 1B thk vert. bk wl in comms. in English bond; c.m (1:3) (iii
2½/	10.00 3.00		(iv

While common bricks (size: 215mm × 102.5mm × 65mm) are used, wall (i) is not half a brick thick and therefore cutting is required and stated in the description. Wall (ii) is 215mm thick, which is exactly one brick thick (102.5 + 102.5 +10 = 215). Walls (iii) and (iv) are multiple of half a brick and > one and half brick thick, as follows:

Wall (iii): 102.5 × 4 +10 × 3 = 440= 2B
Wall (iv): 102.5 × 5 +10 × 4 = 552.5= 2½B

Therefore, both of them are grouped together and measured as reduced one brick walls. Noticed that the multiple (based on one brick thick wall) is put in the timesing column accordingly. In addition, it is recommended to enter the timesing factor as a fraction, rather than a decimal number (e.g. 2.5) to avoid confusing the decimal point as a dotting-on.

Question 14.2

i length of the brick wall: G_m
- according to the SMM, the average length of the brick wall should be measured.

ii length of the extra over for fair face work on the external side: G_e
- fair face should be measured extra over the brick wall on which it occurs (Seeley, 1988). Therefore, the external girth, G_e, should be used as the fair face work applies to the external side only. Noticed that the mean girth should not be used here as it is shorter than the external girth.

iii length of extra over for fair face work on both sides: $2G_m$
- If the fair face work is applied on both sides, the total length of fair face work is $G_i + G_e$, which is equal to $2G_m$.

15 Measurement of wood works

Introduction

Wood works, conventionally described as carpentry and joinery works, cover a wide range of items from structural timber works such as timber roof trusses to joinery works such as furniture.

According to HKSMM4 Rev 2018 Section XIII, Wood Works covers the following major work items:

- Structural timber works
- Flooring
- Raised access floors
- Sheet linings, sheetings and claddings
- Unframed trims
- Timber suspended ceilings
- Roof coverings and wall claddings
- Partitions
- Doors and the like
- Stairs
- Furniture and fittings
- Minor buildings and structures

Due to the broad coverage of this SMM section, only the major non-structural timber work items that are usually required in a local, simple building are discussed in this chapter.

General

- State the thickness, kind and quality of timber used for each item (XIII(b).S.1).
- No deduction should be made in items measured superficially for voids $\leq 0.5m^2$ (XIII(b).M.2).
- Curved works shall be measured separately (XIII(b).M.3).
- All fixings ranging from simple methods such as bolting and screwing to more complicated ones such as the use of holdfasts, patent connectors, drilling and mortices in structure and subsequent grouting are deemed to be included (XIII(b).C.1).

Flooring, partitioning and sheeting

XIII(c).1–4 Timber flooring

- Timber flooring is a popular floor finish in Hong Kong. Different designs of timber flooring have to be measured separately:
  - Block flooring
  - Parquet flooring
  - Parquet flooring in preformed panels
  - Strip flooring

Points to note

Since there are no definition rules given in the HKSMM for the various flooring types, the British Standards is used as a reference. Timber block flooring uses timber blocks that are square edged or with an interlock to join with each other, fixed to a floor screed by bitumen or adhesive. Parquet flooring is manufactured from selected timber cut to form patterns either by laying the boards directly onto a floor/screed (parquet flooring) or pressed onto a backing and laid as panels (parquet flooring in preformed panels). For strip flooring, it is made from timber strips with tongue-and-groove joints. These timber strips are fixed to floor battens or to a plywood sub-base by nailing or adhesive.

- Measured in m^2, including the area of borders (if any).
- State the size of strips, blocks or parquet fingers.
- State the pattern, method of jointing and bedding materials.
- Any sub-bases or battens required for fixing the timber strips are deemed to be included.

XIII(c).6 Borders (example see **Figure 15.1**)

- Measured as extra over timber flooring in m.
- Mean girth shall be measured.
- Fully described.

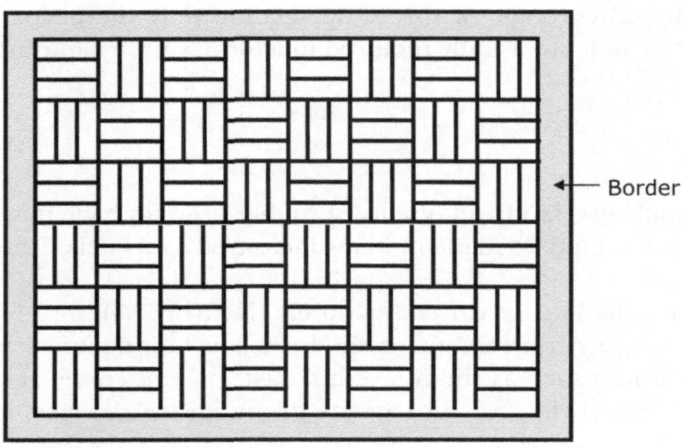

Figure 15.1 Strip flooring in square basket pattern with border.

XIII(c).7 Accessories

- This item covers movement joints (including expansion joints), cover strips and dividing strips. Movement joints are gaps/joints filled with compressible material to accommodate movement in the building components. Cover strips are protecting strips to cover the edge of finishes at, for instance, door openings. Dividing strips are used to separate two different floor finishes.
- Movement joints, cover strips and dividing strips each measured separately.
- Measured in m.
- State the dimension and method of fixing.

Boarding and lining to walls, ceilings, roofs etc.

This section is not limited to continuous boardings and linings made of timber. Other non-metallic sheetings or linings made of fabric, plastic laminate, fire resistant materials, thermal insulation boards, and so forth are also included.

> **Reminder**
>
> Waterproofing to roofs and tanking are not included in this section but in Section X. Other waterproofing layers and vapour barriers (except the non-asphaltic damp proof course in brick work) are all measured under this section (XIII(e).D.1).

XIII(e).2–12 Linings, sheetings, claddings and the like

- The items have to be separately measured when they are:
  - applied to different building elements such as walls, ceiling and beams, roofs etc.
  - applied to different heights of ceilings/beams, which is sub-classified into ≤ 3.50m high and > 3.5m high in further stages of 1.50m (XIII(e).M.2) (for sheet linings and sheet claddings)
  - designed for different functions such as acoustic, fireproof, waterproofing etc.
  - made of different materials such as fabric, plastic laminate etc.
  - applied to external (XIII(e).D.2)
- Measured in m^2.
- All fixing materials such as screws, matching filler, adhesives etc. are deemed to be included (XIII(e).C.1).
- Details like materials, backing of linings/sheetings, fixing method and the like have to be described.

Unframed trims

XIII(e).13–19 Skirtings, architraves, picture rails, dado rails, cornices, cappings and the like

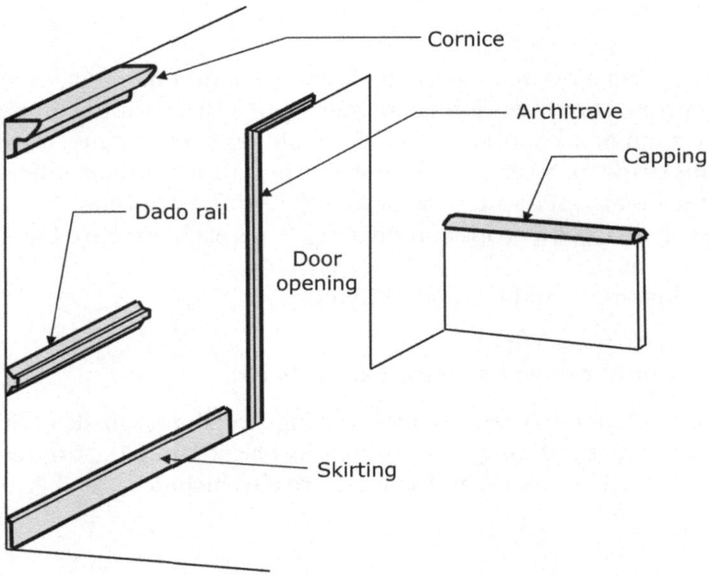

Figure 15.2 Cornice, dado rail, skirting, architrave and capping.

- Measured the nett length as fixed in m.
- Different items (as shown in **Figure 15.2**) have to be measured separately.
- State the sizes.
- State where services are to be located within built-up work.
- State the grounds and/or backings if included in the item.

Question 15.1

If a timber dado is made up from two types of timber, should we measure the dado as two items using different timber?

Question 15.2

Do we need to separately measure the following?

i skirting to wall and skirting to staircase
ii internal corners and external corners of the cornice
iii architrave and door frame

Fixed partitions and demountable partitions

XIII(g).1–2 Fixed partitions and demountable partitions

- Measured in m.
- Mean length shall be measured (XIII(g).M.1).

Question 15.3

Which length should be measured for the timber partition as shown in the drawing?

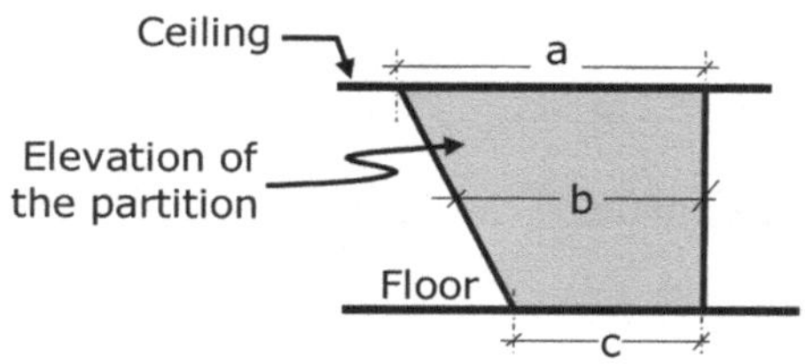

- The following have to be measured separately (XIII(g).M.2):
  - Straight work
  - Curved work
- State the thickness and height, glazed or semi-glazed, factory or site applied finish.
- Describe all details of the partitions, including the kind and quality of materials, method of construction, fire rating (if applicable) and the catalogue number (in case of proprietary partitions) (XIII(g).S.1–S.7).
- Building services to be put inside the partitions shall be stated.
- Deemed to include the following (XIII(g).C.1–C.3):
  - all integral components, finishes and holes preformed at factory;
  - cutting holes for building services and subsequent blocking;
  - all angles, intersections, ends, abutments etc.;
  - glazing and glazing beads; and
  - extra framing members, trimmers and linings around blank openings.

XIII(g).4 Openings on partitions

- Measured as extra over the partitions in number.
- State the dimensions of openings.
- Different types of openings have to be measured separately:
  - Blank openings
  - Doors
  - Windows (state the number of lights)
  - Hatches (small openings equipped with hinged covers for passage of people or ventilation)
  - Glazed panels (state the number of lights)
  - Access panels
- Deemed to include ironmongery, glass, any components filling the openings, glazing beads, linings and the like but excluding trim.

Points to note

Here, openings refer to 'breaks in the general construction of partitions and include the components filling the openings' (XIII(g).D.3).

XIII(g).3 Trims

- Trims are regarded as separate items fixed on site as cover pieces to edges or panel joints (XIII(g).D.2).
  - Separately measured in m.
  - State the dimensions.

Cubicle partitions

XIII(h) Toilet cubicle partitions, office cubicle partitions and fittings

- Measured in sets (or number for proprietary office cubicles).
- Toilet cubicles and proprietary office cubicles have to be measured separately.

> **Reminder**
>
> Demountable partitions (under XIII(g).2) and office cubicles [under XIII(h)] are measured separately.

- Building services to be put inside the partitions shall be stated.
- State the catalogue number, overall size and detailed dimensions (e.g. number of panels with sizes, number of cubicles with sizes, number of doors with sizes etc.).
- All integral components, doors, ironmongery, glazed panels (for office cubicles only) and the like are included but exclude trims that are fixed on site (XIII(h).D.1&D.2).
- Trims fixed on site at junctions of cubicles and at junctions with adjoining constructions are measured in m, with dimensions stated (XIII(h).5).

Doors

XIII(i).1–3 Doors and frames

- Measured in set.
- The following types of doors to be stated and measured separately:
  - flush doors
  - framed and panelled doors
  - fire-rated doors
  - single leaf doors
  - double leaf doors
  - swinging/hinged doors
  - folding doors
  - sliding doors
  - sliding and folding doors

> **Reminder**
>
> Under HKSMM4 Rev 2018, all doors, including non-fire-rated doors, are measured with the door frames, door linings, glazing, glazing beads, louvres and louvre frames.

- All doors measured are complete with door frames, linings, glazing, glazing beads, louvres, louvre frames and necessary fixings such as dowels, holdfasts and the like (XIII(i).M.3).
- Forming openings for glazing panels and louvre panels on doors are deemed to be included (XIII(i).C.1).
- Surface treatment applied as part of the production process should be described (XIII(i).S.1).
- State detailed dimensions of doors; frames and linings; construction, core material and facings of doors; stiles and mouldings; size, number and type of glass panels and louvres (XIII(i).M.1 and XIII(i).1–3.1).
- Ironmongery for doors shall be measured separately following the SMM XIV Ironmongery Section – measured in number (XIII(i).M.3).
- Fire rating to be stated for fire-rated doors (XIII(i).3.1.*.1).
- Fire-rated doors and frames are deemed to include all necessary testing and certification and obtaining approval from relevant authorities (XIII(i).C.4).

XIII(i).4 Louvres and frames

- Louvres are measured with the frames.
- Measured in number.
- State the louvre size, thickness, sectional sizes of frames, construction method and surface treatment applied as part of the production process.

> **Reminder**
>
> This measurement rule applies to louvres (including frame) that do not form part of the doors.

XIII(i).5–6 Service hatches, access doors, trapdoors and frames

- Service hatches, access doors, trapdoors and the like are measured with the frames.
- Measured in number.
- State the door size, thickness, sectional sizes of frames, construction method and surface treatment applied as part of the production process.

Others

XIII(j).1 Stairs

- Measured in number.
- Give full dimensioned description including the number of treads, winders, and so forth.

XIII(j).3–4 Handrails and balustrades

- Measured in m.
- Give full dimensioned description, material and jointing or mounting method.
- Plain ends, ramps, wreaths, bends and ornamental ends (see **Figure 15.3**) are deemed to be included.

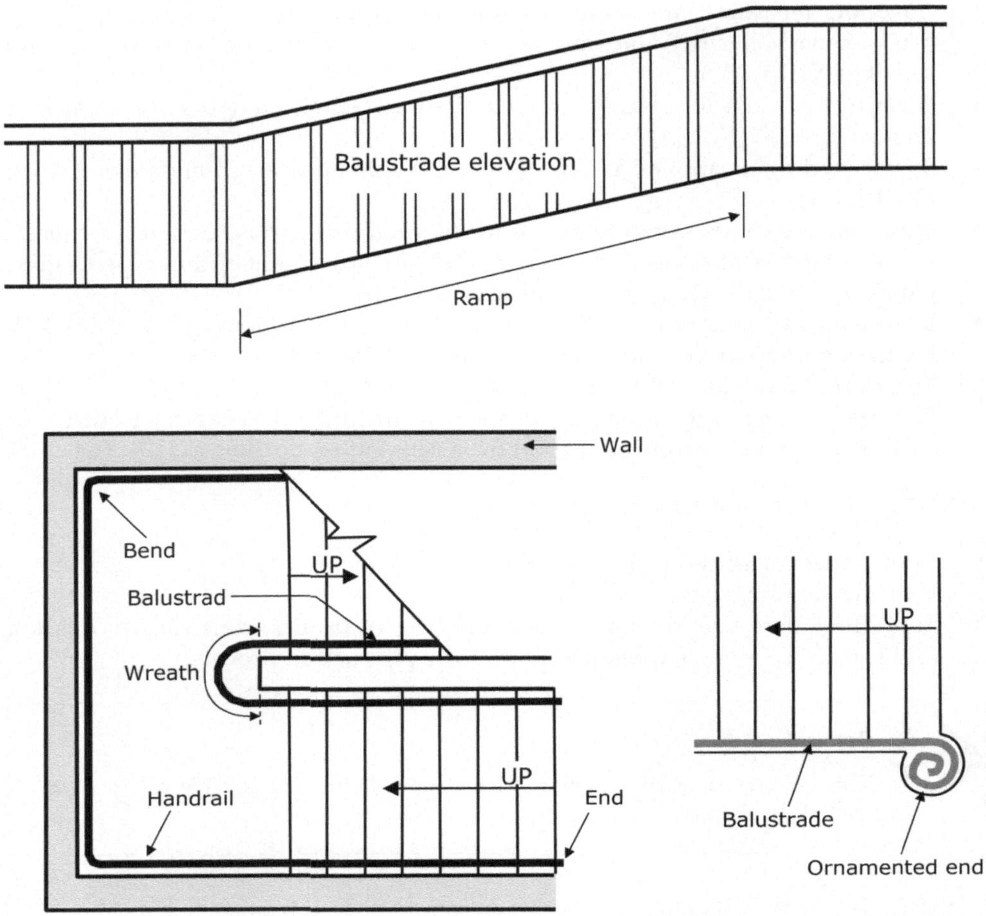

Figure 15.3 Ramps, wreaths, bends and ornamented ends.

- For handrails, state if handrails are hollow and screwed to metal core.
- For balustrades, the spacing of balusters and opening portions should be stated.

> **Reminder**
>
> Under HKSMM4 Rev 2018, all ramps, wreaths, bends etc. are deemed to be included in the handrails or balustrades item.

XIII(k) Furniture, fittings, signboards, playground equipment and the like

- Measured in number.
- All components required to complete the furniture/fitting, such as ironmongery, glazing, mirrors, metalwork, carcassing, doors, shelving, drawers, off-site and on-site finishes are deemed to be included.

- State the drawing reference and/or dimensioned description as appropriate.
- State the background for fixing.
- Describe the ironmongery and/or metal fittings, glass panels, finishes (both off-site and on site) etc. which are to be included in the item.

Suggested answers

Question 15.1

No, the dado should be measured as one item, described it as a 'built-up dado' with full dimensions. Built-up work should be so described (XIII(e).16.1).

Question 15.2

i No, timber skirting to wall and staircase can be measured together under the same timber skirting item. According to SMM, item XIII(e).13, there is no need to classify the skirting item according to the building elements to which it is fixed.

ii No, we do not need to measure. Internal corners and external corners for the cornice are deemed to be included in the cornice item.

iii Yes, according to SMM, item XIII(i).M.3, door frame should be measured together with the door leaf as a set. The architrave should be measured separately in m. Note that although the architrave and the door frame appear to be the same, they are technically different. As shown in **Figure 15.4**, the door frame is an essential member of the door set for fixing the door leaf. The architrave is only a decorative item to improve the appearance by covering the joint between the wall and the door frame.

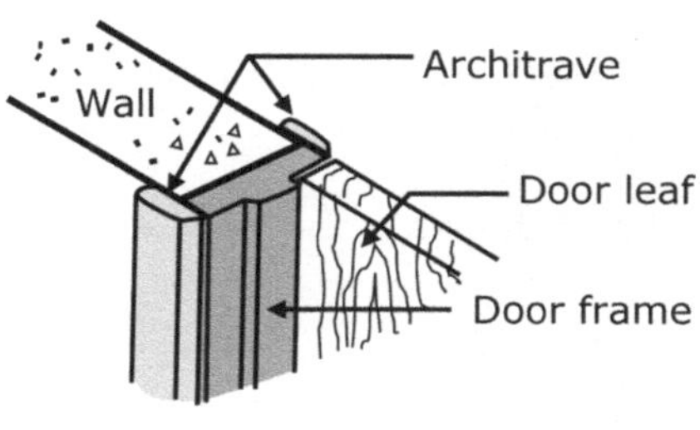

Figure 15.4 Door frame and architrave.

Question 15.3

'b' should be measured as the partition length. According to SMM, item XIII(g).M.1, the mean length as fixed should be measured for all partitions.

16 Measurement of steel and metal works

Introduction

Steel and metal work represent a large sector of work covering heavy steelwork structures and general metal works. Classification of structural steelwork can be summarised as follows:

- Structural members such as columns, beams, purlins and bracings
- Grillages
- Built-up columns, trestles and towers
- Built-up trusses and girders
- Overhead crane rails
- Connections
- Fittings such as tees, plates etc.
- Holding down bolts, special bolts or fasteners
- Surface treatment of steel members

For general metal works, some of the classification is similar to that of the joinery works. Typical metal works include:

- Linings, coverings and claddings
- Roofing and flashings
- Doors, grilles and the like
- Stairs
- Fencing and gates
- Minor buildings and structures
- Raised access floors
- Partitions and cubicle partitions
- Suspended ceilings
- Windows
- Shop fronts
- Curtain walling
- Furniture and fittings
- Sundries

In view of the popularity of structural steelwork in the local industry, this chapter will introduce the measurement rules of structural steel work as well as the measurement rules applied to the major metal work items.

Structural steelwork

Understanding of structural steel members

Before starting any measurements, there must be a general understanding of the structural steel terminology. The major members and structural components are illustrated diagrammatically from **Figure 16.1** to **Figure 16.3**.

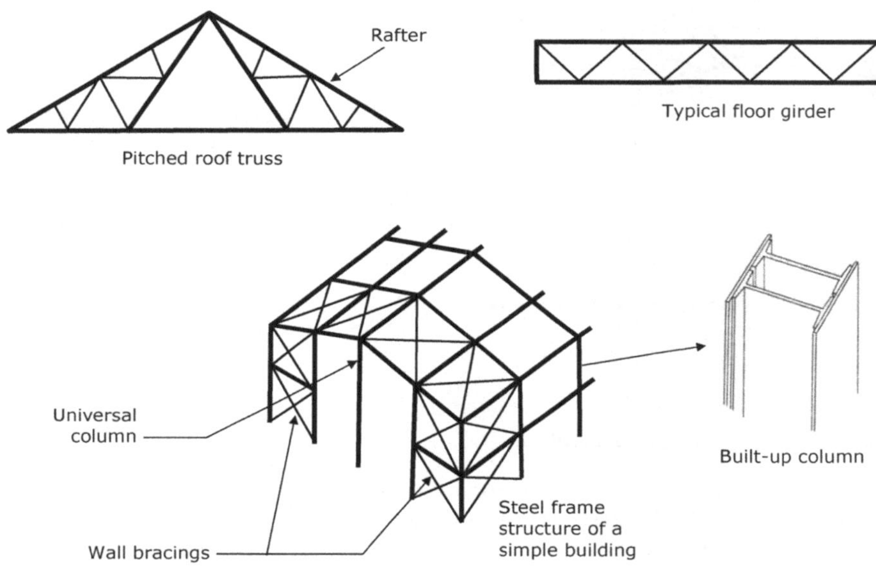

Figure 16.1 Illustration of steel structures.

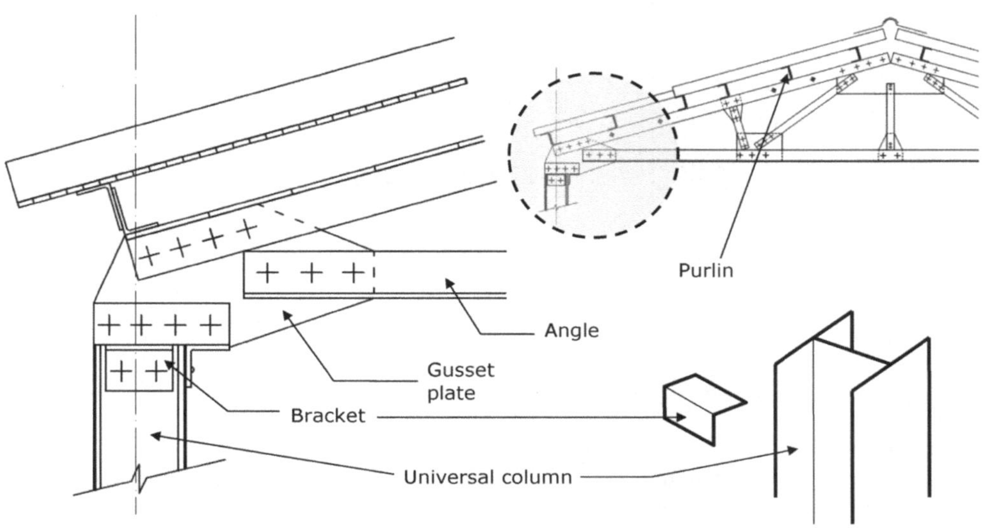

Figure 16.2 Typical roof truss details.

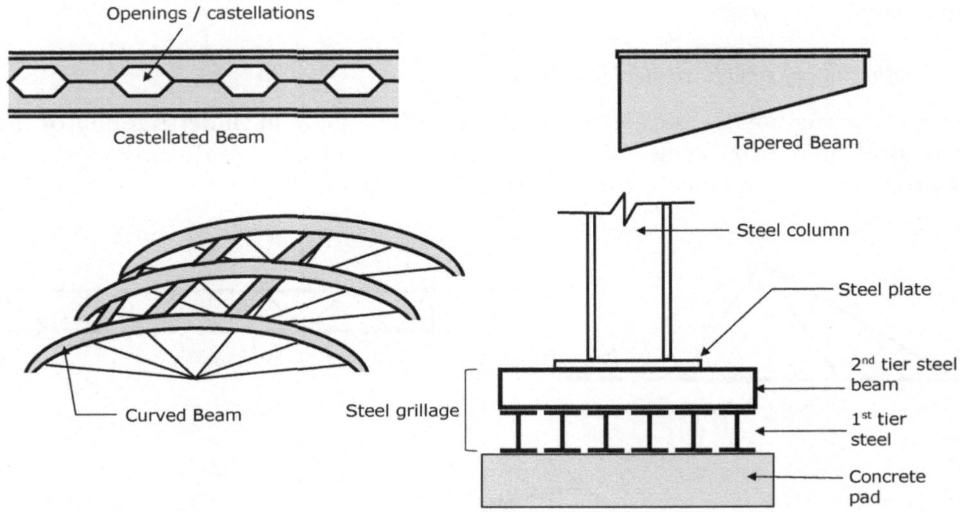

Figure 16.3 Common structural steel members.

Conversion of steel length/area to weight

The SMM rules require most structural steel members to be measured in kg. When taking-off, we measure the length or area of the members and convert the quantities into mass by multiplying the length or area of each member by the unit weight for that member. For example, a 127 × 76 × 13kg universal beam (UB) in 10.00m length will be measured as:

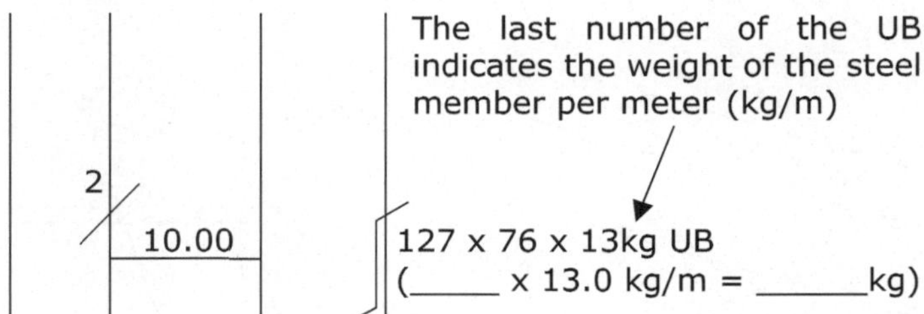

Reminder

The conversion equation is put in brackets following the item description. The multiplication in the bracket will be completed in the squaring process (to be explained in **Chapter 20**), which will become:

(<u>20.00</u> × 13.0 kg/m = <u>260.00</u>kg)

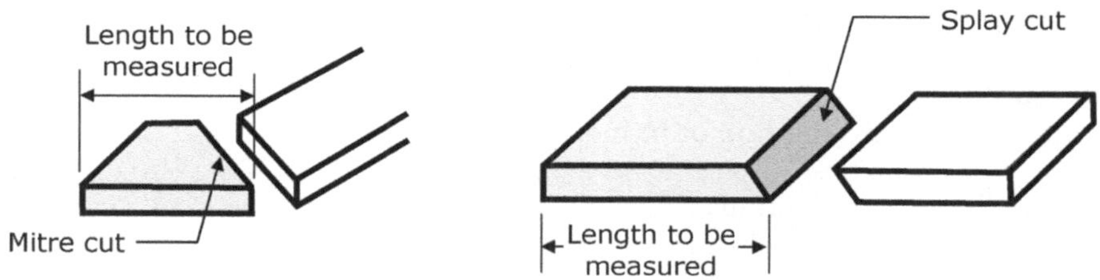

Figure 16.4 Overall length measurement of mitre-cut and splay-cut members.

- The overall lengths should be measured (as **Figure 16.4**) with no deductions for splay cuts or mitred ends or of the mass of metal removed to form notches and holes (XV(a).M.2).
- No allowance (addition) is made for welding, fillets, bolts, nuts etc. (XV(a).M.3).

Steel members and connections

XV(a).2–8 Structural steel framing members

- Steel columns, beams, bracings, purlins, cladding rails, grillages, built-up columns, trestles, towers, trusses and girders are all separately measured.
- Measured in kg.
- State the type, size and weight of sections.
- Where structural steel is not a major part of the structure, it shall be grouped according to the bearing levels for hoisting and fixing which shall be stated in stages of 3m above ground level (XV(a).M.4). For example: 'Steel Structure; hot-rolled steel to BS EN 10025; grade S275; galvanized; bearing level 6.00m – 9.00m above ground level; welded and bolted fabrication'.

XV(a).10 Connections
Connections are essential items used to fix the various steel members to form a complete frame. For instance, cleats, which look similar to angles, are used to hold two members in position. Brackets are for support and gusset plates are for connecting two or more members together (**Figure 16.2**).

- Cleats, brackets, plates, packings etc. are grouped together to arrive at the mass for this item (XV(a).M.5).
- Measured in kg.

Question 16.1

Referring to the diagrams below, should we classify the steel plates as part of the built-up column or as connections?

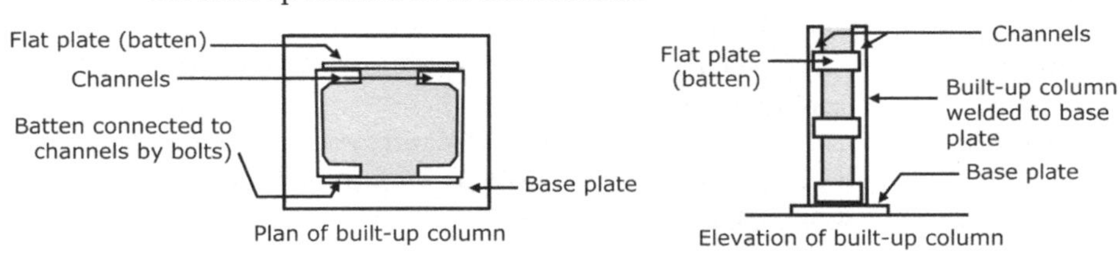

XV(a).12 Proprietary sections

- Measure in m.
- State the type and method of fixing.

XV(a).13 Filling hollow sections

- Measure in number.
- State the size of hollow section, filling material and method of filling.

XV(a).14–15 Bolts, fasteners and the like

- Shop and site rivets, bolts including black bolts (unfinished, ordinary bolts) and high friction grip bolts, nuts and washers are deemed to be included and no need to measure (XV(a).C.1(b)&(c)).
- Holding down bolts or assemblies are measured in number, details such as diameter and length stated.
- Special bolts such as resin anchor bolts or fasteners measured in number, details such as diameter and length stated.
- Welding is held to be included (XV(a).C.1(d)).

Sundries

XV(a).C.1(h) Wedging and grouting under bases

- Wedging and grouting under bases are held to be included.

XV(a).C.1(d),(f) Tests

- Welding tests, non-destructive tests and performance tests are deemed to be included.

Question 16.2

If a base plate of size 1.2m(L) × 1.2m(W) × 30mm thick is used, how is this item measured on the dimension paper?

XV(a).17 Surface treatment

Surface treatment includes painting and fire protection coating.

- Measured in m^2.
- Surface preparation is deemed to be included (XV(a).C.3).
- Painting to additional surface areas of connections such as cleats, brackets etc. are deemed to be included (XV(a).C.2).
- Galvanisation to structural steel members is deemed to be included (XV(a).C.1(e)).
- State the surface preparation, the type of surface treatment (sprayed coating, protective coating or so forth), number and thickness of coats, fire rating and finish.

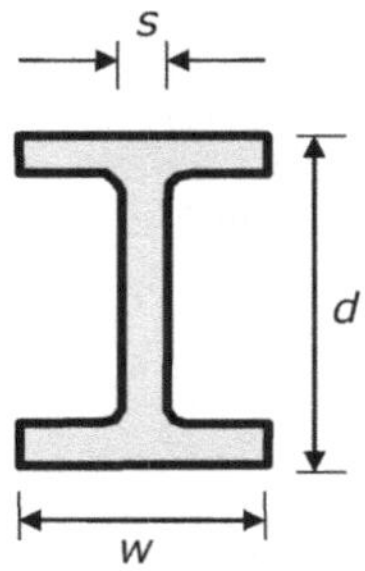

Figure 16.5 Cross-section of a universal column.

For instance, referring to **Figure 16.5**, the area of surface treatment/painting to a universal column is:

> **Points to note**
>
> Universal columns/beams (UC/UB) are described by their serial sizes: depth × width × weight (per metre). For example, UB 127×76×13kg is 127mm (*d*) × 76mm (*w*) weighing 13kg/m. However, the serial size of the UC/UB does not always tell the precise cross-sectional dimensions of the member. For instance, UB 457×152×52kg is 449.8mm(*d*) × 152.4mm (*w*), weighing 52.3kg per m. UC 305×305×198kg is in fact 339.9mm (*d*) × 314.5mm (*w*), which weighs 198.1kg per m. Using the serial size of UC/UB to measure the weight is generally acceptable. However, if you want to calculate the exact girth or cross-sectional size of the UC/UB, it is advisable to check the UC/UB table, which can be found from the internet.

Girth (G) around the column with size $d \times w$

$= 4w - 2s + 2d$

Area of surface treatment/painting $(A) = G \times L$
where
G = girth around the member
L = length of member

Metal works

Linings, coverings and claddings

This section only covers the linings, coverings and claddings to walls, ceilings and fittings. Metal roof coverings and the like are measured in accordance with SMM XV(c).

XV(b).2 Sheet linings

- Measured in m^2.
- State the thickness, quality of material, finish and method of fixing (XV(b).S.1–S.3).
- No deduction for voids $\leq 0.50m^2$ (XV(b).M.4).

- The following works shall be measured separately:
  - curve works (XV(b).M.1)
  - to walls
  - to ceilings, sides and soffits of beams
  - to ceilings and beams that exceed 3.5m above floor, in stages of 1.50m high (XV(b).M.3)
  - to fittings
  - others

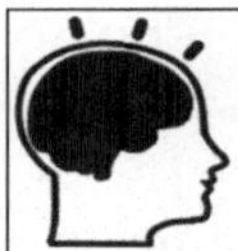

Question 16.3

Do we need to separately measure sheet linings to ceilings and sheet linings to beams?

Doors, gates and the like

XV(d) Doors, gates, shutters, grilles and hatches

- The following items have to be measured separately:
  - Doors
  - Sliding and folding partitions
  - Hatches
  - Strong room doors
  - Grilles, screens and louvres
  - Gates
  - Rolling shutters (**Figure 16.6**)
  - Folding doors
  - Collapsible gates (**Figure 16.7**)
- All measured in number.
- State the dimensions, reference to drawing or manufacturer's catalogue.
- Describe the materials, surface treatment and method of jointing, operating equipment and fire rating (if applicable) (XV(d).S.1–S.7).
- Describe the frames, linings, tracks, rails and shutter hoods (where applicable), ironmongeries required (or alternatively, ironmongeries can be measured separately).

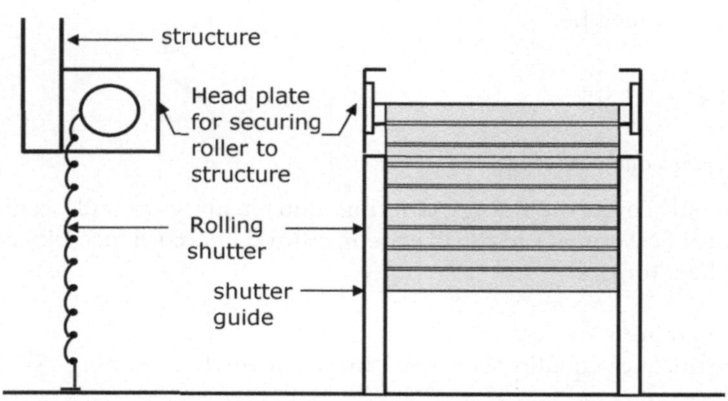

Figure 16.6 Rolling shutter.

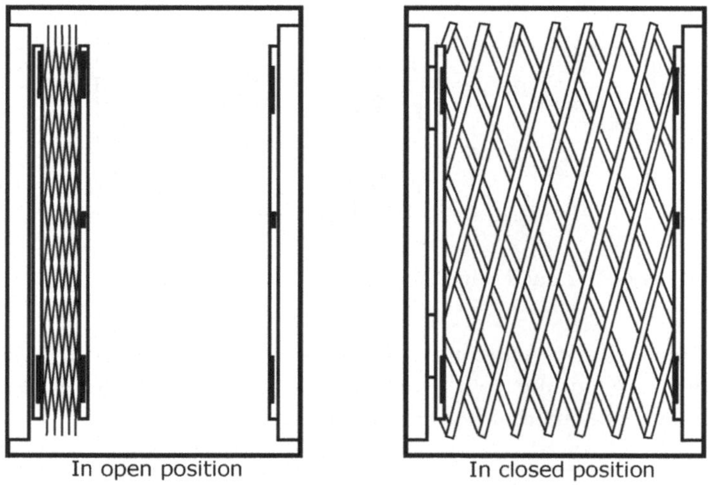

Figure 16.7 Collapsible gate.

Framed work and stairs

XV(e) Framed work, stairs, handrails and balustrades

- Curved works to be measured separately (XV(e).M.2).
- Bars, flats, rods, angles, tees, channels etc. each shall be given separately and state the size (XV(e).M.1).
- Kind, quality of material, surface treatment applied as part of the production process, method of fixing, form of construction etc. shall be described (XV(e).S.1–S.5).

XV(e).2–3 Framed work

- The following works to be measured separately:
  - Gratings
  - Grilles
  - Ladders
  - Balustrades
  - Railings
  - Others (e.g. fencing)

> **Reminder**
>
> Any metal balustrade with glazed panels are regarded as glazed metal balustrades and should be measured in accordance with XV(e).7, not XV(e).2.4.

- Measured in m.
- Work should be described under headings (XV(e).M.3).
- State the number of balusters, standards and ladder rungs (see **Figure 16.8**).
- Brackets to framed work shall be measured in number with dimensions described.

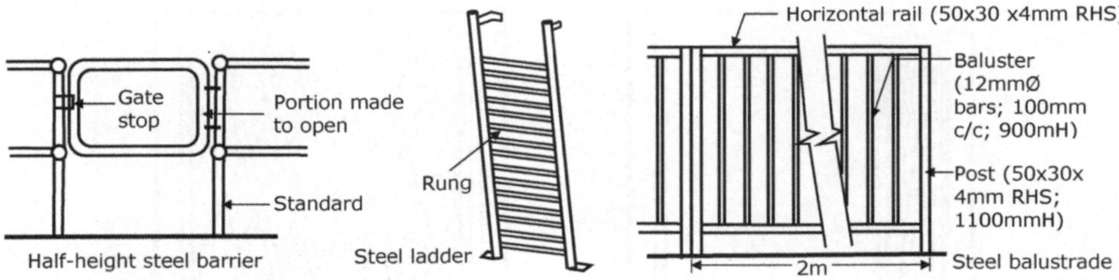

Figure 16.8 Various steel framed work.

- Portions made to open (example see **Figure 16.8**) are measured as extra over framed work in number, with dimensions described (XV(e).4). Any necessary ironmongery is deemed to be included (XV(e).C.2(e)).

Question 16.4

The overall length of the steel balustrade in Figure 16.8 is 6m long. Take-off the quantities.

XV(e).6 Core-rails, handrails, tubular handrails

- Measured in m.
- Bends, angles, ramps, wreaths, junctions, capped ends, flange plates, brackets and the like are deemed to be included (XV(e).C.3).

XV(e).7 Glazed metal balustrades

- Measured in m.
- State the overall dimensions and sizes of each member.
- Give description of glass panels.
- Balustrade in different locations shall be measured separately (XV(e).M.6).
- Metal handrails, glazing panels and supporting fixings (see **Figure 16.9**) are deemed to be included (XV(e).C.4(a)).
- Ends, angles, ramps and the like are deemed to be included (XV(e).C.4(b)).

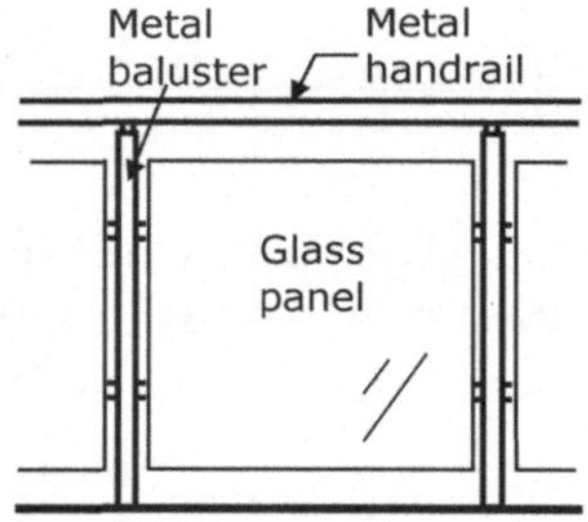

Figure 16.9 Glazed metal balustrade.

XV(e).8 Wire work

- Measured in m^2.
- State diameter of wire and size of mesh.
- Work not rectangular in form shall be measured in square and described as 'in irregular shapes measured square' (XV(e).M.7).

Fencing and gates

XV(f) Fencing and gates

- Where fencing and gates consist of wire mesh, railings or other infill material supported by metal framing, it shall be measured in detail according to Section XV(e). (XV(f).M.1).
- Work is deemed to include excavating holes for supports, backfilling and disposal, earthwork support and making good existing hard pavings (XV(f).C.1).
- Ironmongery shall be measured separately and in accordance with Section XIV Ironmongery (XV(f).6).

XV(f).2–3 Post and wire fencing

- Measured in m.
- State the height of fencing (above ground surface (XV(f).D.1)), spacing, height and depth of supports.
- End posts, angle posts, integral gate posts, straining posts etc. shall be measured in number as extra over fencing, with size, height and depth stated.

XV(f).5 Gates

- Measured in number.
- State the type, height and width.
- Deemed to include gate stops, gate catchers and gate stays.

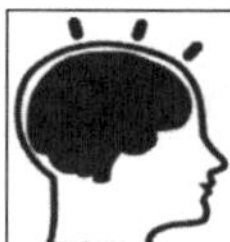

Question 16.5

Which measurement rule should be used to take-off the steel and metal work as shown in cases 1, 2 and 3 respectively?

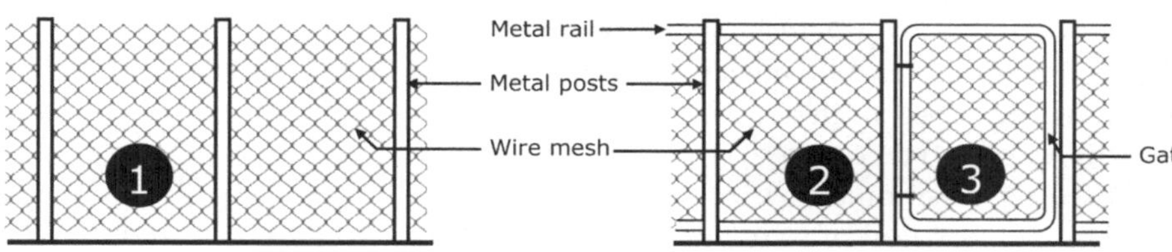

Sundries

XV(h).2–3 Gratings and frames to floor channels and the like

- Floor gratings (see **Figure 16.10**) are measured in m. The width, thickness and general length of each section to be stated.

Figure 16.10 Typical grating to surface channel and air grille.

- Angle frames to floor gratings are measured in m, dimensions and method of fixing to be described.
- All fixing lugs and bedding mortar are deemed to be included.

XV(h).4 Gratings or grilles (see **Figure 16.10**) to openings, ventilators and the like

- Measured in number with dimensions stated.
- Frames to be included in the item.

XV(h)5 Manhole covers and frames

- Measured in number.
- State the dimensions, approximate weight and opening size.
- If comply with BS497 or other Standard, this is stated.
- The following have to be measured separately:
  - rectangular cover and frame
  - circular cover and frame
  - double triangular cover and frame

XV(h)11–12 Tactile studs and strips

Tactile studs and strips are raised truncated domes or bars fixed on the walking surface. They are designed to help vision-impaired pedestrians navigate on the walking surface safely and to provide all pedestrians an anti-slip surface. Stainless steel tactile indicators are durable and popular in Hong Kong.

- Measured in m^2 or m (width stated) or number.
- The following have to be measured separately (see **Figure 16.11**):
  - warning/stop tactiles
  - direction/go tactiles
  - turning/positional tactiles
- Dimensions, method of fixing and background for fixing to be stated.

Partitions and cubicle partitions

This section covers all partitions that are made with a metal framework. The measurement rules for steel partitions are the same as those for the timber partitions. The detailed classification and measurement method can be referred to **Chapter 15** and is not repeated here. Only the general rules are highlighted below.

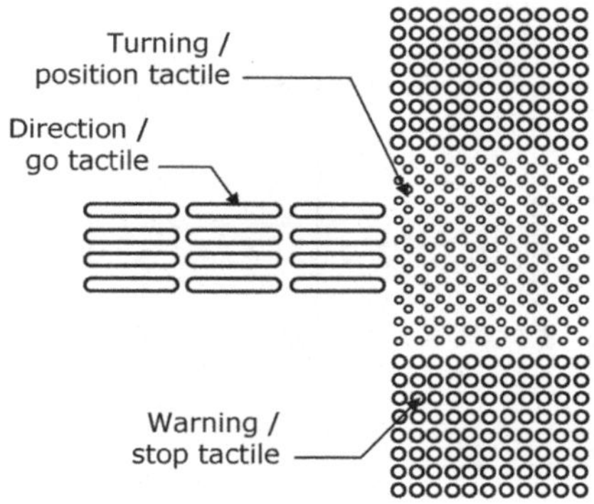

Figure 16.11 An example of tactile guide path.

XV(k).2–3 Fixed partitions and demountable partitions

- Measured in m. Mean length as fixed shall be measured.
- State thickness and height.
- Fixed partition, demountable partition, straight work and curved work are each measured separately.
- Partitions are deemed to include all integral components preformed at factory but exclude trims fixed on site.
- Openings on partitions for doors, glazed panels and the like are measured as extra over the partitions in number. Except for trims fixed on site to be measured in m, ironmongery, components filling the openings such as glass panels, linings and the like are deemed to be included in the openings measurement (XV(k).4–5).

XV(l) Toilet cubicle partitions and office cubicle partitions

- Toilet cubicles and proprietary office cubicles have to be measured separately.
- Measured in sets (or number for office cubicles).
- State the catalogue number and detailed dimensions.
- All integral components, doors, ironmongery, glazed panels (in office cubicles) and the like are included, but trims that are fixed on site are excluded.

Suspended ceilings

- Suspended ceiling of plasterboard or of plastered metal lathing are measured in Section XVI – Plastering and Paving (XV(m).M.2).

XV(m).2–3 Tiled or panelled ceilings, metal strip ceilings

- Measured in m^2.
- Kind, material and catalogue number (if any) should be described.
- Works over 3.50m above floor to the ceiling soffit shall be measured separately, in stages of 1.50m (XV(m).M.3).

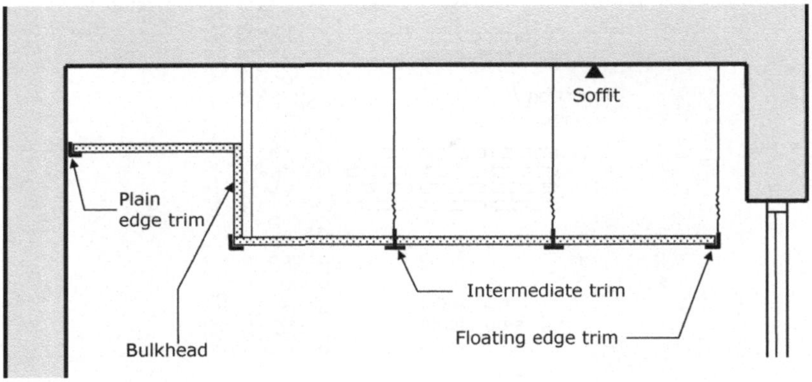

Figure 16.12 Typical section of suspended ceiling system.

- For patterned, sloping and curved works, details to be stated and measured separately.
- Measured separately for different depths of suspension (see **Figure 16.12**):
    - Depth ≤ 1.50m
    - Depth > 1.50m, thereafter in stages of 1.50m

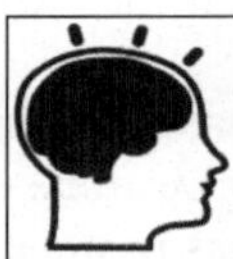

Question 16.6

When measuring the depth of suspension for classification purpose, which of the following should be used?

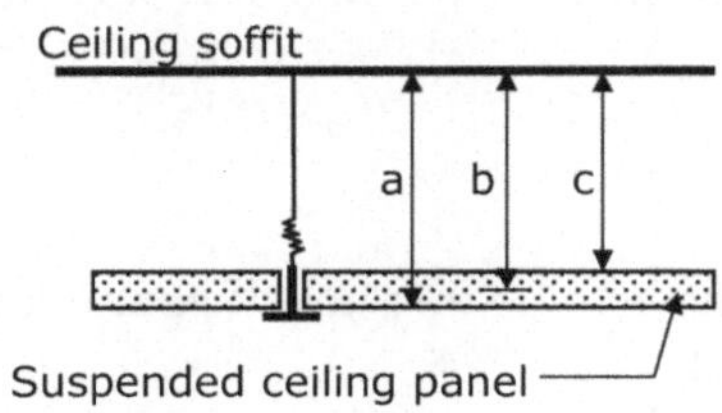

- All work is deemed to be internal unless described as external (XV(m).D.1).
- No deduction is made for voids ≤ 0.50m² (XV(m).M.4).
- All hangers, suspension and framed members, edges and trims (see **Figure 16.12**) are deemed to be included.

Reminder

Under HKSMM4 Rev 2018, all edges and trims of metal suspended ceiling are deemed to be included. Also, the suspension depth range for classification is widened to allow simpler measurement.

XV(m).4 Access panels to suspended ceilings

- Measured as extra over the ceiling in number.
- State dimensions.
- Trimming and necessary fixings shall deemed to be included.

XV(m).5 Upstands, bulkheads

- Measured in m.
- Classified as height ≤ 600 mm and thereafter in 300 mm stages (see **Figure 16.12**).

Windows and glazed doors

XV(n) Windows and Glazed Doors

- Measured in number.
- State the overall size and drawing reference.
- Describe the kind and quality of materials.
- Describe the number of fixed lights, number and type of opening lights.
- Describe the glazing and ironmongery.
- Deemed to include all fixings (such as fixing lugs), glazing, glazing beads and compounds, sealant and ironmongery (XV(n).C.1).

Reminder

According to HKSMM4 Rev 2018, all glazing panels of the windows and doors are deemed to be included in the window/door items.

Points to note

Since windows and doors (in both Wood Works Section and Steel and Metal Works Section) are measured in number, it is more practical and convenient to prepare the take-off by schedules. By counting the number of each type from the drawings, the total quantities can be found easily. For instance, the windows of a ten-storey building can be summarised as:

Location	*W1*	*W2*	*W3*	*W4*	*W5*	*W6*	*W7*	*W8*
East wing		3	2	3	2	1	3	1
West wing	2	3	3	1	1	2	2	2
Stair A	1		1					1
Stair B	1		1					1
Total per floor	4	6	7	4	3	3	5	5
Total for 10 floors	40	60	70	40	30	30	50	50

Assuming a drawing no. WD-10 shows the layout of W1 and W2 as below. The measurement for W1 can be written as:

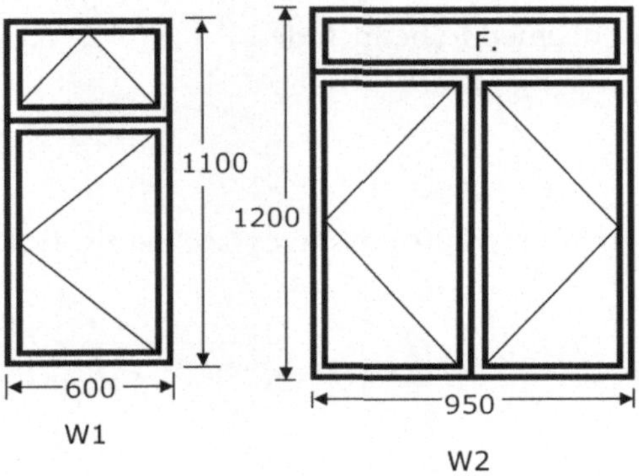

			<u>Designing, supplying and installing windows; hot-rolled steel to BS EN 10025; galvanised; complete with 12mm thick tempered glass; framing, water bars, fittings, fixing lugs, brackets, bolts and ironmongery; glazing beads; PVC weatherstrip; assembling, jointing, cutting and pinning lugs; bedding in water-proof cement grouting; as Drawing no. WD-10</u>
	40		Windows 600 × 1100mm overall (Type W1); one top hung opening light; one side hung opening light

Question 16.7

Complete the measurement for W2 based on the information above.

Furniture, fittings, signage etc.

XV(q) Furniture, fittings, shelving, racks, playground equipment, signage etc.

- Measured in number.
- State the drawing reference and/or dimensioned description as appropriate.
- State the thickness, kind and quality of material, finish and surface treatments.
- State the background for fixing.
- Describe ironmongery and/or metal fittings, glass panels etc. which are to be included in the items.

Suggested answers

Question 16.1

The flat plates (battens) on the sides of the double channels are part of the built-up column. All the battens and the channels shall be measured under the 'built-up column' item. However, the base plate is only a connection piece and should be measured under 'connections'.

Question 16.2

	1.20 1.20		Connections (_____ × 235.3kg/m^2 = _____kg)	*The unit weight of steel is 7,843kg/m^3. The unit weight (kg/m^2) of a 30mm steel plate can be obtained by the following equation:* *30mm thick steel plate unit weight* *= 7,843 × thickness of steel plate* *= 7,843 × 0.03* *=235.3kg/m^2* *Noticed that base plates are measured as connections which will be measured together with other connections such as cleats, angles and gusset plates.*

Question 16.3

No, we don't need to separately measure sheet linings to ceiling and sheet linings to sides and soffits of beams. All of them can be grouped together and measured as 'metal sheet linings to ceilings, and sides and soffits of beams'.

Question 16.4

			<u>Framed balustrades; hot-rolled galvanised steel to BS EN10025; Grade S275; welded and bolted connections</u>	*A heading is required.*
3/	1.94		Top rails; 50 × 30 × 4mm RHS	*2 × 30mm (width of the rectangular hollow section posts) has been deducted from the length of horizontal rails.*
3/	1.94		Bottom rails; 50 × 30 × 4mm RHS	*Individual members that form the framed balustrade are measured separately.*
6/	1.10		Posts; 50 × 30 × 4mm RHS (in 6 no.)	
3/ 19/	0.90		Balusters; 12mm dia. bars (in 57 no.) (Brackets are excluded here)	*No. of balusters and the like has to be stated.*

Question 16.5

Case 1: Item XV(f).2 – Post and wire fencing, measured in m.

Case 2: Item XV(e).2 – Framing measured in m; Item XV(e).8 – Wire mesh measured in m^2.

Case 3: Item XV(e).4 – Portions made to open, measured as extra over framed work in number.

Question 16.6

Dimension 'a' should be used when measuring the depth of suspension. According to the SMM, item XV(m).M.5, the depth of suspension should be measured from the structural soffit to the face of suspended ceiling.

Question 16.7

	60	950 × 1200mm overall (Type W2); one fixed light; two side hung opening lights	*Fixed light is a window or portion of a window that is not intended to be opened; with the glazed panel inserted directly to the window frame.*

17 Measurement of plastering, paving and painting

Introduction

This chapter includes two SMM sections: the Plastering and Paving Section (XVI) and the Painting Section (XXI). Both sections are related to finishes.

The Plastering and Paving Section covers all the finishing work, except for those related to Wood Works, Steel and Metal Works and Painting. Most of the preparation for finishing work such as screeding is also covered in this section. Major work items in this section include:

- Spatterdash
- Lathing and plasterboard
- Screeds
- Tiling and sheeting finish

Besides plastering and tiling, painting is also a popular finish for building works in Hong Kong. The Painting Section covers:

- Painting
- Polishing and clear finish
- Signwriting
- Wallpaper and fabric hanging

Plastering and paving

Spatterdash

XVI(a) Spatterdash

Spatterdash is normally applied to concrete surfaces before plastering to improve the adhesion of plaster.

- Measured in m^2.
- Work to all surfaces such as columns, walls etc. are measured together.

Lathing and plasterboard

XVI(b).1–5 Metal lathing, plasterboard and fire-rated enclosure.

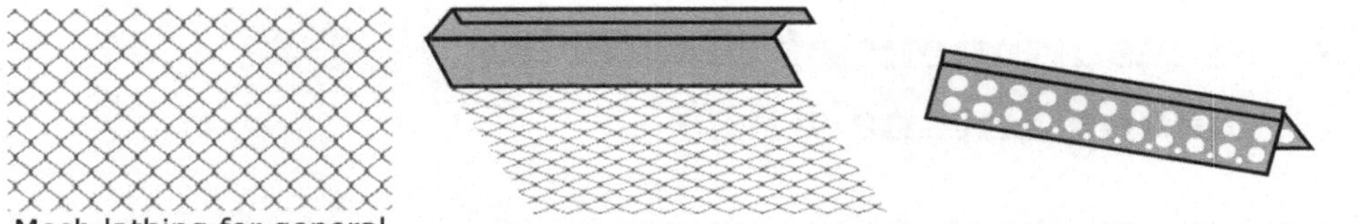

Figure 17.1 Common types of metal lathing and beads.

- Metal lathing (see **Figure 17.1**) is often required in screeding and plasterboard works for various purposes such as:
  - to provide keying for plaster work.
  - to act as corner reinforcement.
  - to reduce cracking of plaster where it is likely to occur such as over the windows and doors.
- Metal lathing, plasterboard and fire-rated enclosures are measured separately in m^2.
- State the kind and quality of material.
- For metal lathing, side and end laps to be stated.
- For plasterboard, thickness of plasterboard and thickness of skim coat (if applicable; XVI(b).S.3) to be stated.
- For fire-rated enclosure, state the materials, thickness and fire resistance period.
- Work to different parts of the building such as walls and columns, ceilings, pipe enclosure etc. should be measured separately.
- Curved work has to be measured separately (XVI(b).M.2).
- Work to ceilings and beams > 3.5m from the floor level has to be measured separately, stating the height in stages of 1.5m (XVI(b).M.4).
- Work is deemed to include narrow widths and small quantities (XVI(b).C.2(a)).

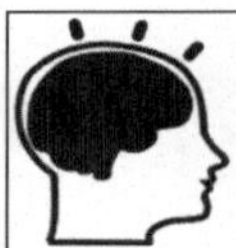

Question 17.1

'Narrow widths and small quantities are deemed to be included' – does it mean that we do not need to measure the narrow widths and small quantities?

- Timber or steel framing to support plasterboard or fire-rated board is held to be included in the item (XVI(b).C.3).
- No deduction is made for voids $\leq 0.5m^2$, nor for voids $\leq$ 300mm wide (XVI(b).M.3).
- Beads for protecting edges (stop beads) or corners (angle beads) (see **Figure 17.1**) are measured separately in m (XVI(b).4–5).

In-situ finishes and screeds, tiled, slab, sheet and carpet finishes

General rules for screeds, tiled and sheet finishes

- The following work should be measured separately:
  - curved work (XVI(c).M.4)
  - confined space work (XVI(c).M.7)

- work in repairs (XVI(c).M.8)
- external work (XVI(c).D.1)
- Work to ceilings and beams > 3.5m from the floor level, stating the height in stages of 1.5m. (XVI(c).M.6)

- Work to floors and ceilings is measured to the area between structural walls (XVI(c).M.2)
- Work to walls and the like is measured to the area of the base, i.e. structural background (XVI(c).M.3, XVI(c).D.2)
- Work is deemed to include preparation of background as specified (XVI(c).C.1(a))

Question 17.2

Do we need to measure the tiling work in the grey area as curved work?

Plan

Question 17.3

The Architect specifies, 'Before applying screeding to concrete ceilings and walls, the surfaces must be prepared with spatterdash'. Is applying the spatterdash here equivalent to the 'preparation of background' as mentioned in item XVI(c).C.1(a)?

- Work is deemed to include narrow widths and small quantities (XVI(c).C.1(b)).
- Work to floors and pavings is deemed to include laying to slopes and falls (XVI(c).C.2).
- No deduction is made for voids $\leq 0.5\text{m}^2$, nor for voids ≤ 300mm wide (XVI(c).M.5).
- Movement joints, expansion joints, cover strips, division strips and non-slip inserts have to be measured in m (or alternatively in number for non-slip inserts).
    - State the dimensions and method of fixing.

Question 17.4

Movement joints and the like are found in various sections and sub-sections of the SMM. Can we group the movement joints applied at the different floor finish, for instance, marble flooring and timber flooring, and measure them under the same item?

In-situ finishes and screeds

XVI(d).1–10 In-situ finishes and screeds

In-situ finishes and screeds cover rendering, plastering, screeding, granolithic terrazzo, bitumen emulsion and asphaltic concrete or macadam and the like (XVI(d).M.1).

Points to note

Bitumen emulsion, asphaltic concrete and macadam are various products that combine bitumen/asphalt with water and emulsifier or aggregates for road construction and paving. They should not be confused with the waterproofing work, Section X.

- Each type of work such as rendering, screeding, terrazzo etc. has to be put under a separate heading (XVI(d).M.1).
- Measured in m^2.
- State the thickness and number of coats for screeds and the like. For bitumen emulsion, asphaltic concrete or macadam to roads/pavings, state the thickness.
- For screeds, the materials to be received e.g. quarry tiles, marble tiles etc. shall be stated and each measured separately.
- Besides the general rules, the following works shall be measured separately:
  - work on plasterboard
  - work on metal lathing

Question 17.5

If screeds and the like applied to plasterboard and metal lathing must be measured separately, how about if the screed is applied to the concrete surface and the brick wall surface (without metal lathing)? Should we measure the work to these two surfaces separately?

- Work to different parts of the building are each measured separately:
  - walls and columns
  - ceilings and sides and soffits of beams (horizontal and slopping measured separately)
  - soffits of stairs (sloping and flewing soffits measured separately)
  - treads and risers
  - floors
  - roofs
  - others
- Coves, grooves, rounded external angles to screeds and the like; and angles, stops, intersections, fair ends etc. on coves, skirtings, channels, curbs and the like are deemed to be included (XVI(d).C.1.(c–d)).

Figure 17.2 In-situ finishes or screeds to various skirting profiles.

XVI(d).13–19 In-situ finishes and screeds to miscellaneous items

- Screeds and the like to the following items shall be measured in m:
    - skirtings; state the thickness of screed (or the like) and the height of skirting above finished floor; each type in **Figure 17.2** is measured separately.
    - wall strings and open strings (**Figure 17.3**); each measured separately, state the thickness of screed (or the like) and the average height of the string.
    - curbs; state the thickness of screed (or the like) and the girth of curb.
    - channels of different finish to floor; state the width and depth of the channel.
    - angle fillets; state the width and depth of the angle fillet.

Reminder

An angle fillet looks like a cove as shown in the diagrams. According to HKSMM4 Rev 2018, angle fillets are measured separately (XVI(d).19) but coves are included in the in-situ finishes and screed items (XVI(d).C.1).

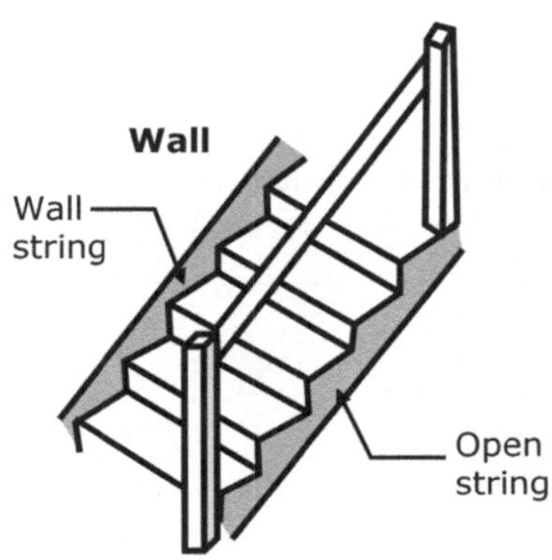

Figure 17.3 Wall string and open string.

Question 17.6

According to the SMM items XVI(d).14–15, wall strings and open strings are measured in m with their average height stated. With reference to the diagram,

i what is the average height of the wall string?
ii what is the average height of the open string?

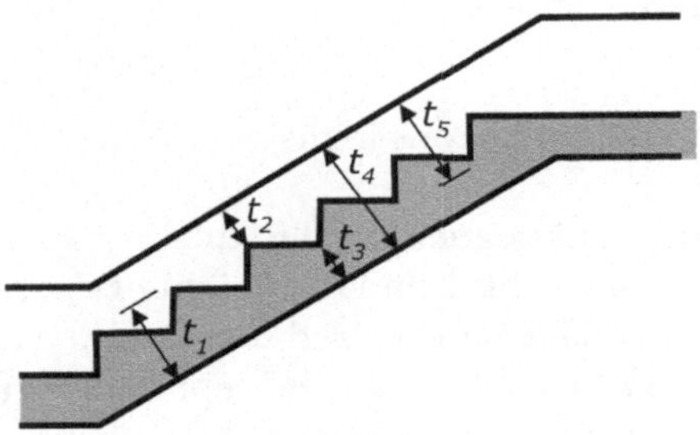

Question 17.7

Referring to the sectional diagram of the curb, how long is the 'girth' to be stated for the cement and sand screed to the curb?

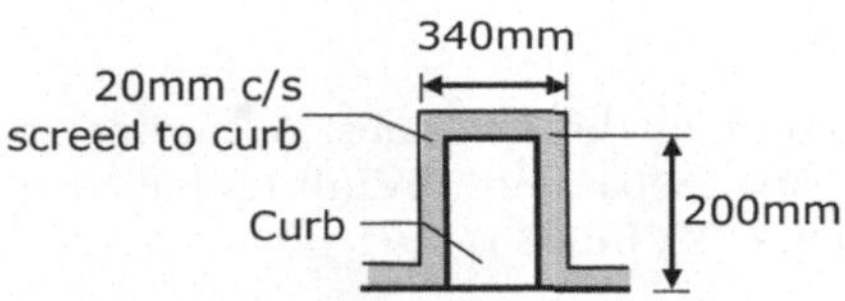

Tiled and slab finishes

This part covers finishes such as ceramic, synthetic rubber, granite, marble, slate and mosaic tiling, precast items and the like. Each type should be grouped under a separate heading (XVI(e).M.1).

Reminder

Large pieces of granite, marble or slate slabs installed to structural backgrounds are measured in accordance with Stone Works. (XVI(e).M.2)

XVI(e) Tiled and slab finishes

- Measured in m^2.
- State kind of the materials, size, shape and thickness of the tiles.
- Describe the nature of the base, underlay, bedding, adhesive and fixing details.
- Details of pattern (for patterned work), size of inserts, colour (in case of colour work) shall be fully described.

- Work to different parts of the building are measured separately:
  - walls and columns
  - treads and risers (measure separately)
  - floors and pavings
  - external cills, copings, horizontal louvres and the like
  - external vertical fins, sunbreakers and the like
  - all surfaces of external small features such as air-conditioner boxes

Tactile tiles XVI(e).24

- Measured in m^2 or m (with the width stated).
- The following types (see **Figure 16.11**) should be measured separately:
  - warning/stop tactiles
  - direction/go tactiles
  - turning/positional tactiles
- Dimensions, method of fixing and background for fixing to be stated.

Reminder

Steel tactile studs and strips are measured under steel and metal works but tactile tiles are measured under plastering and paving works. Tactile tiles can be made of ceramic, rubber and the like.

XVI(e).9-15 Tiled and slab finishes to miscellaneous items

- The following items are measured in m, with any coves or splays, rounded edges; height above finished floor or girth stated:
  - skirtings
  - strings
  - aprons (see **Figure 17.4**)
  - curbs and upstands (tops and sides)
  - channels (sides and bottoms)
- Different profiles such as level, vertical, raking (i.e. work on a slope), raking with stepped base and stepped following profile of treads and risers (see also **Figure 17.2**) are each measured separately.

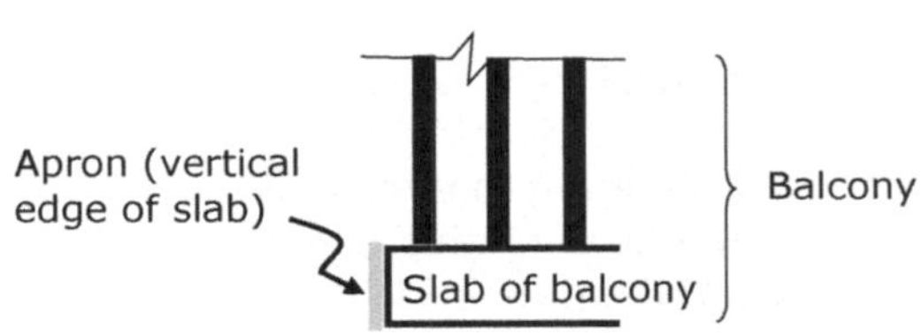

Figure 17.4 Example of an apron.

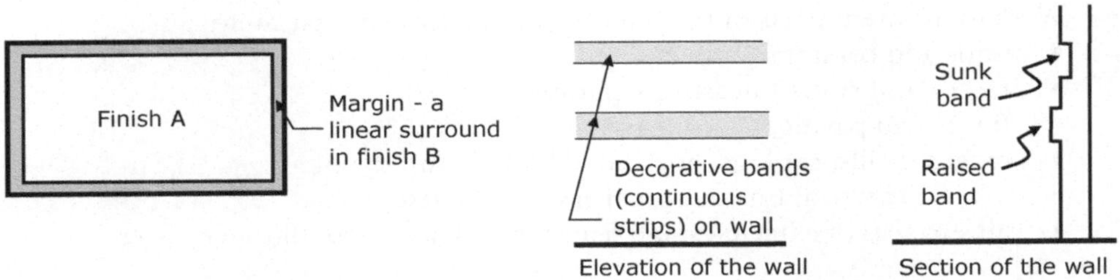

Figure 17.5 Margins and bands.

Non-standard tiles XVI(e).16

- Non-standard and special tiles such as nosing bead tiles, rounded edge tiles, corner tiles, margins and raised/sunk bands in contrasting colour (**Figure 17.5**) etc. shall be measured as extra over the items in which they occur and measured in m (XVI(e).16).

Question 17.8

If ceramic tiles are used to finish a square column, do we need to measure the tiles at the corners separately?

Rubber and plastic sheeting, Carpeting and the like

This part covers synthetic rubber, linoleum, cork, rubber, vinyl sheeting, fibreglass membrane, ball court finishes, acoustic sheeting, carpet, carpet tiles, artificial turfing, and the like. Each type should be grouped under a separate heading.

Points to note

Linoleum and vinyl often get mistaken with each other. Linoleum is made by pressing natural raw materials such as linseed oil, ground cork dust, wood flour, pigments and mineral fillers. Vinyl is usually polyvinyl chloride, a synthetic plastic made from ethylene and chlorine. Both materials are popular for floor finish.

XVI(f) Rubber and plastic sheeting, carpeting and the like

- Measured in m^2 in general.
- Kind of material, pattern, details, size, inserts, and surface treatment shall be fully described.
- State the nature of base, bedding, nature and number of underlay, adhesive, extent of laps, laying direction and fixing details.
- Work to different parts of the building are measured separately:
    - walls and columns
    - treads and risers (each measure separately)

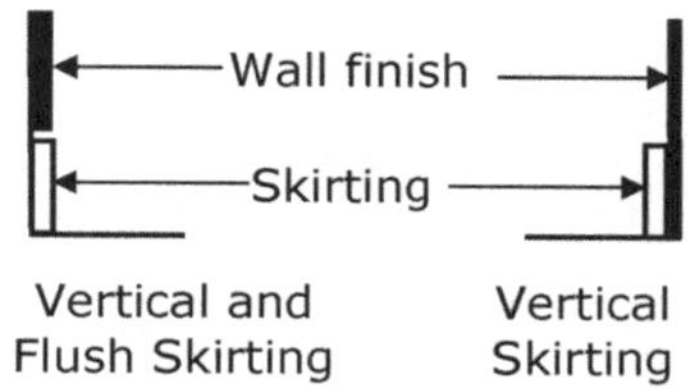

Figure 17.6 Sections of different skirting profiles.

- floors and pavings
- strings
- aprons
- Rounded internal and external angles > 100mm radius are classified as curved work (XVI(f).D.1).
- Measurement rules for rubber tactile tiles are similar to those for ceramic tactile tiles, measured in m^2 or m (XVI(f).19).

XVI(f).8–9 Rubber, carpet etc. to skirtings and curbs

- Measured separately in m:
  - Skirtings; with height stated
  - Curbs; with girth stated
- Finishes to skirtings/curbs in various profiles, such as flush, vertical (see **Figure 17.6**) and raking have to be measured separately.

XVI(f).16–18 Sundries

- Line markings on sports pavings shall be measured in m; width stated (XVI(f).16).
- Letters or figures on sports pavings shall be measured in number; height stated (XVI(f).17).
- Nosings shall be measured in m with dimensions stated.
- Stair rods (see **Figure 17.7**) shall be measured in number with dimensions stated.

Painting

This section covers general painting and polishing work to different types of materials such as masonry work, timber and steel. Fabric and wallpaper hanging are also included in this section.

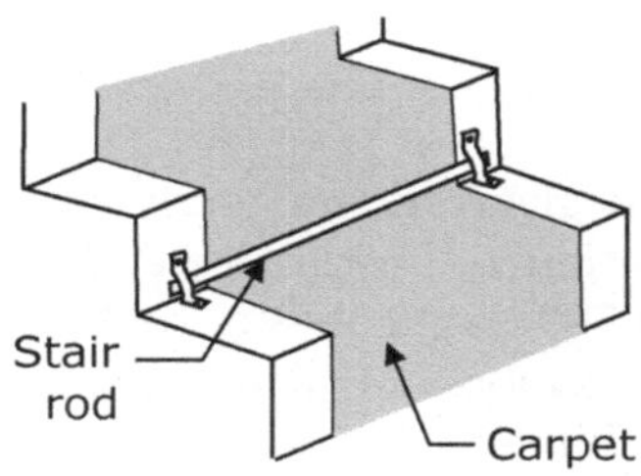

Figure 17.7 Stair rod.

XXI(a) General

- Work on different types of surfaces such as concrete, brick, plaster, stone, metal, wood etc. shall each be given separately (XXI(a).M.1).

Question 17.9

Two coats of cement paint will be applied to fair faced concrete wall surfaces and standard concrete column surfaces. Do we need to measure the cement paint to these two surfaces separately?

- External work to be stated and measured separately.
- Works to ceilings and beams exceeding 3.5m from the floor shall be given separately, in stages of 1.5m (XXI(a).M.5).
- No deduction is made for voids $\leq 0.50m^2$ (XXI(a).M.6).
- The following is deemed to be included:
  - multi-colour work (XXI(a).C.1(a))
  - preliminary preparatory work before painting (XXI(a).C.1(b))
  - removal of ironmongery and hardware before painting and reinstatement after painting or provide protection to the items (XXI(a).C.1(e))
- Describe the type of paint and the method of finish (if any) (XXI(a).S.1–2).
- State the preparatory work such as knotting and stopping, priming and sealing coats, the number of undercoats and finishing coats (XXI(a).M.7).

XXI(b) Painting

- Painting work is generally divided according to girth/surface area, as follows:
  - Girth > 300mm wide: measured in m^2
  - Surfaces with girth $\leq$ 300mm: measured in m
  - Isolated surfaces $\leq 0.50m^2$: measured in number
- Work to different areas such as general surfaces, walls and columns, ceilings and beams, corrugated surfaces and the like are measured separately. Special attention has to be paid for the following items:

Question 17.10

How do we classify painting work as 'general surfaces' work? Give a few examples.

  - Work on windows and glazed doors, mesh-covered fencing, gates, ornamental fencing, grilled or louvred surfaces etc. shall be measured flat overall on both sides (XXI(b).1.7–9) regardless of voids (XXI(b).M.2).
  - Work on individual members of plain, open-type balustrades, fencing (see **Figure 17.8**), gates, railings and the like are described according to the size of individual members, i.e. either in m^2 or m depending on whether the individual member's girth is > or $\leq$ 300mm (XXI(b).1.10) and (XXI(b).M.1).

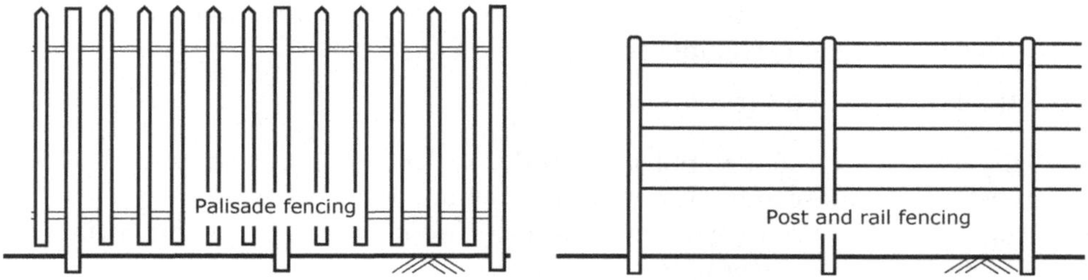

Figure 17.8 Examples of open-type fencings.

Reminder

Work on mesh-covered/ornamental fencing and work on open-type fencing are measured in different ways. In the absence of a definition rule for open-type fencings, it is reasonable to consider the open-type fencing as fences that contain gaps in between the boards or vertical members.

Question 17.11

How do we measure the painting work to the glazed screen as shown in the simplified diagram?

i if it is an internal screen.
ii if it is installed on the external wall.

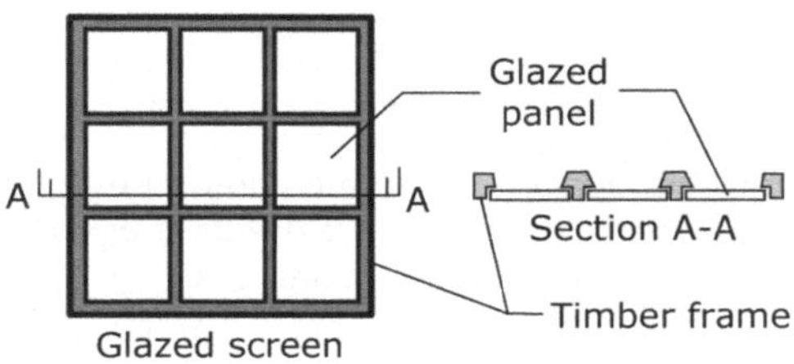

Question 17.12

If the same type of painting and treatment is applied to the following items on site, how do we measure the work? Can we group them together under one item? What is the measurement unit?

i Timber door
ii Timber door lining
iii Timber furniture

- Painting on pipes and conduits are also classified in the same way according to the girth of each pipe.

XXI(c) Polishing and Clear Finishing

- This sub-section refers to polishing and clear finishing such as wax and shellac polishing, cellulose and epoxy clear finishes and the like.
- Work to timber floors, general surfaces of woodwork and general surfaces of marble/granite are measured separately.
- All work is classified as follows:
    - Girth > 300mm: measured in m^2
    - Isolated surfaces with girth ≤ 300mm: measured in m
- State the type of finish and preparatory work (e.g. staining).

XXI(d) Signwriting

- Straight and curved lines/arrows shall be measured separately.
- Lines and broken lines shall be measured in m; width stated.
- Arrow heads, tails, letters, numerals, punctuation marks, Chinese characters, symbols or logos shall be measured in number; height stated.

XXI(e) Paper and fabric hanging

- Wallpaper and fabric linings shall be measured in m^2.
- State the manufacturer's reference and the adhesive to be used.
- Matching pattern and cutting around pipes, switches and the like is deemed to be included.
- Work is classified according to the building parts that receive the paper/fabric:
    - to walls and columns
    - to ceilings and beams
- Border strips and the like shall be measured in m (XXI(e).2).

Question 17.13

A wall is divided into two portions (an upper portion and a lower portion) by a horizontal dado strip. All three parts are finished in different types of wallpaper. Can we measure all the paper hanging work together in one item?

Finishes

As mentioned in the beginning of this chapter, plastering and painting operations are typically related to finishing work. Having examined the measurement rules associated with the two sections, a quick review of the overall procedures to measure internal/external finish should be useful.

Internal finishes

Depending on the range of materials and application methods adopted, measurement of internal finishes can be more complicated than it appears to be. Fixing method, specified treatment of background/base and the required surface finish of the finishing material should be noted with care. Most taker-offs prefer to measure finishes by a schedule, which is introduced in **Chapter 9** and not repeated here.

Although the SMM does not included any rule for measurement sequence, many authors suggested a general approach (Packer, 2014; Seeley and Winfield, 1999), which is also the classical approach used by the local industry for measuring internal finishes. The measurements should be dealt with in a floor by floor, room by room manner. The overall approach can be summarised as below:

1 Measure the floor finish together with the ceiling finish.

Since the floor area and the ceiling area are the same, they are often measured together. Measure the areas between structural walls (XVI(c).M.2). All the isolated beams and columns, fixtures/furniture are ignored at the moment. Adjustments will be made later.

2 Measure the wall finish.

Measure the girths of the room to the area of the base (structural background, metal lathing or plasterboard) (XVI(c).M.3 & XVI(c).D.2). By multiplying the girths with room height, wall areas can be obtained. Similar to the floor/ceiling finish, all the fixtures, furniture, windows, doors and the like are ignored. Adjustments will be made later.

Question 17.14

The actual girth length of the wall tiling should be shorter than that of the wall screed. Therefore, the girth of the wall tiling should be adjusted for the thickness of the screed. Is this correct?

3 Measure the cornices, dadoes, skirtings and the like.

Use the girths of the walls to calculate the length of skirtings, cornices, dadoes and the like. Adjustments will be made later.

4 Make adjustments

Make due adjustments for the finishing areas/lengths with respect to the isolated columns, beams, windows, doors, openings, furniture and fixtures. Here are some reminders for the tasks that require extra caution:

- Deduct the floor areas and ceiling areas for the isolated columns and add the vertical areas to the walls and columns finish.
- Deduct the wall areas for windows only when they are $> 0.5m^2$.
- Although ceilings and sides and soffits of beams are grouped as one item according to XVI(d).2, the finish to the sides and soffit of a beam is often different from the ceiling, but instead the same as the finish to the wall.
- Deduction of the skirting lengths and dado lengths should be made for the door openings. Deduction of the dado lengths and the like may be required for the windows depending on the size and position of the windows.
- Areas of reveals around doors, windows and bay windows should be added accordingly.
- Wall finishes may or may not be applied behind the fixtures such as cabinets. The same may happen in the floor finishes as well.
- Special tiles such as corner tiles may be required at the tiled surfaces.

External finishes

In general, external finish mainly covers the walls, floors, soffits and paving finish. Unlike internal finishes, there is no inter-relationship between the walls, floors and soffits areas, and each of them can be measured independently. Here are some reminders for the tasks that require extra caution:

- All the finishing work is deemed to be internal. A clear heading to state that the work is external must be given.
- Most buildings consist of typical and non-typical floors. It is advisable to divide the external finish into these two portions and measure them separately. A floor-by-floor approach should be adopted.
- When measuring the external wall finish, multiply the girth of the external wall with the floor height. Then adjust the areas for features such as balconies, planters etc. Multiply the area with the number of typical floors accordingly.
- Remember to measure the finishes on all sides of external features.

Suggested answers

Question 17.1

'Narrow widths and small quantities' are deemed to be included in the plastering and tiling work. While there is no separate item for the narrow widths such as the soffit and reveals of a door opening, we normally measure the areas involved in the narrow widths together with the item where the narrow widths occur. Therefore, the area of the soffit and reveals will be added to the overall area of the wall when we measure the lathing, screeding or tiling over there.

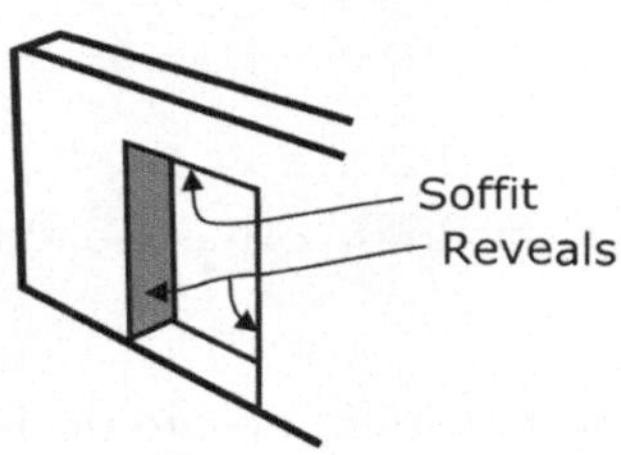

Question 17.2

The spatterdash applied before screeding or plastering is not regarded as 'preparation of background' and therefore must be measured in accordance with item XVI(a). Preparation of background refers to treatments such as removal of dirt, grease and markings, grinding or scrabbling the surface and the like to enable screeding/plastering to stick.

Question 17.3

No, the floor tiling in the grey area is not curved work.

According to SMM, curved work refers to work on curved surfaces such as:

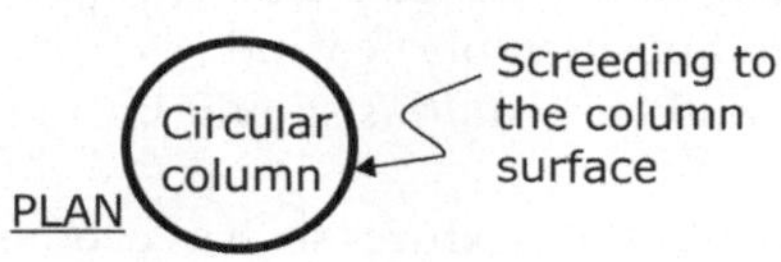

Question 17.4

Movement joints, cover strips and division strips appear in various sub-sections of the Plastering and Paving Trade including XVI(d), (e) and (f), as well as in the XI(b) section (stone facings) and XIII(c) section (timber flooring). However, the movement joints at the marble flooring should not be grouped together with those at the timber flooring.

First, the materials and installation details of movement joints are usually different when applied at different floor finishes. Second, it is easier to check the quantities and to sublet the work to separate specialists if the accessories are measured in each corresponding trade section.

Question 17.5

No, cement and sand screed to concrete surface and brick wall surface are measured together. SMM4 Rev 2018, XVI(d).1.1–3 stated clearly that only the works on plasterboard and on metal lathing are measured separately. In other words, in-situ finishes, screeds and the like are classified into three groups if referred to the base: (1) on general surfaces, (2) on plasterboard and (3) on metal lathing.

Question 17.6

i The average height of wall string is:

$$= \frac{t_2 + t_5}{2}$$

ii The average height of open string is:

$$= \frac{t_1 + t_3}{2}$$

Note that we are not taking the extreme or overall height as we have done in measuring the formwork. When measuring the length of the wall string and open string, the length of soffit (L) can be applied.

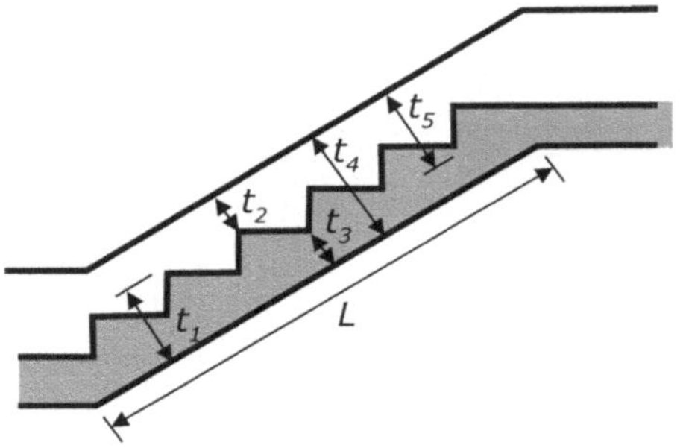

Question 17.7

Face girth of the curb should be measured along the surface of the curb, not the surface of the screeding, which is:

$= a + 2b$
$= 200 \times 2 + 300(\text{mm})$
$= 700(\text{mm})$

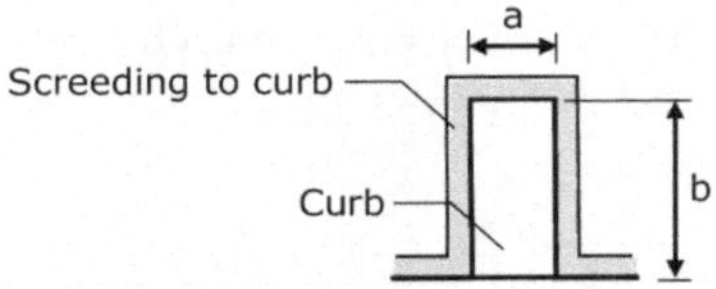

Question 17.8

External corners, unless using corner tiles or rounded edge tiles, are not required to be measured separately. The same applies to internal corners or edges.

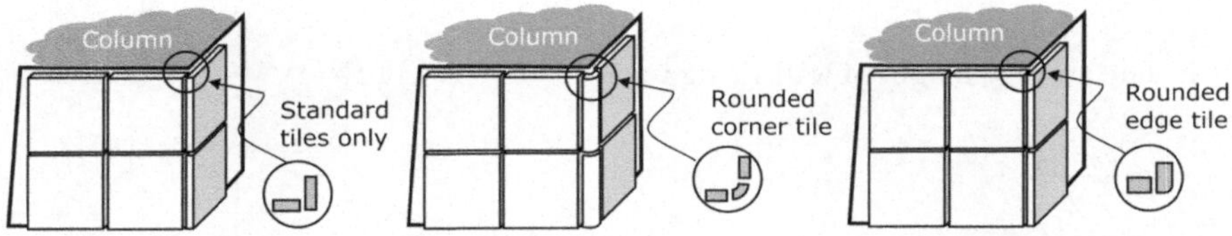

External corner tiled with standard tiles

External corner tiled with rounded corner tiles

External corner tiled with rounded edge tiles

Question 17.9

SMM4 item XXI(a).M.1 requires painting and paperhanging on concrete, brick, plaster, stone, metal, wood or other surface to be measured separately. This means that the base must be identified and painting to different bases has to be measured separately. Although painting to walls and columns can be grouped together (XXI(b).1.2), the work to the two areas needs to be measured separately here, one as cement paint to concrete walls and columns and the other as cement paint to fair faced concrete walls and columns.

Question 17.10

General surfaces work covers painting to areas other than items XXI(b).1.2–10, for example floors, flush doors and frames, skirtings etc.

Question 17.11

According to SMM4 Rev 2018, XXI(b).1.7, glazed screens and the like are measured flat on both sides. The painting work should be measured in m^2. Note that work to opening edges of the frame is deemed to be included (XXI(b).C.2).

i If the glazed screen is installed internally, it should be measured as:

2/	L W		<u>Knotting; applying one coat primer, stopping, painting two undercoats and one finishing coat of synthetic paint; on wood</u> Windows, glazed screens and glazed doors; both sides measured flat overall; over 300mm girth

ii If the screen is installed on the external wall, half of the above area should be measured as internal work, and half of the area should be measured as external work. Note that internal work and external work are measured separately (XXI(a).D.1).

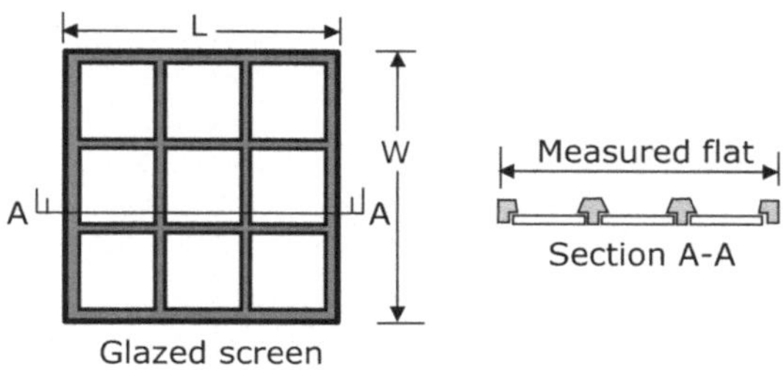

Question 17.12

(i) and (ii) Depending on the girth of the door lining, the door lining can be measured together with the door in m² under 'General surfaces; over 300mm girth' if its girth length exceeds 300mm. Otherwise, the door lining should be measured separately as 'General surfaces; not exceeding 300mm girth' in m (XXI(b).1.1.1–2). Note that only surface treatment that was applied during the production process is included in the door item (XIII(i).S.1).

(iii) Both on-site and off-site finishes (and surface treatments) applied to timber furniture should be included in the timber furniture item and not measured here (XIII(k).S.2–3).

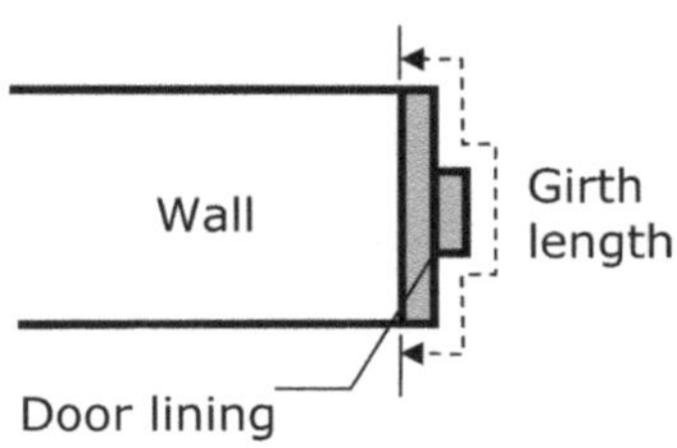

Question 17.13

While the manufacturer's reference of the wallpaper has to be stated in the item description (XXI(e).1.1), the upper portion and lower portion should be measured separately in m^2. For the dado strip, it should be measured separately in m as border strips (XXI(e).2).

Question 17.14

No. According to SMM item XXI(c).M.3, plastering and tiling work to walls is measured to the area of the base. Therefore, the girth length of wall tiling is the same as that of the wall screed underneath. As suggested in the text, the girth length of wall is used as the girth lengths for screed, tiling, skirting and the like.

18 Measurement of waterproofing

Introduction

Waterproofing does not apply to roofs only, but also to other locations such as basements, swimming pools, kitchen floors and bathroom floors. Major work items covered in Section X Waterproofing include:

- Liquid membrane waterproofing work using asphalt or liquid membrane products such as epoxy, polyurethane and the like.
- Sheet membrane waterproofing work using sheet-based membranes such as bituminous membrane, rubber sheet and the like.

General

Some general rules apply to both types of waterproofing works, which include:

- Roof waterproofing system and waterproofing work to non-roof locations (i.e. tanking and damp-proofing) shall be measured separately.

> **Reminder**
>
> Roof coverings:
>
> - waterproofing work to roofs
>
> Tanking:
>
> - tank-like waterproofing against ingress of water, applied to basements, swimming pools and the like
>
> Damp-proofing:
>
> - waterproofing to areas such as kitchens and toilets

- Details of the waterproofing materials, underlays, preparation of underlying surface, fixing method and the like shall be described.
- No deduction shall be made for voids $\leq 0.50\ m^2$. (X(a)&(b).M.2 and X(c)&(d).M.1).
- Work is deemed to include working at any heights, laid to falls, narrow widths and laps. (X(a)&(b).C.1 and X(c)&(d).C.1).

- Solar protection layer, reflective paint and the like shall be measured in m², state the details such as size of chippings and number of coats. (X(a)&(b).19, 20 and X(c)&(d).13, 14).
- Testing of waterproofing shall be measured as an item. (X(a)&(b).21 and X(c)&(d).15).

Liquid membrane waterproofing

X(a)&(b).2 Asphalt and liquid membrane waterproofing to surfaces

- Measured in m².
- Measure the nett area covered.
- All waterproofing work to general surfaces shall be measured separately and in accordance with the following classification:
  - horizontal
  - sloping (work inclined to horizontal > 15°; X(a)&(b).D.2)
  - on soffits
  - vertical
  - curved
  - executed overhand (refer to **Figure 14.5**)
  - executed in confined situations where working space < 0.6m (see **Figure 18.1**) (X(a)&(b).D.1)

X(a)&(b).3–11 Asphalt and liquid membrane waterproofing to linear items

- Waterproofing to smaller areas such as skirtings, gutters, valley gutters, curbs and channels, each are measured separately in m (X(a)&(b).5–11).
- If height/girth > 300mm, the work shall be measured separately with the height/girth stated. Height/girth is measured on face (as shown in **Figure 18.2**).
- Work to the following profiles are measured separately:
  - horizontal
  - stepped
  - raking (i.e. declined > 15° to the horizontal (X(a)&(b).D.3))
  - curved

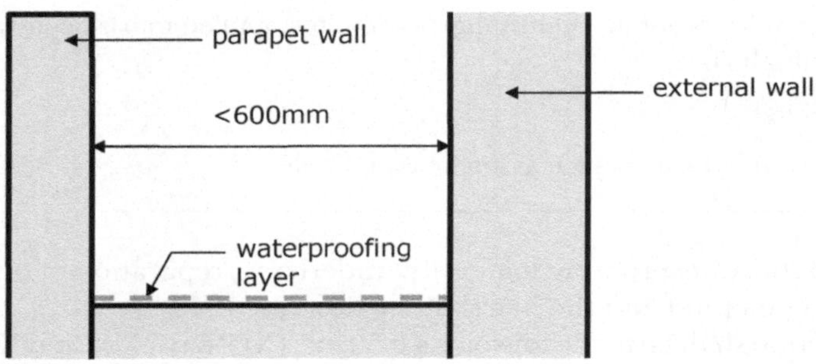

Figure 18.1 Waterproofing work to confined space.

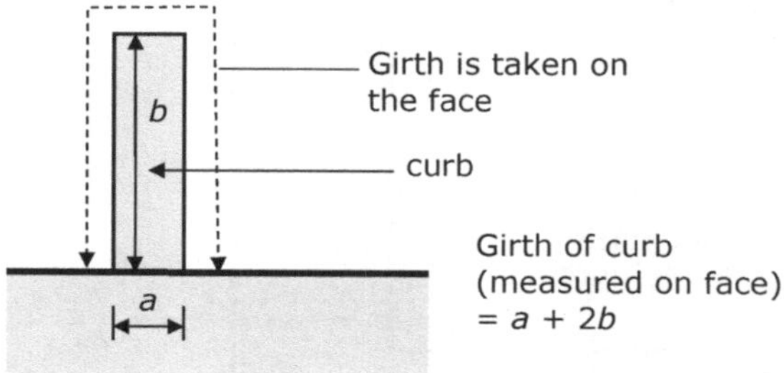

Figure 18.2 Girth measured on face.

Question 18.1

Prepare sketches to show the girth length of waterproofing to the following:

i gutter
ii valley gutter
iii channel

- Work is deemed to include edges, arrises, internal angle fillets, dressing over tilting fillets, turning into grooves **(Figure 18.3)**, stop ends, fair ends etc.
- Angle fillets are measured in m (X(a)&(b).3) when they are not in association with items such as skirtings, curbs, collars etc.

Question 18.2

35mm thick mastic asphalt is applied along the skirting of a circular column. Calculate the length of the asphaltic skirting.

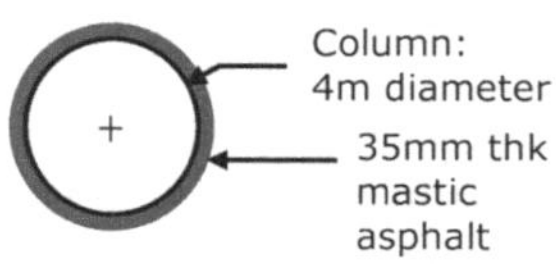

Sheet-based waterproofing

X(c)&(d).2 Felt roofing and sheet membrane waterproofing to surfaces

- Measured in m^2.
- Measure the nett area covered.
- Sheet-based waterproofing for roofs is sub-classified into:
    - single layer felt roof covering (using single layer of bituminous felt for roof waterproofing, insulation and roof finish)

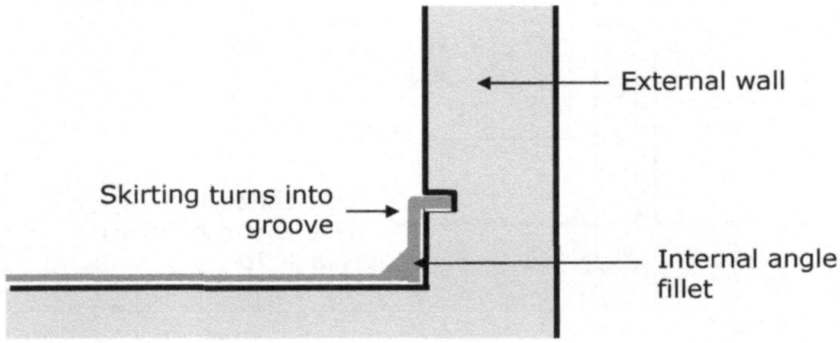

Figure 18.3 Turning waterproofing material into groove.

- built-up felt roof covering (using multiple layers of bituminous felt or similar laid and jointed in bitumen or similar compounds including vapour barrier and insulation)
- flexible sheet roof covering (using impermeable flexible sheet material as roof covering)

Question 18.3

The architect specifies that a minimum lapping of 300mm should be provided for the roofing felt installation. Should we add extras to the length and the width of the roofing felt when taking-off?

- Waterproofing work to general surfaces shall be measured separately according to the following classification:
    - horizontal
    - sloping (work inclined to horizontal > 15°); (X(c)&(d).D.1)
    - vertical
    - curved

Reminder

Unlike liquid membrane waterproofing, overhand work and work in confined space is not required to be measured separately if sheet-based waterproofing membrane is used.

X(c)&(d).4–8 Felt roofing and sheet membrane waterproofing to linear items

- Waterproofing to skirtings and aprons are measured separately in m.
    - State the height (for skirting) or girth (for aprons), measured on face.
    - Work to horizontal, sloping (i.e. declined > 15° to the horizontal (X(c)&(d).D.2)), vertical and curved profiles are measured separately.
    - Work is deemed to include edges, angles, dressing over tilting fillets and turning into grooves.

- Waterproofing to curbs, linings to channels and linings to gutters are measured separately in m.
  - State the girth, measured on face.
  - Work is deemed to include edges, angles, dressing over tilting fillets and turning into grooves.

Reminder

Sheet-based waterproofing work to curbs, channels and gutters are measured irrespective to their profiles (such as curved, sloping etc.).

Suggested answers

Question 18.1

The girth length of waterproofing to gutters, valley gutters and channels should be measured as below:

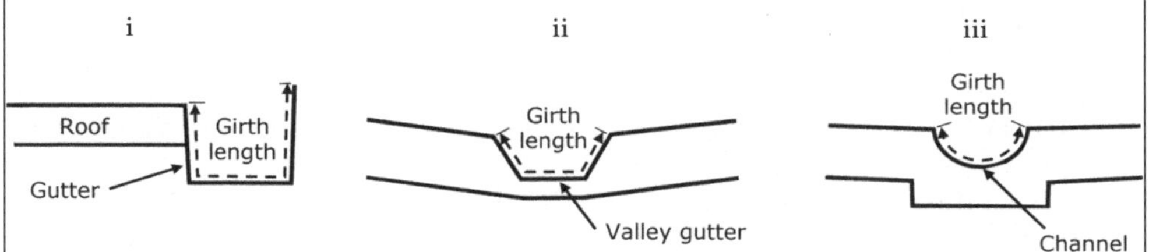

Noticed that all the lengths are measured on face.

Question 18.2

The length of asphaltic skirting should be:

$= 4\pi(\text{m})$

Although there is no measurement rule in the SMM, it is a normal practice to measure the skirting along the perimeter of the base where the skirting is fixed, not along the face of the skirting. (As in the case of screed or tiled skirting,

> work to walls and the like is measured to the area of the base
>
> (XVI.(c).M.3).)

Question 18.3

No additional lengths should be added for the laps. Nett area should be measured (X(c)&(d).M.2). This applies to the liquid membrane as well.

19 Worked examples

Introduction

Having reviewed the SMM measurement rules, a few worked examples are prepared to illustrate the take-off techniques for the typical items found in simple building projects. Each example covers more than one trade or SMM section to demonstrate the planning of take-off tasks in a practical context. Outstanding information required for taking-off is tabulated in the query list, with answers given aside. **Table 19.1** is a summary of the SMM sections covered in each worked example.

General notes for the drawings

To simplify the drawing presentation, some design information is listed below as general notes for all examples.

1. All drawings are not to scale. Unless otherwise stated, all dimensions are in mm.
2. Existing ground is levelled.
3. Unless otherwise stated, no surface excavation is required.
4. All excavated soil to be removed from site, deposited to a tip provided by the contractor.
5. All backfill material to be material arisen from the excavation.

Table 19.1 Coverage of Worked Examples.

	Worked Example 19.1	*Worked Example 19.2*	*Worked Example 19.3*	*Worked Example 19.4*
General description	Simple garden works	Steel frame structure	Roofing system to concrete flat roof	Simple warehouse building
Excavation	✓	✓		✓
Concrete works	✓	✓		✓
Brickwork & blockwork	✓			
Waterproofing			✓	✓
Wood works			✓	✓
Steel & metal works		✓		✓
Plastering & paving	✓		✓	✓
Painting				✓

Readers who are interested in 3D modelling can visit the publisher's website (www.routledge.com/9780367862329) for the Autodesk Revit models of the worked examples.

Example 19.1 – Simple garden works

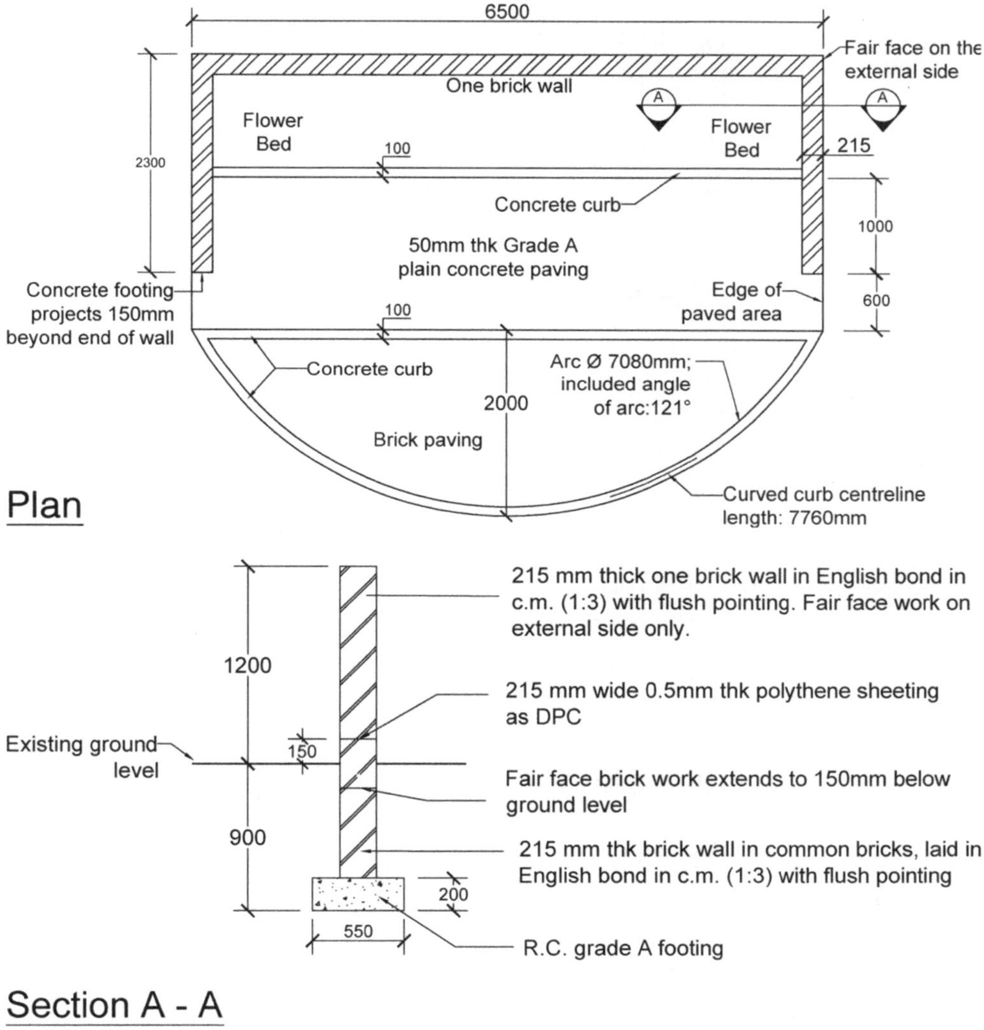

Figure 19.1a Plan and section A-A of garden works.

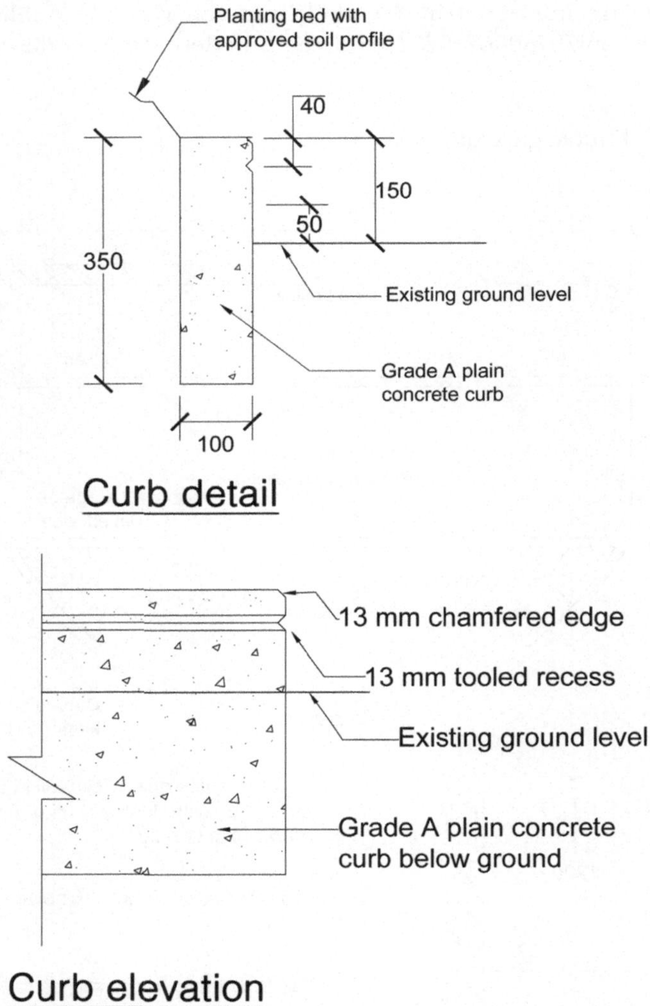

Figure 19.1b Concrete curb detail and elevation.

Query list Questions:	 *Answers:*
1. What is the thickness and type of soil in the flower bed?	*1. To be done by specialist contractor. No need to measure.*
2. Brick paving details missing.	*2. Specification of bricks to be confirmed. Bricks to be laid directly on compacted existing ground, bedding and jointing with cement and pour sand in herringbone pattern.*
3. Reinforcement details of the footing is missing.	*3. To be confirmed later, no need to measure.*
4. Is formwork required for casting the curb?	*4. Yes, sawn formwork is required.*

			Simple Garden Works Figures 19.1a & 19.1b Excavation	*Take-off list* *1. Excavation, backfill & removal* *2. Concrete footing* *3. Concrete curb* *4. Formwork to in-situ concrete* *5. Brick wall* *6. Damp proof course* *7. Concrete paving* *8. Brick paving*
			Centerline of Excavation 6500 2/2300 4600 fdn proj. 2/150 300 11400 Less 2/2/½/215 430 fdn centreline 10970	*Note : Take-off list should be prepared prior to measurements. Any misisng information or ambiguities should be raised in the query list.*
			Add w/s 2/250 500 11470	*HKSMM4 Rev 2018, VI(a).M.2(a)* *250mm working space has to be allowed where the formwork < 0.6m in height. Here, the height of formwork required for the foundation is 200mm.*
			Width of excavation fdn 550 w/s 2/250 500 1050	*Abbreviations are commonly used in taking-off. A list of abbreviations for use in taking-off is shown in the Appendix.*
	11.47 1.05 0.90		Trench excavn for fdn commencing at existing grd level, not ex. 1.50m dp	*HKSMM, VI(a).3.5.1.1* *Commencing level has to be stated, and the excavation item is measured in successive stages of 1.50m. The overall depth of the foundation is 900mm, which only requires the 1st stage of excavation.*
			&	
			Backfilling with sel. exc. matl	*HKSMM, VI(a).7.0.1* *Here, we assumed all excavated soil is backfilled immediately after excavation. The soil volume occupied by the footing can be deducted later when we measure the concrete for footing. Potential errors due to miscalculation of the nett backfill volume can be avoided.*
			Curb L 6500 Less wl 2/215 430 6070	

	6.07		Trench excavn for 100 x 350mm curbs without beds comm. at existing grd lev., n.e. 0.25m av. depth	*HKSMM, VI(a).3.9.1* *Excavation for curb should be measured in m. Average depth of excavation for curbs shall be described in multiples of 0.25m deep. Here, the excavation depth of the curb is 200mm and thus the average depth of excavation should be "not exceeding 0.25m deep". The depth range, size of curb, with/without beds and commencing level has to be stated in the item description according to SMM.*
			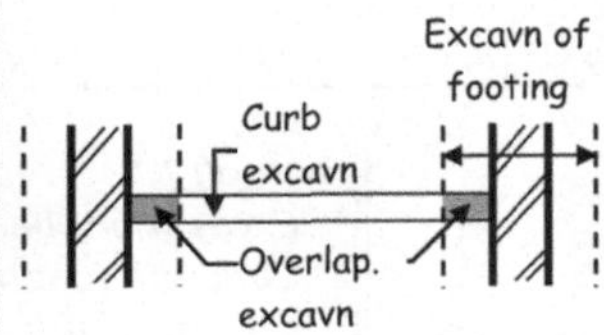	*Note that the item includes backfilling and disposal. Thus, no soil adjustment is required.*
			<u>overlapped exc</u> proj 167.5 w/s <u>250</u> <u>417.5</u>	*Part of the curb excavation has been measured in the trench excavation for footing and therefore deducted here.*
2/	0.42		<u>Ddt</u> Ditto (overlap. excavn	*"Ditto" (or "do.") means "the same as what is said just before". This abbreviation is often used by taker-off so as to reduce the description of the next item.*
	7.76 6.40		Trench excavn for 100 x 350mm curbs without beds a.b.d. (curved (straight	*HKSMM, VI(a).3.9* *Trench excavation for curbs that are curved on plan are not required to measure separately.*
			<u>Concrete</u>	
	10.97 0.55 0.20		R.C. Grade A in foundation & <u>Ddt</u> Backfilling with sel. exc. matl & <u>Add</u> Disposal of exc. matl from site to a tip provided by Contractor	*HKSMM, VII(a).8* *Concrete to foundation to be measured in* m^3. *Centreline of foundation is calculated earlier and therefore not repeated here.* *HKSMM, VI(a).7.0.1 &* *HKSMM, VI(a).6.1.3* *When we measured the excavation item, we have assumed that the excavated soil was backfilled entirely. Now, the concrete volume replaces the backfilled soil, and therefore, disposal of soil should be measured and the volume of backfill should be adjusted. This is a customary way to deal with soil adjustment.*

	6.07 0.10 0.35	Grade A plain concrete in curbs, with 13mm tooled recess and 13mm chamfered edge	HKSMM, VII(a).12
	7.76 0.10 0.35	(curved	
	6.40 0.10 0.35		
2/	0.42 0.10 0.20	Ddt Backfilling with sel. exc. matl (overlap. excavn & Add Disposal of exc. matl from site to a tip provided by Contractor	Soil adjustment for foundation backfill and disposal quantities where the concrete curb locates.
		Formwork	
2/	10.97 0.20	Sawn fwk to vert. sides of fdn	There is no specific classification in the SMM for formwork to foundations. The rules for measuring vertical surfaces for walls, columns and like are applied here. Note that for foundation, ground slabs and the like, no formwork to soffit is required.
2/	0.55 0.20	(ends	Sawn formwork refers to untreated timber formwork which is used to produce general concrete surfaces. If fair face concrete is required, wrought (wrot) formwork should be used.
2/	6.07 0.35	Sawn fwk to vert. sides of curbs exceeding 300mm wide	HKSMM, VII(d).9.1.1 Formwork to curb exceeding 300mm width (i.e. height of curb) has to be measured in m^2 and described.
2/	6.40 0.35		
2/	7.76 0.35	Ditto, curved on plan	HKSMM, VII(d).9.2.1 Formwork curved on plan has to be measured separately. VII(d).C.4 states that formwork is deemed to include chamfered edges and recesses. Therefore, no separate measurement is required.

			<u>Brickwork</u>	
			<u>Centerline of brk wl</u> 6500 2/2300 <u>4600</u> extl girth: 11100 <u>Less</u> wl 2/2/½/215 <u>430</u> <u>10670</u>	
			<u>Height of wl</u> above grd 1200 below grd 900 <u>Less</u> fdn <u>200</u> <u>1900</u>	
	10.67 1.90		1B wls, vert., in comms in Eng. bond in c.m. (1:3) with flush pointing	*HKSMM, VIII(a).2.2.1* *Brick wall ≤ one and a half brick thick will have the thickness stated. (Here, the brick wall is one brick thick.) Quality and size of brick, bond, mortar mix, pointing should be stated. (VIII(a).S.1–S.4)*
			<u>Height of fair face</u> 1200 <u>150</u> <u>1350</u>	
	11.10 1.35		E.o. ditto for fair face in Eng. bond in c.m. (1:3) with flush pointing	*HKSMM, VIII(a).6* *Fair face brickwork is measured as extra over the brick work on which it applies. Therefore, the external face area of the brick wall is measured. Details of the bond, mortar mix, etc. should be stated (VIII(a).S.5-S.7).*
			<u>Height of brkwl under grd lv</u> 900 <u>Less</u> <u>200</u> <u>700</u>	
	10.67 0.22 0.70		<u>Ddt</u> Backfilling a.b.d. (brk & <u>Add</u> Disposal of exc. matl a.b.d.	*HKSMM, VI(a).7.0.1* *Since the brick wall is measured in* m^2 *whereas disposal is in* m^3*, soil adjustment has to be dealt with separately. Note that soil adjustment should only be applied to the brick wall below ground level.* *HKSMM, VI(a).6.1.3*
	10.67		215mm wide and 0.5mm thk polythene sheeting DPC, laid horiz.	*HKSMM, VIII(a).20.1.2.1* *DPC ≤ 225mm wide has to be measured in m. Orientation (horizontal), material and thickness has to be stated. Note that the brick wall centerline is used to calculate the length of DPC.*

	6.50 1.60	50mm thk grade A plain conc. paving, laid on existing grd	*HKSMM, XVI(d)* *The paving should be measured in m^2, thickness stated.* *Measure the large rectangular area and then deduct the brick wall area using "Ddt" is one of the usual ways to deal with odd shapes.* *Alternatively, you can divide the area into sub-areas and then add them up.*
2/	1.00 0.22	<u>Ddt</u> Ditto (brk wl	

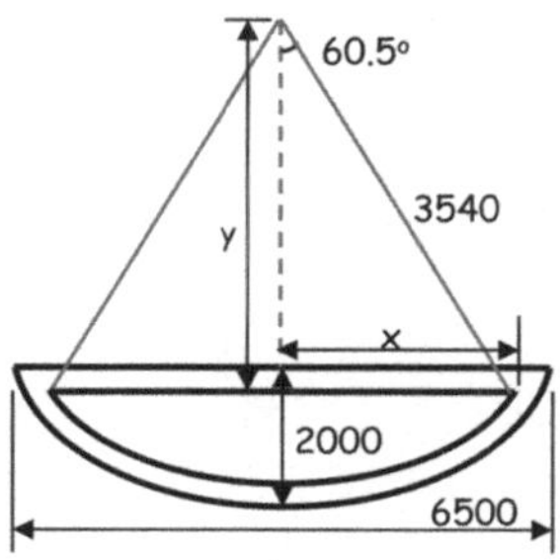

$\frac{121}{360}$/ π/	3.54 3.54	Pavings in bricks (P.C. rate: \$180 per m^2); bedding and jointing with cement and pour sand on compacted existing grd; laid on face in herringbone pattern	*HKSMM, XVI(e).2* *Since the specification of bricks is unknown, a P.C. rate is applied here. Details of P.C. rate and its pricing adjustment please refer to* ***Chapter 23***.
		<u>y</u> 3540 Less: <u>1800</u> <u>1740</u> <u>x</u> 6500 Less: curbs <u>200</u> ½ / <u>6300</u> <u>3150</u>	
2/ ½/	3.15 1.74	<u>Ddt</u> Ditto (triangle	
		<u>End of Simple Garden Measurement</u>	

Example 19.2 – Steel frame structure

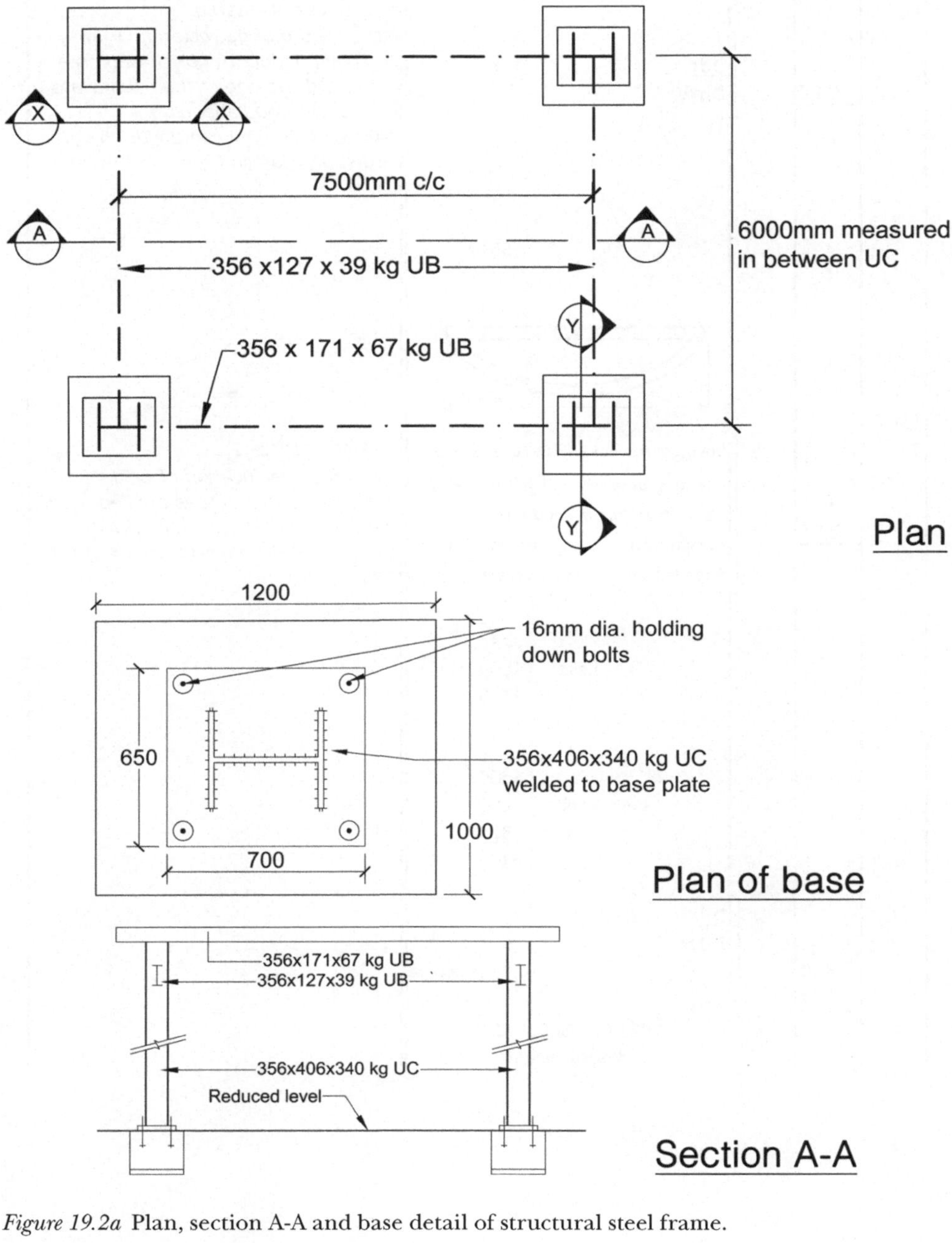

Figure 19.2a Plan, section A-A and base detail of structural steel frame.

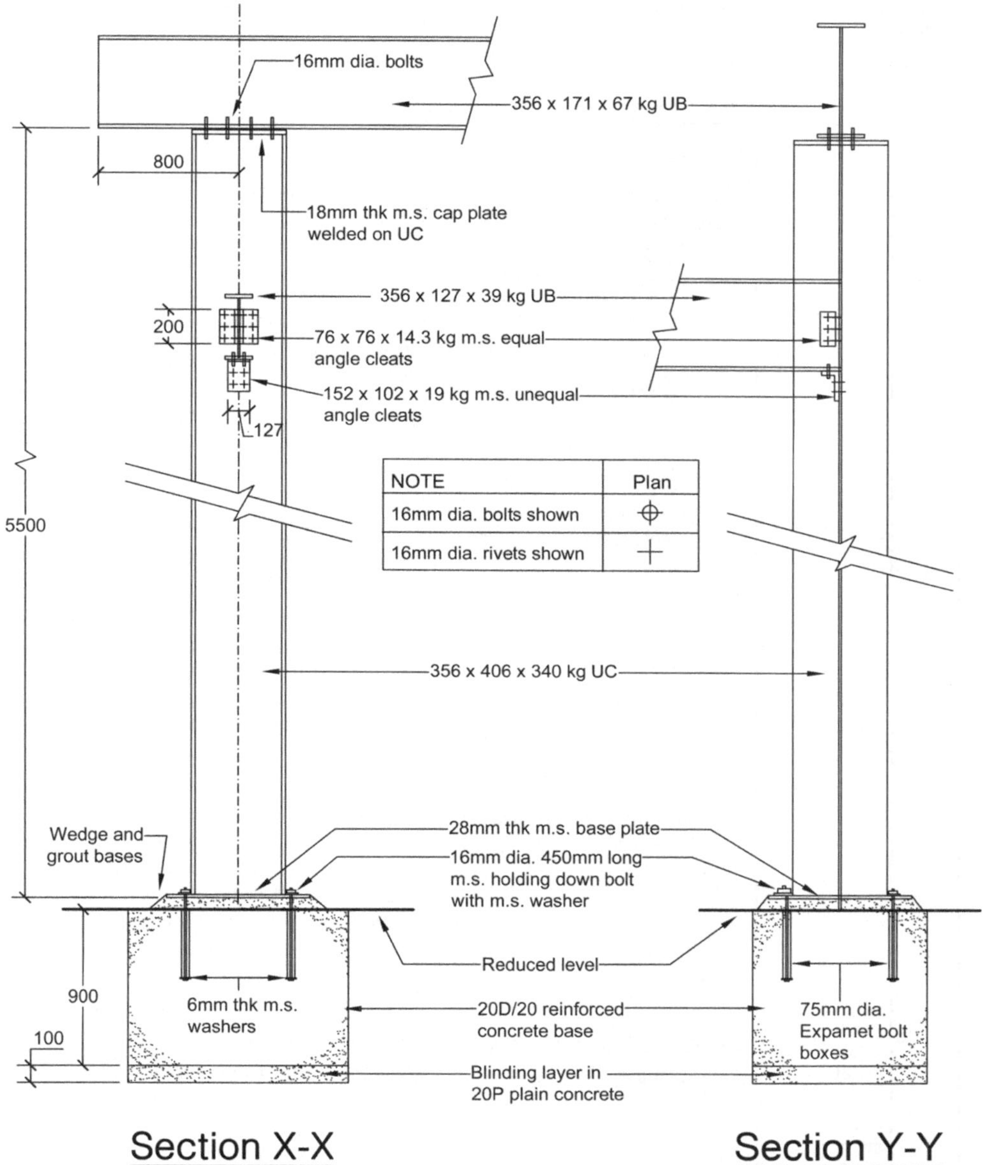

Figure 19.2b Section X-X and section Y-Y of structural steel frame.

Query list Questions: 1. Reinforcement details of the footing is missing.	Answers: 1. To be confirmed later, no need to measure.

		Description	Notes
		Steel Frame Structure Figures 19.2a & 19.2b Excavation	Take-off list 1. Excavation, backfill, removal 2. Blinding 3. Concrete column bases 4. Formwork 5. Universal columns 6. Universal beams 7. Connections 8. Special bolts
		Length of excavation 1200 w/s 2/600 1200 2400	
		Width of excavation 1000 w/s 2/600 1200 2200	HKSMM4 Rev 2018, VI(a).3.7.M.2(b) 600mm working space has to be allowed.
4/	2.40 2.20 0.90	Exc. for col. bases commencing at red. level, n.e. 1.50m deep (base	HKSMM, VI(a).3.7.1.1 Measurement should be taken in successive stages of 1.50m and commencing level should be stated.
4/	1.20 1.00 0.10	(blinding	Since no formwork will be provided for plain concrete (unless otherwise specified by the engineer), working space is not allowed for the blinding layer.
		& Backfilling with sel. exc. matl	HKSMM, VI(a).7.0.1 Similar to Worked Example 19.1, backfilling is assumed for the entire excavation. The adjustment will be subsequently dealt with when we measure the concrete works.
		Concrete	
4/	1.20 1.00 0.10	20P plain conc. in 100mm thk blinding under col. bases	HKSMM, VII(a).11.1 Blinding under foundations should be measured in m^3 with thickness described.
		& Ddt Backfilling a.b.d.	HKSMM, VI(a).7.0.1 Since backfilling has been measured for the entire excavation earlier, volume of concrete should be deducted from the quantity of backfilling.
		& Add Disposal of exc. matl from site to a tip provided by Contractor	HKSMM, VI(a).6.1.3 The volume of soil displaced by concrete should be disposed. Designation of disposal should be stated accordingly.

Timesing	Dimensions	Description	Notes
4/	1.20 1.00 0.90	20D/20 R.C. in col. bases & Ddt Backfilling a.b.d. & Add Disposal of exc. matl from site to a tip provided by Contractor	*HKSMM, VII(a).9* *Concrete to column bases should be measured in m^3.* *HKSMM, VI(a).7.0.1* *Soil adjustment as before.* *HKSMM, VI(a).6.1.3*
		Formwork Girth of column base 1.20 1.00 2/ 2.20 4.40	
4/	4.40 0.90	Sawn fwk to vert. surf. of col. bases	
		Supply, fabricate and erect grade S355 steel structure to BS EN 10210-2, approx. plan area 45m^2; single storey; 5.9m overall height above grd; incl. all necessary welding and bolting	*HKSMM, XV(a).S.1*
		Columns Height of columns 5500 Less Cap plate 18 Base plate 28 5454	
4/	5.45	356 x 406 x 340kg universal columns (____m x 340kg/m = ____kg)	*HKSMM, XV(a).2* *SMM requires steel work to be measured by kg. When taking-off, the lengths of steel members are taken in the dimension column, followed by conversion to mass in the squaring process.*
		Beams Length of beams 7500 proj 2/800 1600 9100	
2/	9.10	356 x 171 x 67kg universal beams (____m x 67kg/m = ____kg)	*HKSMM, XV(a).3* *Steel columns and beams are measured separately, even if they are using the same type of structural steel members.*

2/	6.00	356 x 127 x 39kg universal beams (____m x 39kg/m = ____kg)	*HKSMM, XV(a).3*
		Connections <u>Mass per area (base plates)</u> = 7843 x 0.028 = 219.6kg/m^2 <u>Mass per area (cap plates)</u> = 7843 x 0.018 = 141.2kg/m^2	*Unit weight of steel is 7,843 kg/m^3. Since steel plates of different thickness are used, the conversion factor for each has to be calculated separately.* *Mass/m^2 = 7,843 x thickness of plate (kg/m^2)*
4/	0.70 0.65	Connections including cleats, plates and the like (base plates (____m^2 x 219.6kg/m^2 = ____kg)	*HKSMM, XV(a).M.5* *Connections including cleats, brackets, plates, etc. are grouped together to obtain the total mass for this item.*
4/	0.40 0.41	(cap plates (____m^2 x 141.2kg/m^2 = ____kg)	*The sectional width and depth of 356 x 406 x 340 UC is 403mm and 406.4mm respectively. Therefore, the cap plate required is 403 x 406.4mm^2.* *Since the mass per unit (length / area) is different for each connection, the relevant conversion should be stated clearly and aligned with its dimensions.*
4/ 2/	0.20	(equal cleats (____m x 14.3kg/m = ____kg)	*Refers to 76 x 76 x 14.3kg equal cleats.*
4/	0.13	(unequal cleats (____m x 19kg/m = ____kg)	*Refers to 152 x 102 x 19kg unequal cleats.*
		<u>Special bolts</u>	*HKSMM, XV(a).14*
4/	4	16mm dia. x 450mm long holding down bolts casting into conc. with 6mm thk washers and 75mm dia. expamet bolt boxes	*Holding down bolts have to be measured in no. with the details stated. Note that the quantity entered in the dimension column should not contain any decimal places, or otherwise the quantity will be interpreted as a linear measurement.*
		<u>End of Structural Steel Frame Measurement</u>	*Site rivets and black bolts are deemed to be included in the fabrication and erection of steel frame and are not required to measure (XV(a).C.1(b)-(c)).* *Wedges and grouting between the concrete foundation and base plate (as shown in* ***Figure 19.2b****) are deemed to be included and therefore not measured. (XV(a).C.1(h))*

Example 19.3 – Roofing system to concrete flat roof

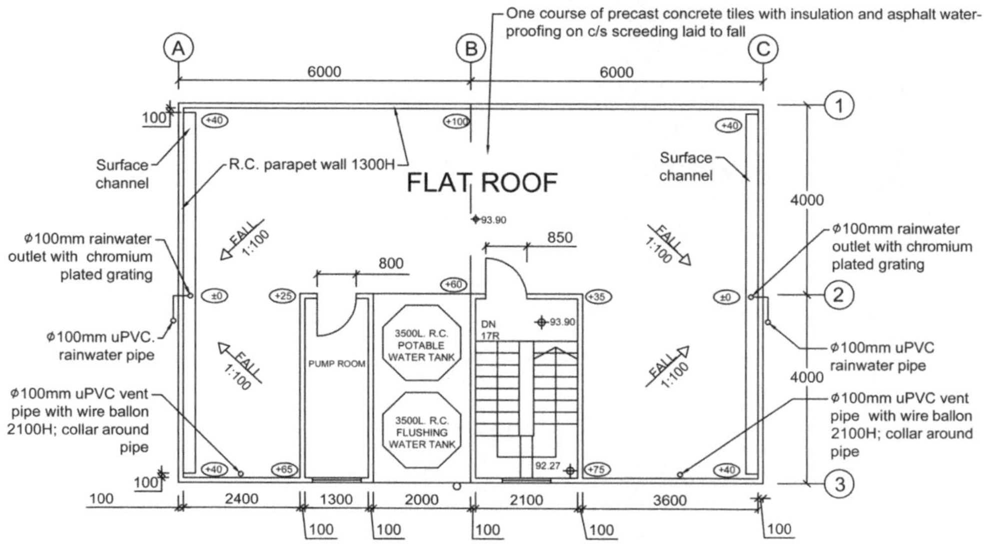

Figure 19.3a Roof plan.

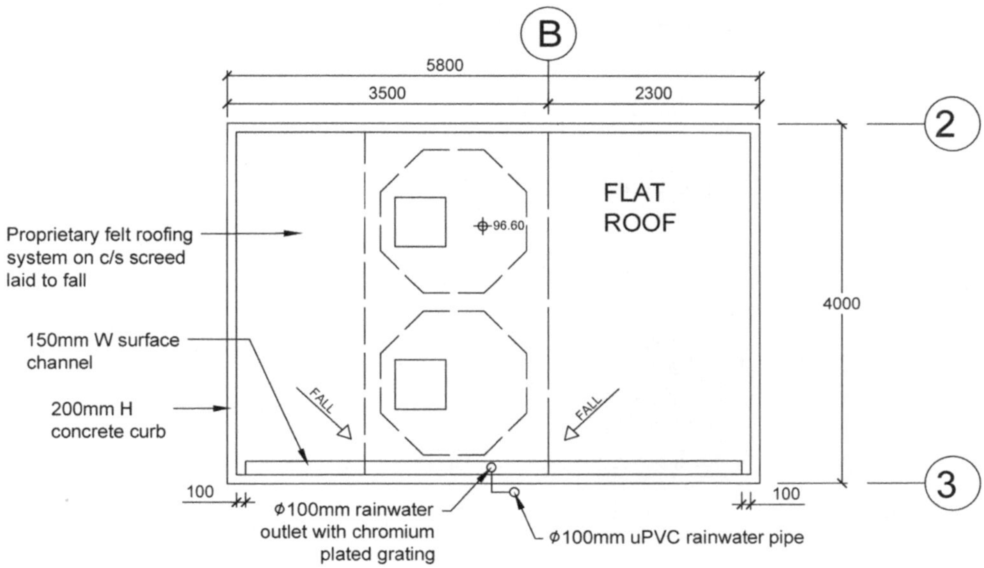

Figure 19.3b Upper roof plan.

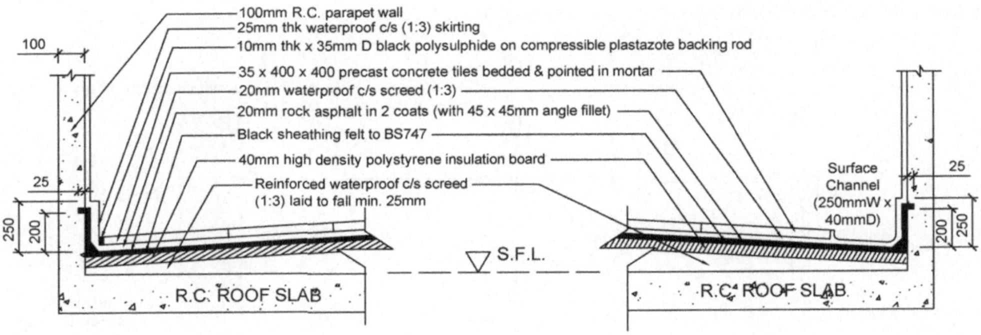

Figure 19.3c Details of main roof.

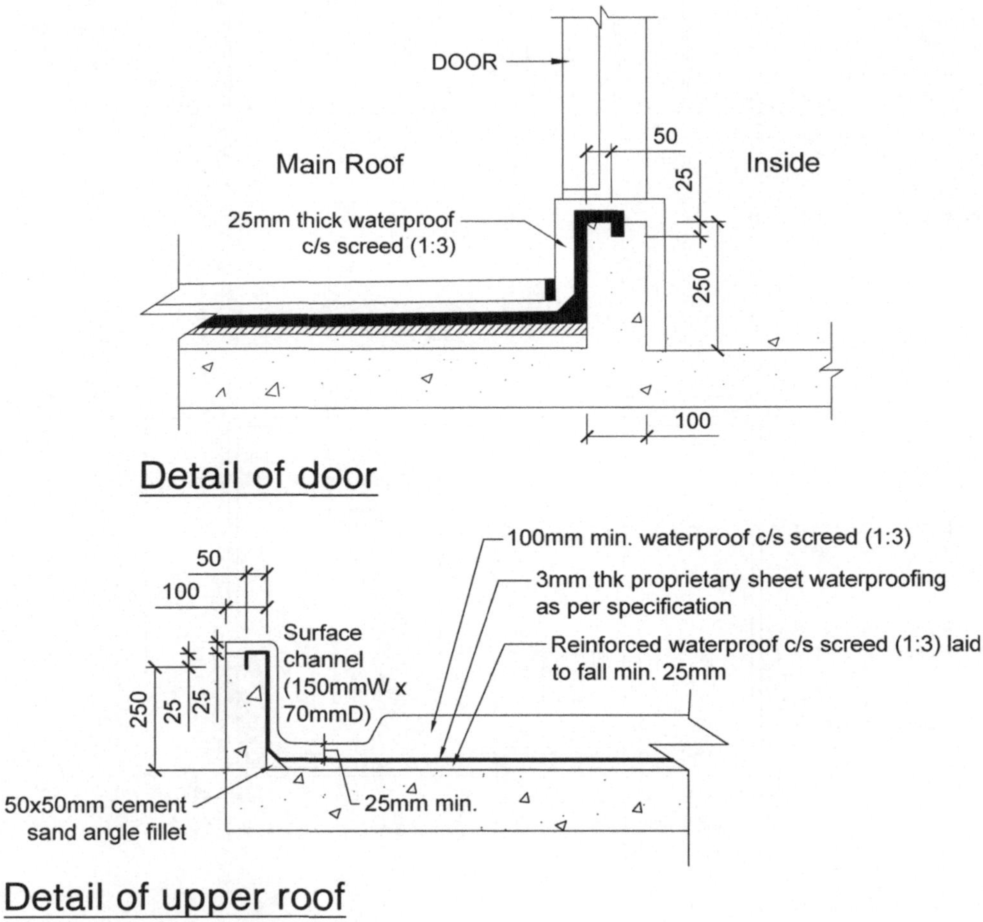

Figure 19.3d Details of door and upper roof.

Query list Questions: 1. Besides along the perimeter, any expansion joint required in the precast concrete tiling ?	Answers: *1. Yes, apply 10mm thick polysulphide sealant on compressible plastazote backing rod, 3m c/c maximum; both ways.*

		Roofing system to concrete flat roof Figures 19.3a, 19.3b, 19.3c and 19.3d **Main roof** Roof: Length / Width 6000 / 4000 6000 / 4000 12000 / 8000 Less Wl 2/100 200 / 200 11800 / 7800	*Take-off list* *Main roof (from bottom to top)* *1. Reinf. screed to roof* *2. Insulation boards* *3. Waterproof coverings to roof* *4. Ditto to skirtings and curbs* *5. Outlets to channels* *6. Working to outlets pipes* *7. Collar around vent pipes* *8. Screed to receive conc. tiles* *9. E.o. for surface channels* *10. Screed to skirtings and curbs* *11. Precast concrete tiling* *12. Expansion joint* *13. Testing* *14. Repeat for upper roof*
		EXTERNALLY	*External plastering work must be so described (XVI(c).D.1).*
11.80 7.80		25mm thk reinforced waterproof c/s (1:3) screed to roof in 1 ct, laid to fall, to receive high density polystyrene insulation board &	*HKSMM4 Rev 2018, XVI(d).7.1.1* *Screeds to roofs have to be measured separately.* *Thickness, number of coats and material to be received (in this case, the insulation board) should be stated.* *Also, no deduction is made for voids ≤ $0.50m^2$, nor for voids ≤ 300mm wide (XVI(c).1.M.5). For the pipes with a diameter of 100mm, the cross-sectional area is only 0.008 m^2. Therefore, deductions for pipes are not necessary.*
		40mm thk high density polystyrene insulation boards to hori. plain areas	*HKSMM, XIII(m).4.1.1* *Insulation boards, sheets, quilts and the like are covered in the Wood works Section (XIII(m).D.1).* *SMM classifies the insulation boards into horizontal, vertical and to soffits, which means sloping work is not required to be measured separately.* *Any surface ≤ 45° from horizontal is considered as horizontal; > 45° is considered as vertical (XIII(m).D.2 & D.3).* *State if the work is applied to plain areas or across members or between members. Here, the insulation is applied to plain areas.*

Dimensions	Description	Notes
	Width of upper roof for Ddt 4000 Less: Wl 100 3900	*No deduction for voids ≤ 0.5m^2 (XIII(b).M.2). No need to deduct the insulation area for pipes.*
5.80 3.90	Ddt Last two items (upper roof	*Deduct both reinforced c/s screeding and insulation.*
	Rock asphalt roofing to BS747 in 2 ct on an underlaying of black sheathing felt to 20mm thk	*A heading can be used to save repeating the same part of description in sevearl items.*
11.80 7.80	Asphalt roof coverings on horiz. surf.	*HKSMM, X(a).S.1* *Details of the asphalt work, including materials, thickness, no. of coats and underlays have to be described.* *HKSMM, X(a).2.2.0.1* *If the work is applied to sloping surface (> 15°), it has to be measured separately and stated (X(a).D.2). As shown in Figure 19.3a, the slope of the roof is 1:100 (i.e. 0.6°) which is designed for clearing surface water effectively. Therefore, the work here should be treated as horizontal.*
5.80 3.90	Ddt Ditto (upper roof	*No deduction is required for area ≤ 0.5m^2 (X(a).M.2). No need to deduct the asphalt area for pipes.*
	Length of skirting L 2/11800 W 2/7800 upper roof 2/3900 47000	*Measure the internal girth of walls where the skirting is fixed.*
47.00	Asphalt coverings to sktg n.e. 300mm high, incl. angle fillet at bottom and turning into grooves	*HKSMM, X(a).5.1* *Asphalt skirtings are classified as: ≤ 300mm high; or state the height if > 300mm high. Height is measured on face (X(a).M.4). Angle fillets are deemed to be included (X(a).C.3).*
0.80 0.85	Ddt Ditto (doors	
	Face girth of curb asphalt 250 50 300 Less: Screed 25 Insulation 40 Asphalt 20 85 215	*HKSMM, X(a).11.1* *The exposed face girth of the curb that receives asphalt is calculated (refer to* ***Figure 19.3d***). *Since the calculated girth length is 215mm, it should be classified as "girth n.e. 300mm".*

	0.80 0.85	Asphalt coverings to curb n.e. 300mm girth, incl. angle fillet at bottom and turn in at top (doors	HKSMM, X(a).11.1
			Waterproofing needs to be fully dressed into the drainage outlets to ensure a watertight seal. Work to the outlets and pipes is measured here.
	2	Outlets to channels	*HKSMM, X(a).15* *Different sizes of the outlets can be grouped together and measured in number.*
	2	Working to outlets pipes & Collars around vent pipes	*HKSMM, X(a).16* *HKSMM, X(a).18* *Collars are required around the vent pipes to ensure watertightness, as shown below:*

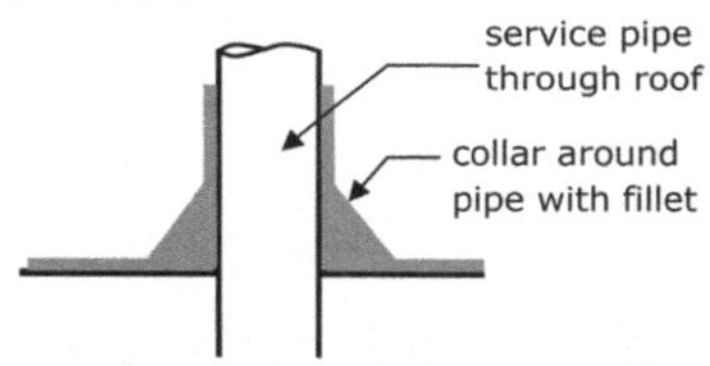

			Internal angle fillet is deemed to be included (X(a).C.6).
	Item	Infra-red thermographic test of asphalt waterproofing system; prepared and executed by an independent specialist contractor	*HKSMM, X(a).21* *Details of the test should be described and measured as an item.*
		- End of Asphalt coverings measurement -	*Where items are measured under a heading, it is a good practice to put a footnote after the last item in the take-off to which the heading refers, stating "End of ..."*
	11.80 7.80	20mm thk waterproof c/s (1:3) screed to roof in 1 ct, to receive precast conc. tiles	*HKSMM, XVI(d).7.1.1* *The material to be received in this case is precast concrete tiles.*
2/	7.60	E.O. ditto for forming 250mm wide x 40mm deep surface channle in floor screed (channels	*HKSMM, XVI(d).11.4* *No need to deduct the area of channel from the floor screed area.*
	5.80 3.90	<u>Ddt</u> 20mm thk waterproof c/s (1:3) screed to roof in 1 ct, to receive precast conc. tiles (upper roof	

Timesing	Dimensions	Squaring	Description	Notes
			<u>Height of skirting</u>	
			250	
			<u>Less:</u> screed <u>20</u>	
			<u>230</u>	
	47.00		25mm thk x 230mm high waterproof c/s (1:3) sktg in 1 ct, incl. laying over angle fillets	*HKSMM, XVI(d).13.1* *Screed to skirting is measured in m. Full description (including thickness, material and the mix) as well as the height above finished floor have to be stated.*
	0.80		<u>Ddt</u> (door	
			Ditto	
	0.85		(door	
			<u>Face girth of curb screed</u>	
			height 2/250 500	
			width <u>100</u>	
			600	
			<u>Less:</u> screed 25	
			insulation 40	
			asphalt 20	
			screed 20 <u>105</u>	
			<u>495</u>	
	0.80		25mm thk x 495mm girth waterproof c/s (1:3) screed in 1 ct to curb, incl. laying over angle fillets	*HKSMM, XVI(d).16.1* *Thickness and girth should be stated.*
	0.85		(doors	
	11.80 7.80		400 x 400 x 35mm thk precast conc. tiles bedded and pointed in c/s (1:3) mortar to roof	*HKSMM, XVI(e).2*
	5.80 3.90		<u>Ddt</u> Ditto (upper roof	
2/	7.60 0.25		(channel	
			<u>Length of movement joint</u>	
			<u>x-axis</u> <u>y-axis</u>	
			12000 8000	
			<u>Less:</u> W1 2/100	
			asphalt 2/20	
			screed sktg 2/25 <u>290</u> <u>290</u>	
			full length: <u>11710</u> <u>7710</u>	
			<u>Less:</u>	
			channel 2/250 <u>500</u>	
			(less channel L) <u>11210</u>	

		<u>No. of movement joint</u>	
		Along <u>y-axis</u> <u>x-axis</u>	Based on the calculated no. of bays in x and y directions, the movement joint layout will be like this:
		7710 11210	
		divided by 3000	
		2.57 3.74	
		say: 3 4	
		Less: <u>1</u> <u>1</u>	
		<u>2</u> <u>3</u>	
		<u>Mean girth of joint along perimeter</u>	
		L 11710	
		W 7710	
		upper roof wl <u>3900</u>	
		2/ <u>23320</u>	
		46640	
		<u>Less</u>: corners 4/10 <u>40</u>	
		<u>46600</u>	
	46.60	10mm wide x 35mm dp movement joint; filled with black polysulphide on compressible plastazote backing rod	HKSMM, XVI(e).17.1 Dimensions have to be described.
		(along perimeter	
2/	11.21	(along X-X; less channel	
3/	7.71	(along Y-Y	
		<u>Upper roof ddt</u>	
		<u>L</u> <u>W</u>	
		5800 3900	
		asphalt + screed 2/45 <u>90</u> <u>0</u>	
		<u>5890</u> <u>3900</u>	
2/	7.60	<u>Ddt</u> Ditto (channels	
	5.89	(along X-X; roof	
2/	3.90	(along Y-Y; roof	
		<u>Upper roof</u>	
		<u>Length</u> <u>Width</u>	
		5800 4000	
		<u>Less</u>	
		Curbs 2/100 <u>200</u> <u>200</u>	
		<u>5600</u> <u>3800</u>	

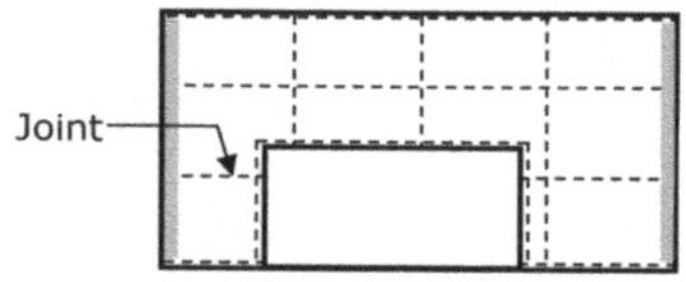

	Dimensions	Description / Workings	Notes
	5.60 3.80	25mm thk reinforced waterproof c/s (1:3) screed in 1 ct to roof, laid to fall, to receive sheet waterproofing membrane	*HKSMM, XVI(d).7.1.1* *The material to be received is the sheet waterproofing membrane.*
		<u>Mean girth of angle fillet</u> L 5600 W <u>3800</u> 2/ <u>9400</u> 18800 <u>Less:</u> corners 4/50 <u>200</u> <u>18600</u>	*Centreline of the angle fillet is taken as the length.*
	18.60	50 x 50mm c/s (1:3) angle fillet	*HKSMM, XVI(d).19.1* *Width and depth should be stated.*
		<u>waterproofing membrane</u> <u>Length</u> <u>Width</u> 5600 3800 <u>Less:</u> fillet approx. 2/25 <u>50</u> <u>50</u> <u>5550</u> <u>3750</u>	
		<u>Proprietary sheet waterproofing material; designed, supplied and fixed by a specialist contractor approved by the manufacturer; laying in accordance with the manufacturer's instructions</u>	
	5.55 3.75	3mm thk sheet coverings on horiz. roof	*HKSMM, X(d).4* *Nett area should be measured for the sheet roof coverings. The angle fillet is deemed to be included in the curbs and therefore deleted from the roof covering area.*
		<u>Exposed face girth of curb</u> curb vert. face 250 curb vert. face 25 curb top 50 fillet approx. <u>35</u> 360 <u>Less</u>: screed <u>25</u> <u>335</u>	
		<u>Total length of curb (intl girth)</u> L: 5600 W: <u>3800</u> 2/ <u>9400</u> <u>18800</u>	
	18.80	Ditto, but to curbs 335mm girth, incl. laying over angle fillet and turn in at top	*HKSMM, X(d).8* *Face girth should be stated.*

1	Cutting & fitting to outlet	
Item	Infra-red thermographic test of sheet waterproofing system; prepared and executed by an independent specialist contractor	*HKSMM, X(d).15*
	- End of Sheet waterproof coverings measurement -	
5.60 3.80	100mm thk waterproof c/s screed (1:3) in 1 ct to roof	*HKSMM, XVI(d).7*
	length of channel 5600 Less: 2/100 200 5400	
5.40	E.o. ditto for forming 150mm wide x 70mm (average) deep surface channel in floor screed	*HKSMM, XVI(d).11.4* *No need to deduct the area of channel from the floor screed area.*
18.80	25mm thk x 325mm girth waterproof c/s (1:3) screed in 1 ct to curb End of Roof Coverings Measurement	*HKSMM, XVI(d).16.1* *Face girth* *= 250+25+100-25(screed)-25(screed)* *= 325mm*

Example 19.4 – Simple warehouse building

GENERAL NOTES

1 Unless noted otherwise, concrete shall be designed mix with the following specified grade strengths and maximum size of aggregate 20mm:

Element	*Concrete grade*
Blinding layer	10P
Underground structures up to floor level of ground slab	40D/20 waterproof concrete
Ground slab	40D/20 waterproof concrete
Walkway	20D/20 waterproof concrete
Column, wall, roof	40D/20

2 Calling up of reinforcement:

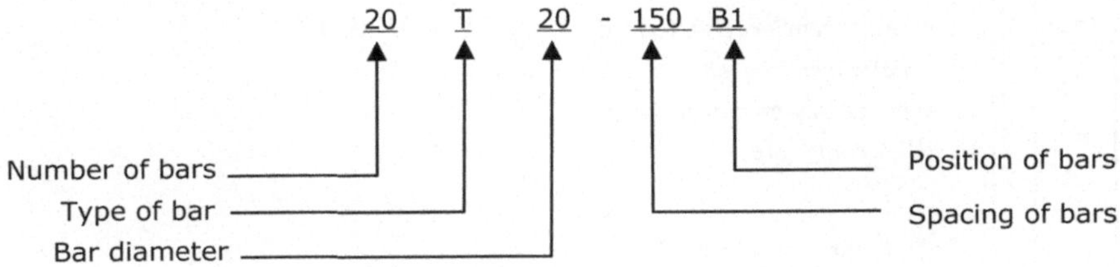

3 Unless noted otherwise, minimum lap and anchorage lengths shall be as follows:

Reinforcement	*High yield bars*
Concrete grade	40/20
Tension anchorage length (T.A.L.)	32D
Tension lap length (T.L.L.)	32D

4 Spatterdash should be applied to all concrete surfaces except floor surface before receiving screed/plaster.

5 The following finishes should be applied unless noted otherwise:

Element	*Backing*	*Finish*
External		
Walkway	-	25mm thick waterproof cement / sand (1:3) screed in one coat; steel trowelled finish
Wall	20mm thick cement / sand (1:3) render in one coat	200 × 100 × 10mm thick light green ceramic tiles bedded in cement mortar (1:4), pointed in white cement
Roof	T.B.A.	T.B.A.
Internal		
Floor	40mm thick waterproof cement / sand (1:3) screed in one coat	300 × 300 × 12mm thick non-slip homogeneous tile bedded in cement mortar (1:4), pointed in white cement
Skirting	20mm thick x 100mm high waterproof cement / sand (1:3) screed in one coat	10mm thick × 100mm high coved homogeneous skirting tile bedded in cement mortar (1:4), pointed in white cement
Wall	20mm thick waterproof cement / sand (1:3) screed in one coat	150 × 150 × 10mm thick white glazed ceramic tiles bedded in cement mortar (1:4), pointed in white cement
Ceiling	10mm thick cement / sand (1:3) plaster in one coat	One coat of water-based lime resistant primer and two coats of emulsion paint

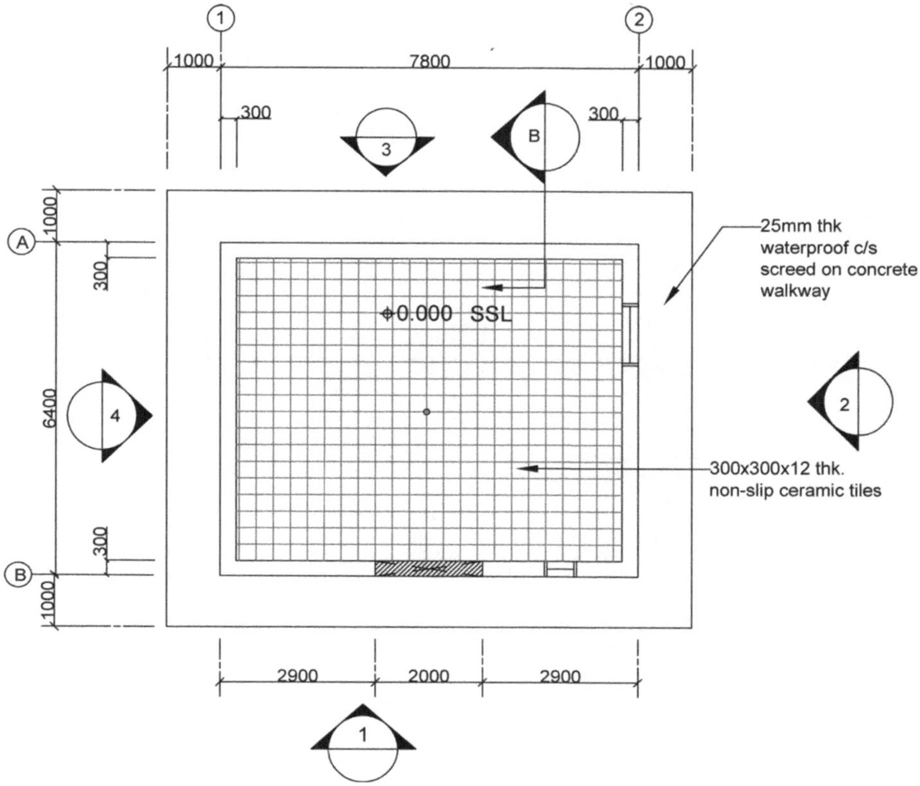

Figure 19.4a Floor plan of warehouse.

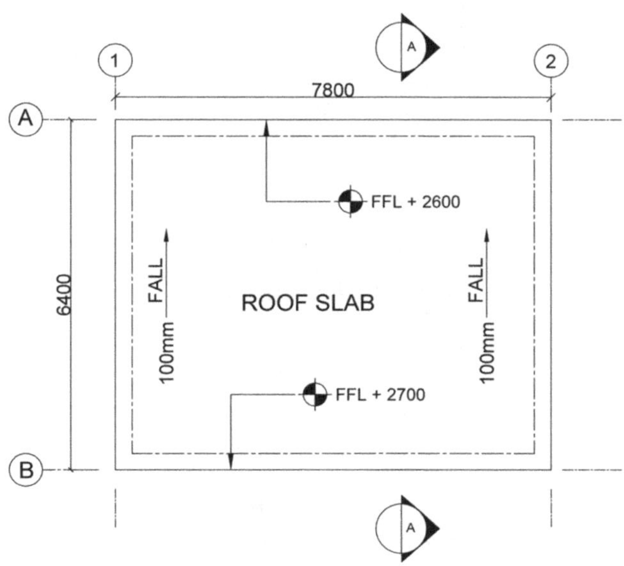

Figure 19.4b Roof plan of warehouse.

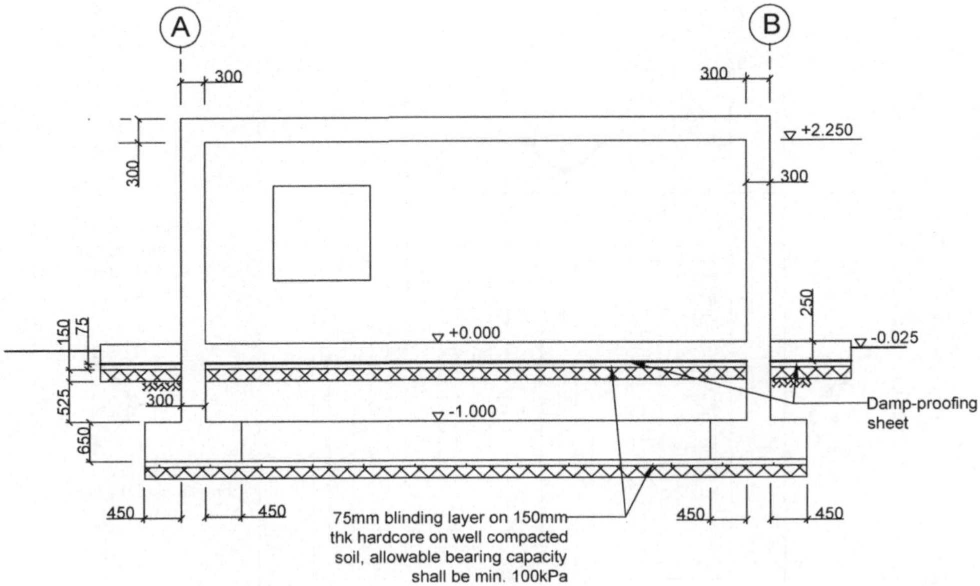

Figure 19.4c Section A-A of warehouse.

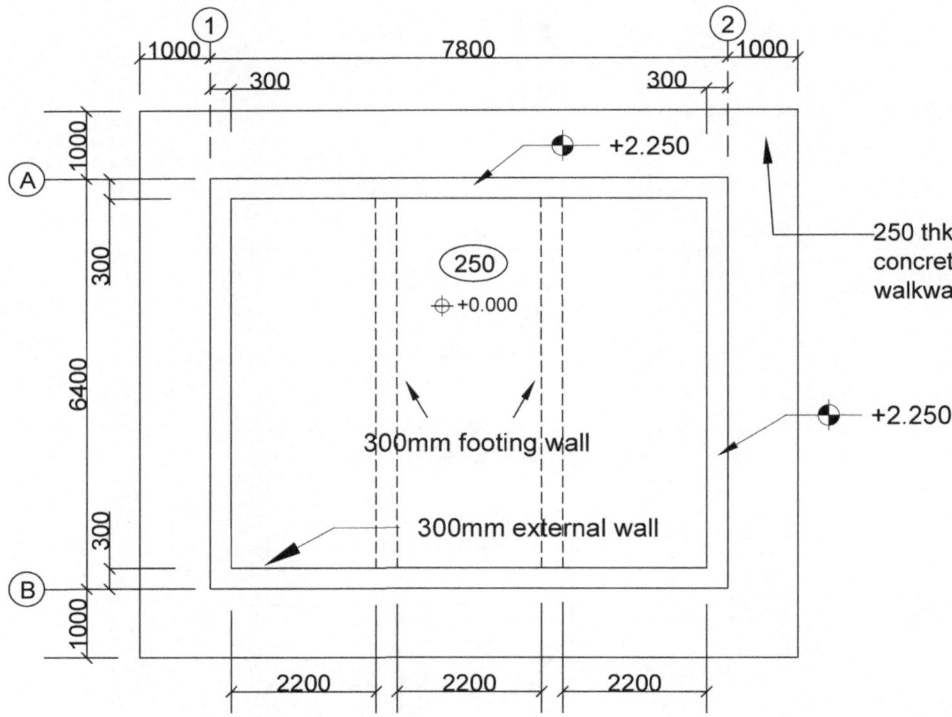

Figure 19.4d Ground floor structural plan of warehouse.

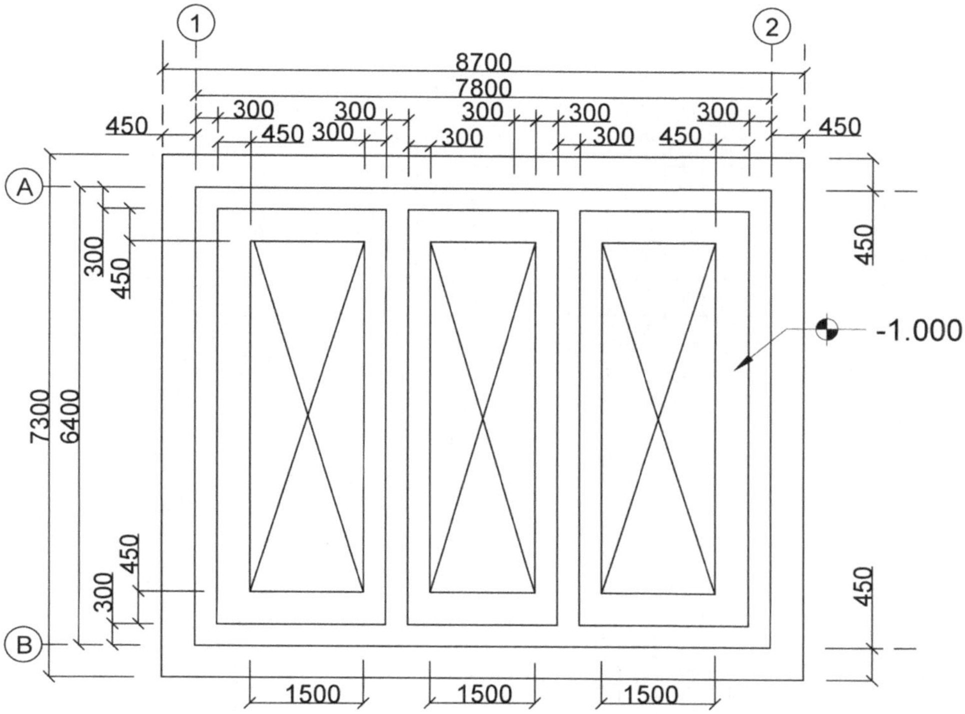

Figure 19.4e Footing plan of warehouse.

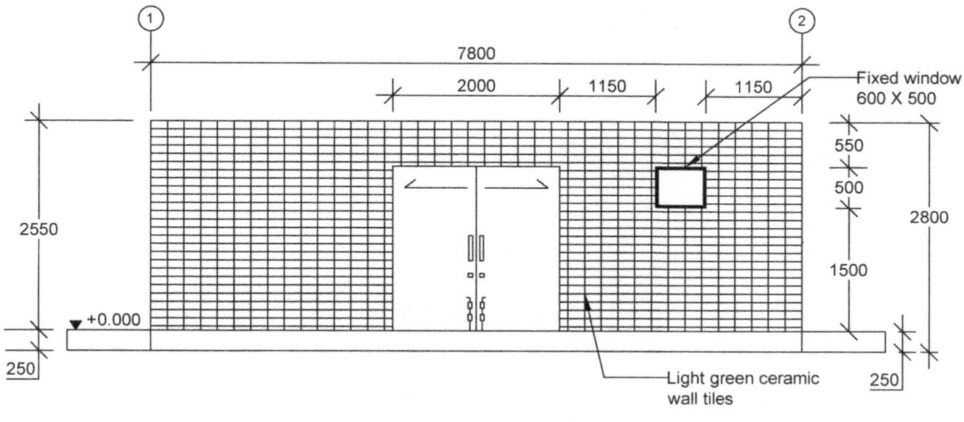

Figure 19.4f Elevation 1.

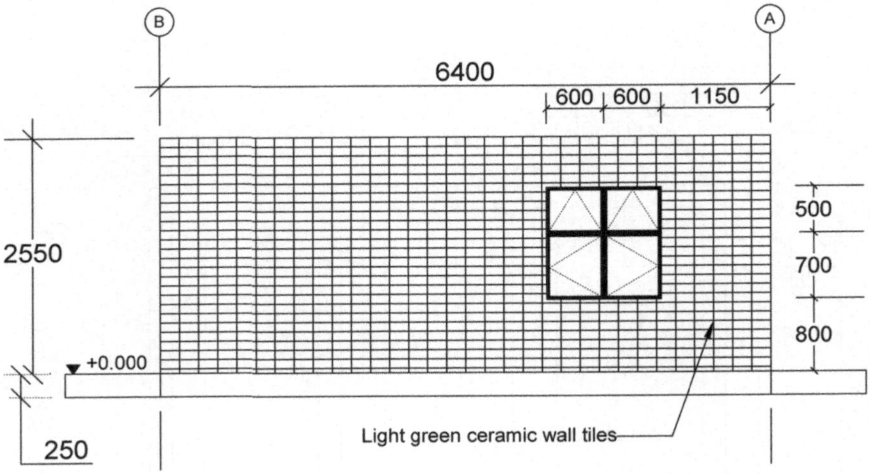

Figure 19.4g Elevation 2.

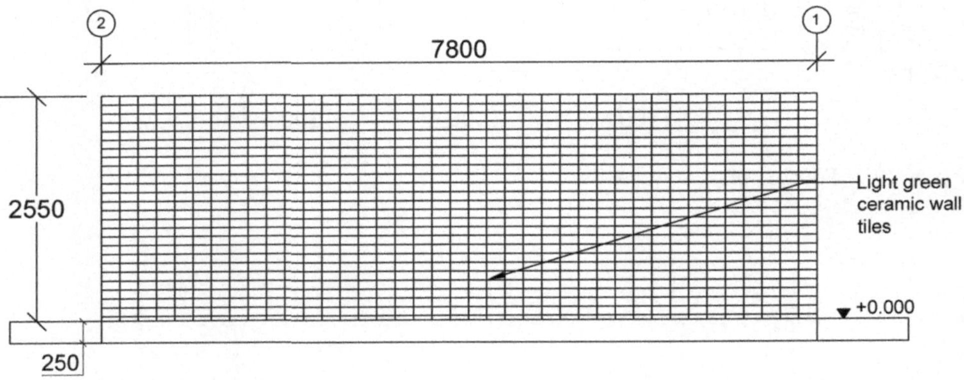

Figure 19.4h Elevation 3.

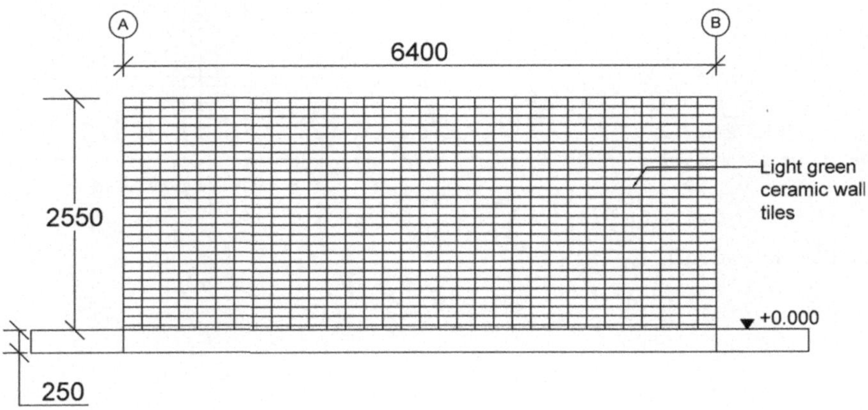

Figure 19.4i Elevation 4.

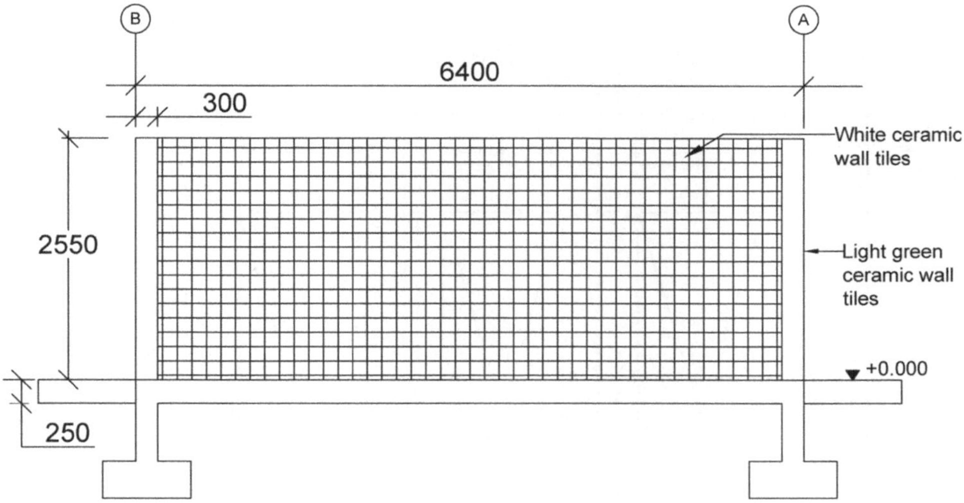

Figure 19.4j Section B-B.

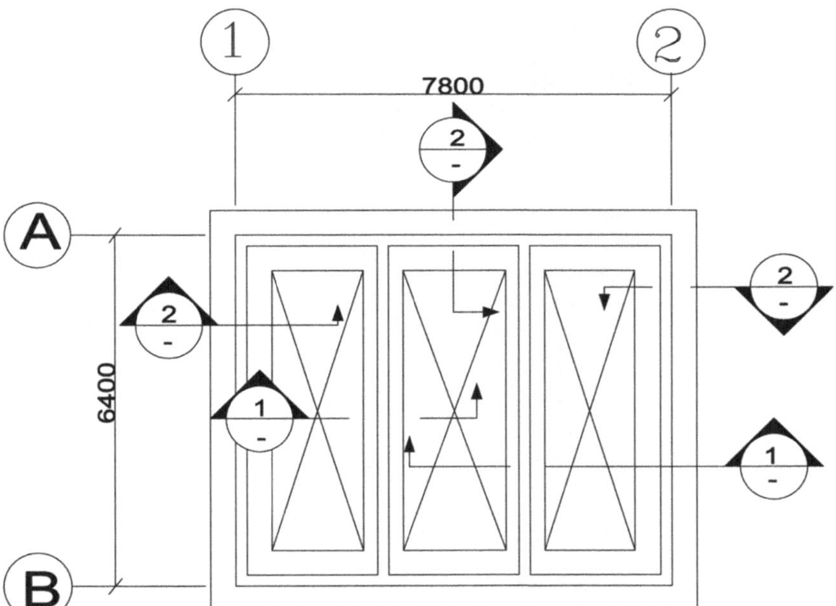

Figure 19.4k Footing plan.

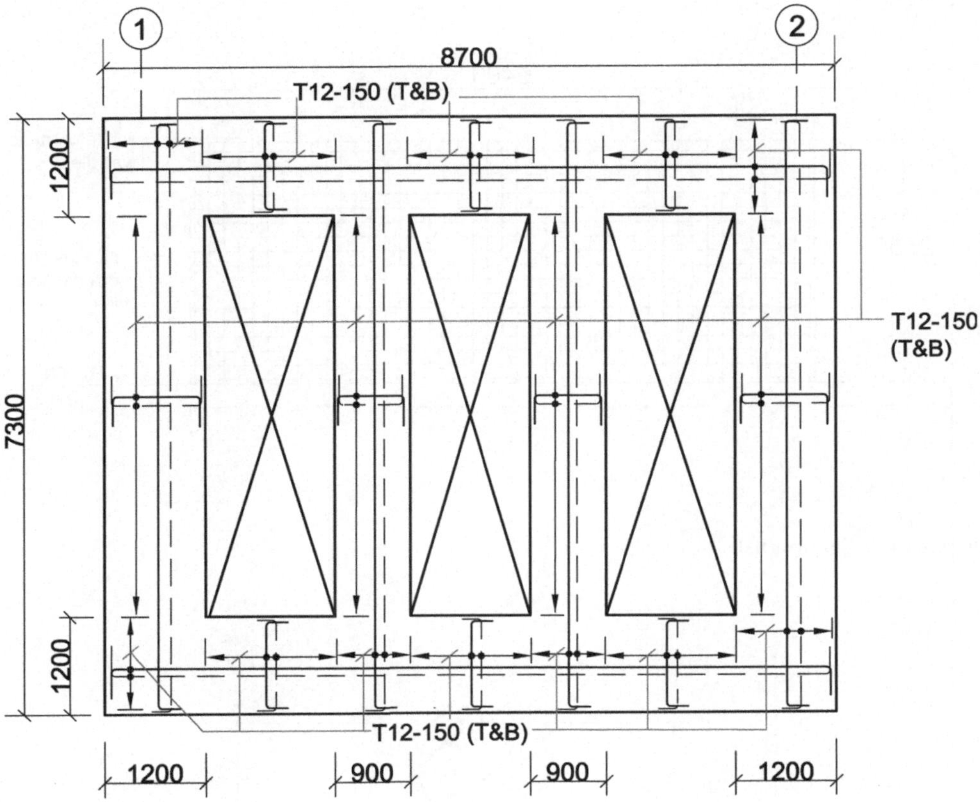

Figure 19.4l Footing rebar plan.

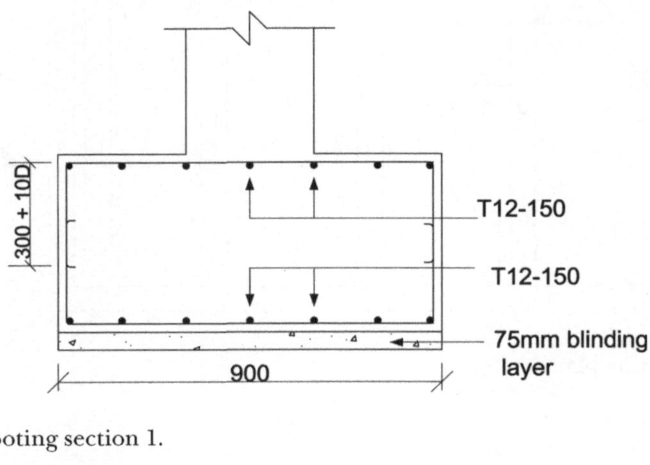

Figure 19.4m Footing section 1.

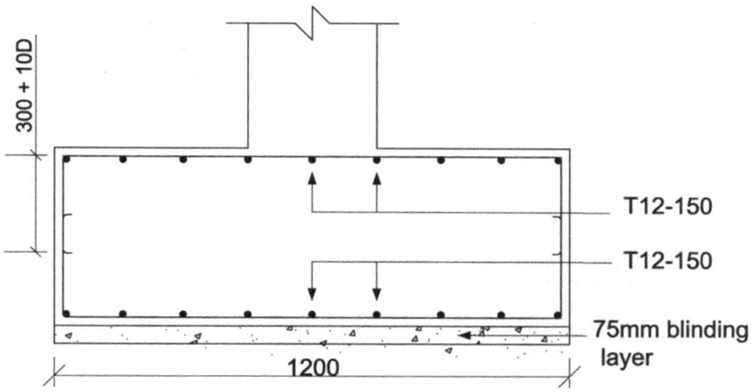

Figure 19.4n Footing section 2.

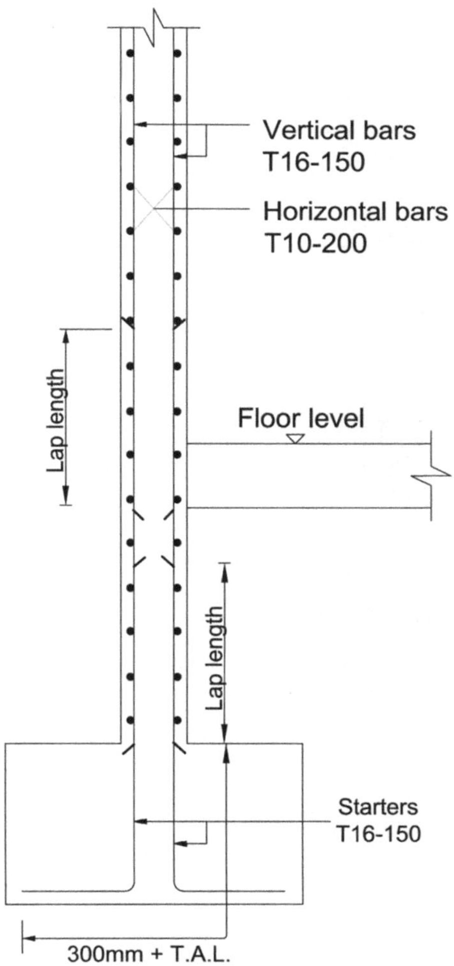

Figure 19.4o Wall base rebar details.

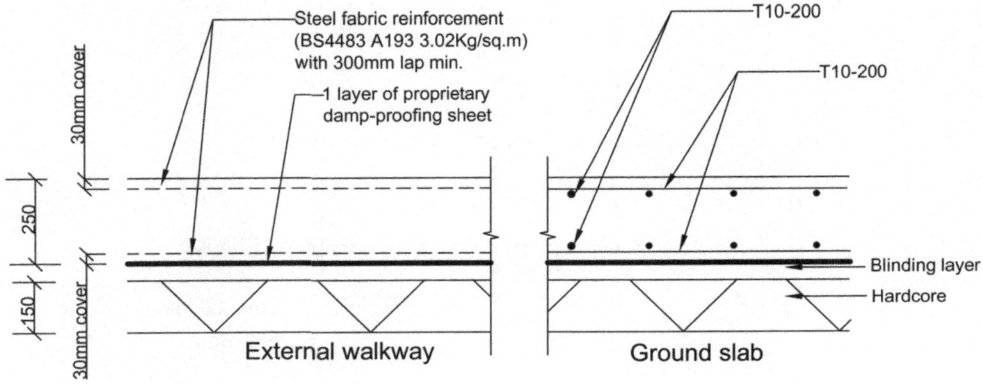

Figure 19.4p Typical rebar details of walkway and ground slab.

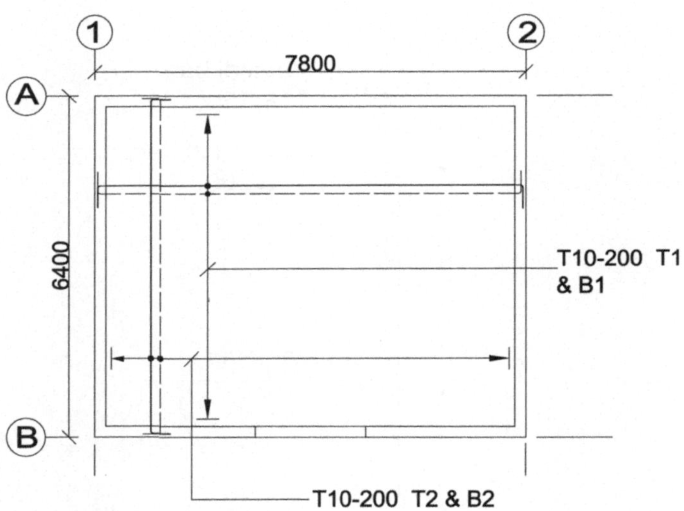

Figure 19.4q Ground floor slab rebar details.

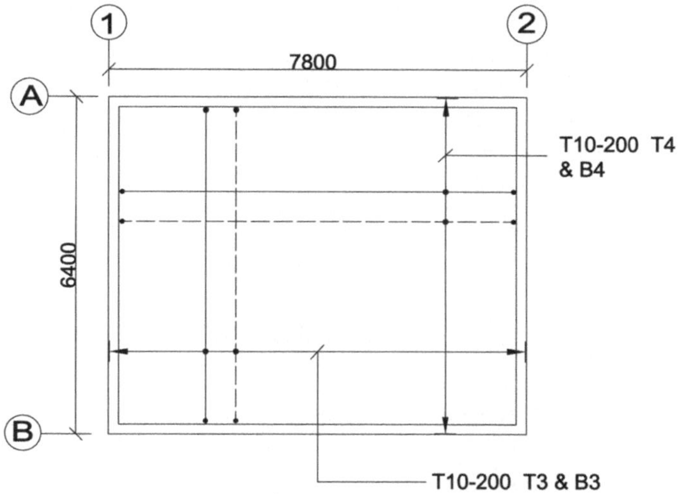

Figure 19.4r Roof slab rebar details.

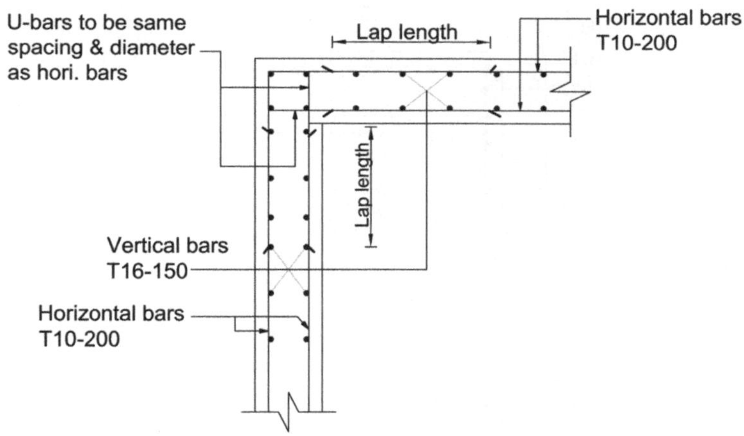

Figure 19.4s Sectional plan of footing wall or wall junction.

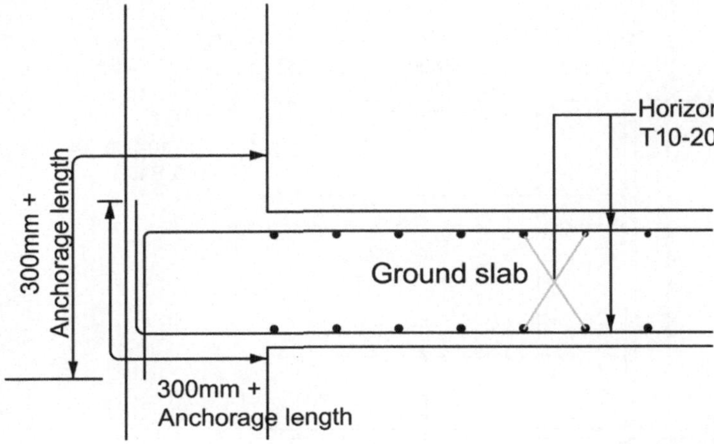

Figure 19.4t Ground slab and wall junction.

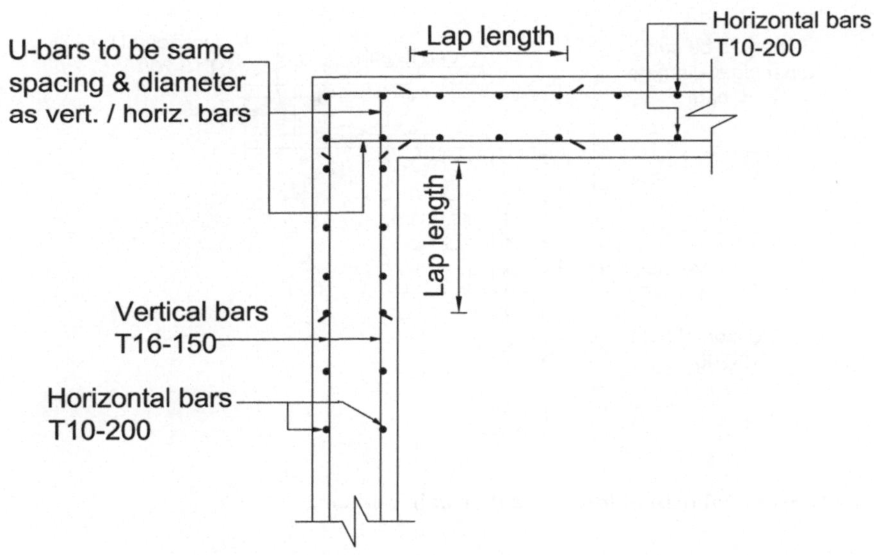

Figure 19.4u Section of wall and roof junction.

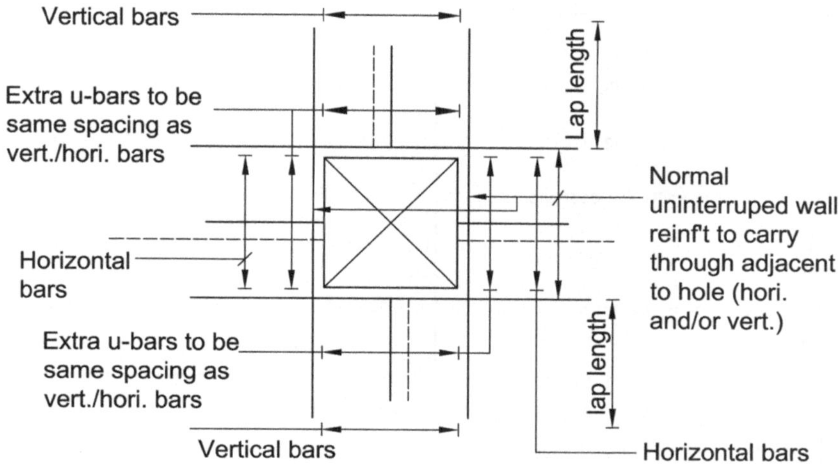

Figure 19.4v Typical rebar detail at wall opening.

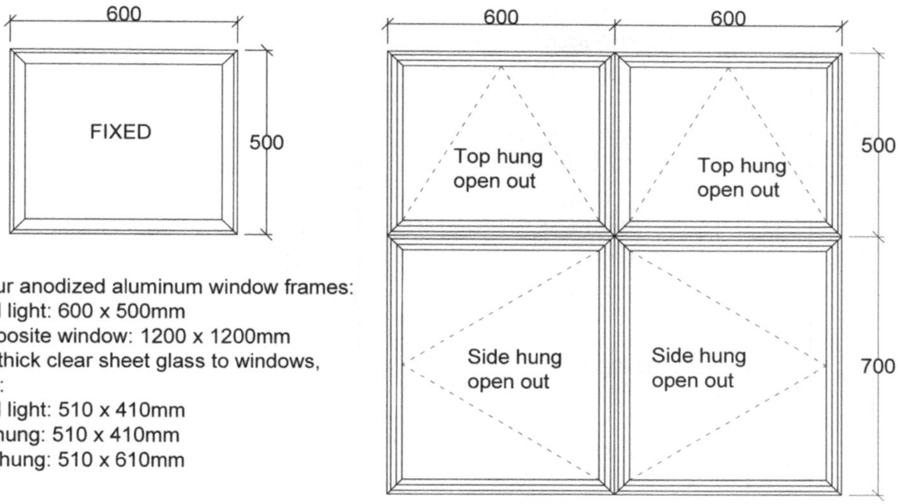

NOTES:
1. Colour anodized aluminum window frames:
   Fixed light: 600 x 500mm
   Composite window: 1200 x 1200mm
2. 6mm thick clear sheet glass to windows, sizes:
   Fixed light: 510 x 410mm
   Top hung: 510 x 410mm
   Side hung: 510 x 610mm

Figure 19.4w Windows layout.

Query list Questions:	Answers:
1. What is the minimum spacing between the rebar?	*1. Same as the minimum concrete cover.*
2. Details of the roof covering and finish missing.	*2. To be confirmed and measured later.*
3. Any details of the entrance door including ironmongeries?	*3. Details to be confirmed later. Allow a 1-hour fire rated, galvanised mild steel door. Ironmongeries to be confirmed and measured later.*
4. Finish details at door and window reveals are outstanding.	*4. Details to be confirmed and measured later.*

Simple warehouse building

Figures 19.4a to 19.4w

Take-off list

Substructure:
1. Oversite excavn to reduced level
2. Soil adjustment
3. Trench excavation for footings
4. Hardcore under footings
5. Blinding under footings
6. Concrete footings
7. Concrete footing walls
8. Hardcore under slab
9. Waterproofing
10. Soil adjustment
11. Formwork
12. Reinforcement

Superstructure:
1. Concrete slab
2. Concrete walkway
3. Concrete walls
4. Adjustment for openings
5. Formwork
6. Adjustment for openings
7. Reinforcement
8. Adjustment for openings
9. Finishes (intl) - spatterdash, screed, tiling
10. Finishes (extl) - repeat as intl finishes
11. Windows
12. Door

Substructure

Excavation

slab:	Length	Width
walkway	9800	8400
w/s 2/250	500	500
	10300	8900

Oversite excavation to the bottom level of the walkway hardcore. HKSMM4 Rev 2018, VI(a).3.2.M.2(a) Working space of 250mm is required for the walkway.

blinding & h.c.:	Length	Width
footing	8700	7300
w/s 2/600	1200	1200
	9900	8500

Excavation for the blinding and hardcore does not require any working space. However, to excavate the trench for the footing below, a slightly larger excavation area is required (as shown in the waste calculation aside and the diagram below). Therefore, 9800x8400mm (dimensions of the walkway without working space) are not taken.

Dimensions	Description	Notes
	Depth blinding 75 h.c. 150 225	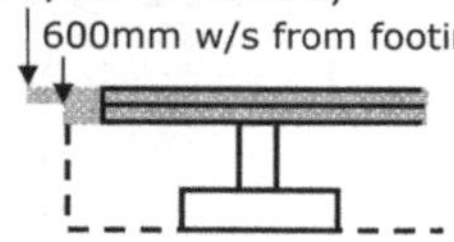
10.30 8.90 0.23	Oversite excavn to red. lev., over 200mm av. depth (slab	*HKSMM, VI(a).3.2* *For excavation to reduced level, it is not necessary to state the commencing level. Average depth has to be stated if oversite excavation is ≤ 0.2m deep.*
	&	
9.90 8.50 0.23	Disposal of exc. matl from site to a tip provided by Contractor (blind & h.c.	*HKSMM, VI(a).6.1.3* *Normally, backfilling will be added right after excavation measurement. Here, soil disposal is added instead of backfilling as disposal dominates.*
	Mean girth of w.s. ard walkway / ard h.c. 9800 / 9800 8400 / 8400 2/ 18200 / 18200 36400 / 36400 w/s 4 corners 1000 / 200 37400 / 36600	*Excavation to the working space has to be backfilled entirely. Soil adjustment is measured here.* *The width of the working space around blinding layer and hardcore is 50mm (i.e. (9900mm - 9800mm) x 0.5).*
37.40 0.25 0.23	Ddt Disposal of exc. matl a.b. (soil adj: slab	*HKSMM, VI(a).6.1.3*
	&	
36.60 0.05 0.23	Add Backfilling with sel. exc. matl (soil adj: blind & h.c.	*HKSMM, VI(a).7.0.1*
	Depth of excavation for footings Fdn wall 525 Footing 650 1175	*As before, excavation for blinding layer and hardcore below footing does not require any working space. Therefore, the two layers are not included in the excavation depth calculation here.*
	Mean girth of perimeter footing L 8700 W 7300 2/ 16000 32000 Less: corners 4/2/½/1200 4800 27200	*The overall excavation for ground slab and hardcore has been measured above already. Here, the trench excavation starts from the bottom of the hardcore (i.e. the reduced level).*

Timesing	Dimensions	Description	Notes
		Length of excavation (900w footing)	
		7300	
		Less:	
		Footing 2/1200 2400	HKSMM, VI(a).3.5.M.2(b)
		w/s 2/600 1200	600mm working space has to be allowed.
		3700	
		Footing width: peri. 900w	
		1200 900	
		w/s 2/600 1200 1200	
		2400 2100	
		Length of excavn (900w blind & h.c.)	
		7300	
		Less: 2/1200 2400	
		4900	
	27.20 2.40 1.18	Trench excavn for footings commencing at red. lev., n.e. 1.50m deep (peri. footing	HKSMM, VI(a).3.5.1.1 Commencing level has to be stated, and in stages of 1.50m deep.
		&	
2/	3.70 2.10 1.18	Backfilling with sel. exc. matl (900w footing	HKSMM, VI(a).7.0.1 As in the previous examples, backfilling is added first. The volume occupied by the concrete footing can be adjusted later.
	27.20 1.20 0.23	(peri. blind & h.c.	
2/	4.90 0.90 0.23	(900w blind & h.c.	
	27.20 1.20	150mm thk hardcore beds (perimeter footing	HKSMM, VI(a).10.1.1 The mean girth of hardcore (at the perimeter) is the same as the mean girth of foundation. Note that hardcore has to be measured in super when thickness ⩽ 300mm.
2/	4.90 0.90	(900w footing	
	27.20 1.20 0.15	Ddt Backfilling with sel. exc. matl (h.c. perimeter	HKSMM, VI(a).7.0.1 Backfilling has been added when measured the trench excavation. Therefore, soil adjustment for hardcore is made here. While hardcore is measured in super but soil disposal and backfilling have to be measured in volume, the soil adjustment is measured separately.
		&	
2/	4.90 0.90 0.15	Add Disposal of exc. matl from site to a tip provided by Contractor (h.c. 900w footing	

		Concrete	
	27.20	Grade 10P plain conc. in 75mm thk	*HKSMM, VII(a).11.1*
	1.20	blinding under footings (peri.	*Blinding layer has to be measured in m^3*
	0.08	&	*with the thickness stated.*
		Ddt	
		Backfilling with sel. exc. matl	*HKSMM, VI(a).7.0.1*
2/		&	
	4.90	Add	
	0.90	Disposal of exc. matl a.b.d.	*HKSMM, VI(a).6.1.3*
	0.08	(900w footing	
	27.20	Grade 40D/20 waterproof R.C. in fdn	*HKSMM, VII(a).8*
	1.20	(perimeter	
	0.65	&	
		Ddt	
		Backfilling with sel. exc. matl	*HKSMM, VI(a).7.0.1*
		&	
2/		Add	
	4.90	Disposal of exc. matl a.b.d.	*HKSMM, VI(a).6.1.3*
	0.90		
	0.65	(900w footing	
		Height of foundation walls	
		525	
		hardcore 150	
		blinding 75	
		750	
	27.20	Grade 40D/20 waterproof R.C. in	*HKSMM, VII(a).22*
	0.30	300mm thk fdn walls	*Thickness of wall to be stated.*
	0.75	(perimeter	
2/	5.80		
	0.30		*Length: 7300 - 2 x (300 + 450) = 5800*
	0.75	(900w footing	
	27.20	Ddt	
	0.30	Backfilling with sel. exc. matl	*HKSMM, VI(a).7.0.1*
	0.53	&	
		Add	*HKSMM, VI(a).6.1.3*
2/		Disposal of exc. matl a.b.d.	*The depth for soil adjustment is 525mm*
	5.80		*instead of 750mm because soil*
	0.30	(soil adjustment	*adjustment for oversite excavation has*
	0.53	for fdn wls	*covered the hardcore and blinding layer already.*
	9.80	150mm thk hardcore beds	*HKSMM, VI(a).10.1.1*
	8.40		*No need to make adjustment for backfilling or disposal as soil adjustment has been done in oversite excavation.*

	9.80 8.40	Flexible sheet damp-proofing in 1 ct proprietary sheet coverings laid horiz. on blinding layer surf.; designing and fixing by a specialist contractor; laying in accordance with the manufacturer's instructions.	*HKSMM, X(d).2.3.0.1* *Profile of the flexible sheet damp-proofing should be stated, which is horizontal in this case.*
	27.20 0.30	<u>Ddt</u> last 2 items (perimeter wl	
2/	5.80 0.30	(900w footing	
		<u>Formwork</u>	
2/	27.20 0.65	Sawn fwk to vert. sides of fdn, grd beams and the like (perimeter footing	*Formwork to vertical sides of footings, foundation walls, ground beams and the like are normally grouped together.*
2/ 2/	4.90 0.65	(900w footing	*Length: 7300 - 2 x 1200 (footing) = 4900*
4/	0.90 0.65	<u>Ddt</u> Ditto (intersections	*VII(d).M.1 – Formwork is measured to each face which requires support, except otherwise stated. Intersections of the formwork should therefore be deducted.*

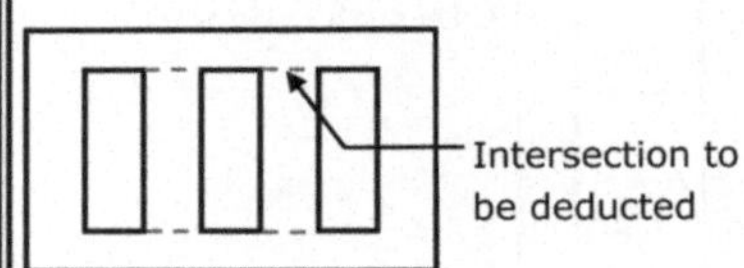

2/	27.20 0.75	Sawn fwk to vert. sides of fdn, grd beams and the like (fdn wl; perimeter	
2/ 2/	5.80 0.75	(fdn wl; 900w footing	*Length: 7300 - 2 x (450 + 300) = 5800*
4/	0.30 0.75	<u>Ddt</u> Ditto (intersections	

Reinforcement

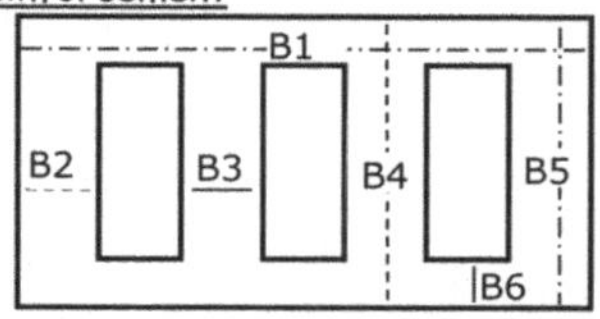

Length of bars in footings (B1-B3)

	B1	B2	B3
length	8700	1200	900
bends 2/420	840	840	840
	9540	2040	1740
Less: cr 2/30			
r 36			
2d 2/12	120	120	120
	9420	1920	1620

*Refer to **Figure 19.4k, 19.4l, 19.4m** & **Figure 19.4n**:*

Formula used: BS4466, code 38

Length – 2 concrete cover + 2 x bends - r - 2d

Bend length: 300 + 10d = 420

*Minimum concrete cover (cr) taken as 30mm based on **Figure 19.4p**.*

According to BS4466 Table 3, r = 36mm when d is 12mm.

Similarly, length of bar B4 & B5:
7300 + 840 - 120 = 8020 mm
Length of bar B6 = same as B2 = 1920mm

No. of bars in footings (B1-B3)

	B1	B2	B3
	1200	4900	same as B2
Less:			
cr (or b) 2/30	60	60	
	1140	4840	
divided by 150	7.6	32.27	
say	8	33	
+ 1	9	34	34
2 identical bays 2/	18	68	68

B2 span for rebar: 7300 - 2 x 1200 = 4900

Although no concrete cover is required to allow for B2, minimum bar spacing (b) has to be kept in between the B2 and B1 bars. As shown in the query answer, 30mm should be allowed.

*B1: Same no. as shown in **Figure 19.4n**. If there is any difference between the calculated vlaue and the number shown on the drawing, a query should be raised.*

No. of bars in footings (B4-B6)

	B4	B5	B6
	900	same as B1	1500
Less:			
cr (or b) 2/30	60		60
	840		1440
divided by 150	5.6		9.6
say	6		10
+ 1	7	9	11
2 identical bays 2/	14	18	22
3 segments 3/			66

cr = minimum concrete cover; b = minimum bar spacing

*B4: Same no. as shown in **Figure 19.4m**.*

Timesing	Dimension	Description	Side notes
2/ 18	9.42	12mm dia. high yield steel bars in general reinforcement (____m x 0.888kg/m = ____kg) (B1	*HKSMM, VII(c).2.1.1* *As per **Figure 19.4m** and **Figure 19.4n**, top and bottom bars are required and therefore each set of measurement is multiplied by 2.*
2/ 66 • 68	1.92	(B2, B6	

2/ 68/	1.62		Ditto (____m x 0.888kg/m = ____kg) (B3	
2/ 18 • 14/	8.02		(B4, B5	

Footing wl L:	Starters	Vert bar	
	300	750	Refer to ***Figures 19.4o***, ***19.4q*** and ***19.4s***.
T.A.L. 32/16	512	0	*The lap length of the vertical bars allows concreting work to the footing and the superstructure to be done separately. In practice, whenever a contractor needs to place concrete to a structure in more than one phase, sufficient laps must be allowed to avoid cold joints.*
Lap L 32/16	512	512	
	1324	1262	

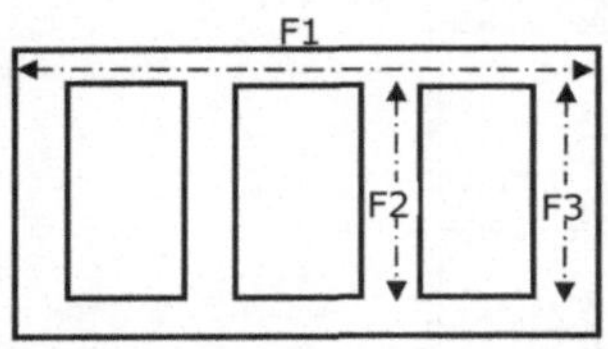

No. of starters/vert. bars in footing walls

	F1	F2	F3	
				Based on the vertical bars layout in ***Figure 19.4s****, starters or vertical bars in F2 or F3 will not overlap with those in F1.*
L	7800	5800	same as F2	*Length of F2: 6400 - 2 x 300 = 5800mm*
Less:				
cr (or b) 2/30	60	60		
	7740	5740		
divided by 150	51.6	38.27		
say	52	39		
+ 1	53	40	40	
2 identical bays 2/	106	80	80	

Footing walls, Length of u-bars

	length	240	*U bar: As shown in* ***Figure 19.4s****, U-bars are required at each wall junction.*
bends	wl 2/270	540	
lap length	2/320	640	*U-bar length: formula based on BS4466, code 38*
		1420	
Less:	r 30		*U-bar length = (wl thk - 2 concrete cover) + 2(wl thk – concrete cover + L.L.) – r – 2d*
	2d 2/10	50	
		1370	*According to BS4466 Table 3, r = 30mm when d is 10mm.*

Footing walls, Length of horzi. bars

	F1	F2/F3	
L:	7800	6400	*Hori. bar length = length – 2(wl thk)*
Less: wl 2/300	600	600	
	7200	5800	

Timesing	Dimensions	Squaring	Description	Side notes
			<u>No. of horiz. bars in footing wls</u> <u>(F1/F2/F3)</u> 525 H.C. 150 blinding <u>75</u> Height of footing wl 750 divided by 200 <u>3.75</u> say 4 + 1 <u>5</u> 2 identical bays 2/ <u>10</u>	*No. of horizontal bars: Same as the number shown in* ***Figure 19.4o****.*
2/ 106/	1.32		16mm dia. high yield steel bars in general reinforcement (____m x 1.579kg/m = ____kg) (F1 starters	*HKSMM, VII(c).2.1.1* *As shown in* ***Figure 19.4o****, the vertical bars, horizontal bars and starter bars for walls are in pairs. Therefore, each set of measurement is multiplied by 2.*
2/ 80 • 80/	1.32		(F2, F3 starters	
2/ 106/	1.26		(F1 vert. bars	
2/ 80 • 80/	1.26		(F2, F3 vert. bars	
2/ 3/ 10/	1.37		10mm dia. high yield steel bars in general reinforcement (____m x 0.617kg/m = ____kg) (u-bars	*HKSMM, VII(c).2.1.1* *U-bars are required at both ends (junctions) of the walls. Therefore, the overall quantity is multiplied by 2.*
2/ 10/	7.20		(F1 horiz. bars	
2/ 10 • 10/	5.80		(F2, F3 horiz. bars	
			<u>Superstructure</u> <u>Concrete</u>	
	7.80 6.40 0.25		Grade 40D/20 waterproof R.C. in 250mm thk beds (grd slab	*HKSMM, VII(a).3.1. Thickness to be stated.* 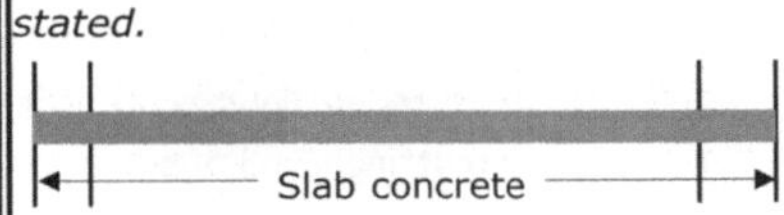

Dimensions	Description	Notes
	mean girth of walkway L: 7800 W: 6400 2/ 14200 28400 4 corners: 4/2/½/1000 4000 32400	
32.40 1.00 0.25	Grade 20D/20 waterproof R.C. in 250mm thk beds (walkway	*HKSMM, VII(a).3.1* *In view of the size and nature, the concrete walkway is measured as concrete bed.*
7.80 6.40 0.30	Grade 40D/20 R.C. in 300mm thk horiz. suspended slab (roof	*HKSMM, VII(a).17* *Concrete to suspended slab (including roof) has to be measured separately.* *HKSMM, VII(a).M.12* *Measurement of suspended slab is taken across the beams and columns if they are of the same mix.* *HKSMM, VII(a).M.16* *Concrete walls are measured up to soffit of slab.*
	Height of concrete wall Refer to Fig. 19.4c ceiling level 2.25 floor level 0.00 2.25	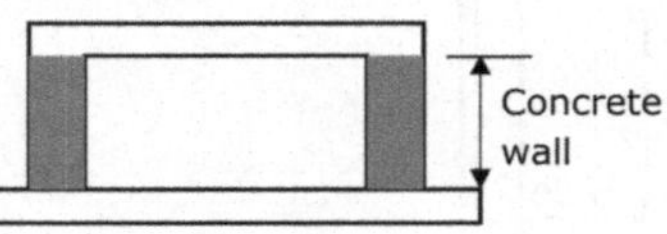
27.20 0.30 2.25	Grade 40D/20 R.C. in 300mm thk walls	*HKSMM, VII(a).22.1* *Thickness has to be stated. Mean girth of the wall is the same as the mean girth of the perimeter footing.*
2.00 0.30 2.00 1.20 0.30 1.20	Ddt Ditto (door (composite wind.	*HKSMM, VII(a).1.M.10(c)* *No deduction of openings in walls if ≤ $0.5m^2$. Therefore, no deduction of the fixed light opening (600 x $500mm^2$) is required.*
	Formwork	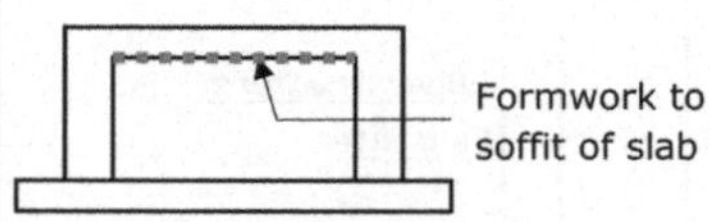
7.20 5.80	Sawn fwk to horiz. soffits of suspended slabs; 200-300mm thk, strutting n.e. 3.50m	*HKSMM, VII(d).2.2.1.1; If slab thickness > 200mm, thickness should be stated In further stages of 100 mm.* *Height to soffit has to be stated: not exceeding 3.50m.*

Timesing	Dimensions	Description	Notes
2/	27.20 2.25	Sawn fwk to vert. sides of walls (b/s	*HKSMM, VII(d).5.0.1* *Mean girth of the wall is used to measure the formwork to both sides.*
		Extl girth of grd slab mean girth 27200 corners 4/2/150 1200 28400	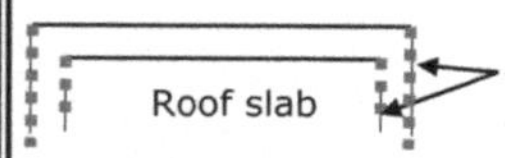
	28.40 0.25	Ditto (edges of grd slab	*Formwork is required at the edge of ground slab as the ground slab is casted separately from the concrete paving. This part of formwork is measured as the wall formwork.*
	28.40 0.30	(edges of roof	
2/	2.00 2.00	Ddt Ditto (door	*HKSMM, VII(d).1.M.3* *No formwork deduction for the fixed light as the area is < 1m²; but deduction has to be made for the door and composite window. Remember to measure 2 vertical sides of formwork in each deduction.*
2/	1.20 1.20	(composite wind.	
2/	2.00	Sawn fwk to vert. surfaces of edges and breaks in walls n.e. 300mm wide (dr open	*HKSMM, VII(d).9.1.2*
2/	1.20	(composite wind.	
	2.00	Sawn fwk to soffits of openings in walls; n.e. 300mm wide (dr open	*HKSMM, VII(d).9.1.2* *According to HKSMM, VII(d).16, boxings for opening are only required when the opening is > 0.5m² and ≤ 1.0m². Therefore, no boxing is measured for the fixed light.*
	1.20	(composite wind.	
		extl girth of walkway 9800 8400 2/ 18200 36400	*HKSMM, VII(d).9.1.2* *Only the external side of the walkway slab requires formwork and therefore external girth of the walkway is measured here.*
	36.40	Sawn fwk to vert. surfaces of edges and breaks in slabs n.e. 300mm wide (walkway	*The item is measured by m if ≤ 300mm wide.*
		Reinforcement	*HKSMM, VII(c).5. Description should include lap length and weight.*
2/	32.40 0.94	Steel fabric reinf. square mesh; BS4483 Ref. A193; weighing 3.02 kg/m²; 300mm minimum side and end laps; measured nett in concrete (walkway	*As shown in* ***Figure 19.4p****, top and bottom fabric reinforcement is required. 30mm concrete cover is allowed at the two ends and thus the width of fabric reinforcement is 0.94m instead of 1.00m.*

Length of bars in grd slab — Refer to **Figure 19.4q** & **Figure 19.4t**.

		1-2	A-B
along G.L.			
	L	7800	6400
Less:			
	wl 2/300	600	600
		7200	5800
bends: 32D			
	300 2/620	1240	1240
		8440	7040

No. of bars in grd slab — *No. of bars in the roof slab is the same as the no. of bars in the ground slab.*

		1-2	A-B
along G.L.			
	L	6400	7800
Less:	wl 2/300	600	600
		5800	7200
	divided by 200	29.00	36.00
	say	29	36
	+ 1	30	37

Length of bars in roof slab — Refer to **Figure 19.4r** & **Figure 19.4u**.

		1-2	A-B
along G.L.			
	L	7800	6400
Less:			
	wl 2/300	600	600
		7200	5800

No. of bars in roof slab

		1-2	A-B
along G.L.			
	L	6400	7800
Less:	cr 2/30	60	60
		6340	7740
	divided by 200	31.70	38.70
	say	32	39
	+ 1	33	40

Timesing	Dimension	Description	Notes
2/30/	8.44	10mm dia. high yield steel bars in general reinforcement (____m x 0.617kg/m = ____kg) (grd slab	HKSMM, VII(c).2.1.1
2/37/	7.04		
2/33/	7.20	(roof	
2/40/	5.80		

Timesing	Dimension	Squaring	Description	Notes
2/ 33/	1.37		Ditto (____m x 0.617kg/m = ____kg) (roof horiz. u-bars	*Length of horizontal u-bars are the same as those in the footing walls.*
2/ 40/	1.37		(ditto	

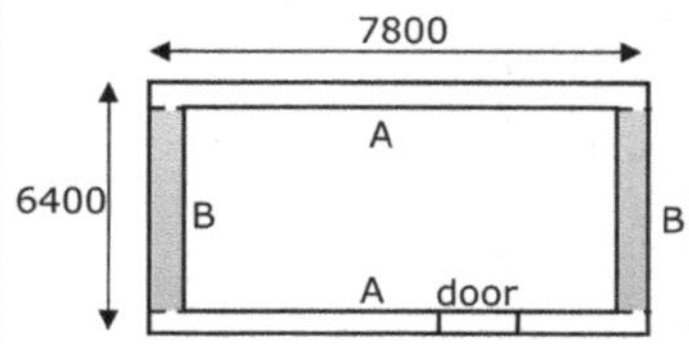

*Refer to **Figure 19.4o**, **Figure 19.4s** and **Figure 19.4u**.*

<u>Length of vert. u-bars</u>

length		240
bends	wl 2/270	540
lap length	2/512	<u>1024</u>
		1804
<u>Less:</u> r	48	
2d	2/16	<u>80</u>
		<u>1724</u>

*Vert. u-bar: As shown in **Figure 19.4u**, u-bars are required at the junction between the roof and walls.*

U-bar length: formula based on BS4466, code 38

U-bar length = (wl thk - 2 concrete cover) + 2(wl thk – concrete cover + L.L.) – r – 2d

According to BS4466 Table 3, r = 48mm when d is 16mm.

<u>Length of vert. bars</u>

	2250
grd slab	<u>250</u>
	<u>2500</u>

<u>No. of vert. bars in walls</u>

	A	B
Length	7800	6400
<u>Less:</u>		
cr (or b) 2/30	60	60
<u>Less</u>: wl 2/300	<u>0</u>	<u>600</u>
	7740	5740
divided by 150	<u>51.6</u>	<u>38.27</u>
say	52	39
+ 1	<u>53</u>	<u>40</u>
2 identical bays 2/	<u>106</u>	<u>80</u>

Vertical bars at the wall intersection have been measured in wall A. Therefore, the length of wall B is based on 5.8m (instead of 6.4m) when calculate the no. of vertical bars.

Timesing	Dimension	Squaring	Description	Notes
2/ 106 • 80/	2.50		16mm dia. high yield steel bars in general reinforcement (____m x 1.579kg/m = ____kg) (wl A&B, vert.	*HKSMM, VII(c).2.1.1* *As the length of vertical bars are constant throughout all the walls, they are added by "dotting on" in the dimension column. Remember to multiply the number of bars by 2 as there are 2 layers.*

Timesing	Dimension	Description	Notes
106 •80/	1.72	Ditto (____m x 1.579kg/m = ____kg) (wl A&B, top u-bar	
		<u>Length of horiz. bars</u>	
		A / B	
		7800 / 6400	
		Wl 2/300: <u>600</u> / <u>600</u>	
		<u>7200</u> / <u>5800</u>	
		<u>No. of horiz. bars</u>	
		grd slab 250	
		height <u>2250</u>	
		2500	
		Less: cr 2/30 <u>60</u>	
		2440	
		divided by 200 <u>12.20</u>	
		say <u>13</u>	
		+ 1 <u>14</u>	
		2 identical bays 2/ <u>28</u>	
2/ 28/	7.20	10mm dia. high yield steel bars in general reinforcement (____m x 0.617kg/m = ____kg) (wl A, horiz.	
2/ 28/	5.80	(wl B, horiz.	
2 /2/ 28/	1.37	(wl A&B, u-bar	*The length of horizontal u-bars are the same as those in the footing walls.*
		<u>No. of bars ddt around openings (vert.)</u>	*Refer to* ***Figure 19.4v*** *.*
		fixed win / c. win / door	*Size of fixed window: 600 x 500mm*
		Length: 600 / 1200 / 2000	*Size of composite window: 1200 x 1200mm*
		divided by 150: <u>4.00</u> / <u>8.00</u> / <u>13.33</u>	
		say: <u>4</u> / <u>8</u> / <u>14</u>	*Size of door: 2000 x 2000mm*
		<u>No. of bars ddt around openings (horiz.)</u>	
		fixed win / c. win / door	
		Width: 500 / 1200 / 2000	
		divided by 200: <u>2.50</u> / <u>6.00</u> / <u>10.00</u>	
		say: <u>3</u> / <u>6</u> / <u>10</u>	

		Ddt	
2/4/	0.56	16mm dia. high yeild steel bars in general reinforcement (____m x 1.579kg/m = ____kg) (fixed win. vert.	Note that 30mm concrete cover has to be allowed to all rebars around the openings.
2/8/	1.26	(comp. win. vert.	
2/14/	2.03	(dr vert.	Concrete cover is only allowed at the door head.
		Ddt	
2/3/	0.66	10mm dia. high yeild steel bars in general reinforcement (____m x 0.617kg/m = ____kg) (fixed win. horiz.	
2/6/	1.26	(comp win. horiz.	
2/10/	2.06	(dr horiz.	
2/4•8/	1.72	16mm dia. high yield steel bars in general reinforcement (____m x 1.579kg/m = ____kg) (wins, vert. u-bar	Vertical u-bars are required for the head and sill of windows.
14/	1.72	(dr, vert. u-bar	Only the door head requires vertical u-bars.
2/3•6/	1.37	10mm dia. high yeild steel bars in general reinforcement (____m x 0.617kg/m = ____kg) (wins, horiz. u-bar	
2/10/	1.37	(dr, horiz. u-bar	
		Finishes; Internally Intl girth of slab L: 7200 W: 5800 2/ 13000 26000	
	7.20 5.80	Spatterdash to all surf. of conc. (ceiling	HKSMM, XVI(a).1.1 Spatterdash to all locations can be grouped together.

26.00 2.25	Spatterdash to all surf. of conc.	(walls	
2.00 2.00	Ddt Ditto	(door	*No deduction rule is defined for spatterdash. Therefore, the area is measured nett with all voids deducted.*
0.60 0.50		(fixed light	
1.20 1.20		(composite wind.	
7.20 5.80	10mm thk c/s (1:3) plaster to ceiling in 1 ct; steel trowelled finish; to receive emuls. paint; n.e. 3.50m above floor		*HKSMM, XVI(d).2.1* *&* *XVI(c).1.M.6* *The height of ceiling has to be stated – not exceeding 3.5m high.*
	&		
	1 ct of water-based lime resistant primer & 2 full ct of emuls. paint on plastered ceilings over 300mm girth; n.e. 3.50m above floor		*HKSMM, XXI(b).1.3.1* *&* *XXI(a).1.M.5* *The height of ceiling has to be stated – n.e. 3.5m high.*
7.20 5.80	40mm thk waterproof c/s (1:3) screed to floor in 1 ct; wood floated finish; to receive non-slip homogeneous floor tiles		*HKSMM, XVI(d).6.1* *Screed to different locations has to be measured separately.* *Thickness, no. of coats and material to be received has to be described.*
	&		
	300 x 300 x 12mm thk non-slip homogeneous floor tiles laid on screed; bedding and jointing in c/s mortar (1:4); pointing in white cement		*HKSMM, XVI(e).2* *Tiling to different locations has to be measured separately. Details of material, method of fixing and treatment of joints have to be described.*
26.00	20mm thk x 100mm high waterproof c/s (1:3) screed to sktgs in 1 ct; wood floated finish; to receive coved homogeneous sktg tiles		*HKSMM, XVI(d).13* *Measured in m along the wall (using the internal girth of the wall), with the height above finished floor stated.*
	&		
	10mm thk x 100mm high coved homogeneous sktg tiles laid on screed; bedding in c/s mortar (1:4), pointing in white cement		*HKSMM, XVI(e).9.1* *Details of material, coved edges and height above finished floor have to be stated.*

Dimensions	Description	Notes
2.00	<u>Ddt</u>	
2.00	Last 2 items	
	(door	
26.00	20mm thk waterproof c/s (1:3) screed to walls in 1 ct; wood floated finish; to receive glazed ceramic tiles	*HKSMM, XVI(d).1.1*
2.25		*Work to walls and the like is measured to the base (XVI(c).M.3).*
	&	
	150 x 150 x 10mm thk glazed ceramic tiles laid on screed; bedding and jointing in c/s mortar (1:4); pointing in white cement	*HKSMM, XVI(e).5.1* *Similar to floor tiling work, details of material, as well as bedding and fixing method, have to be described.*
2.00	<u>Ddt</u>	*HKSMM, XVI(c).1.M.5*
2.00	Last 2 items	*No deduction shall be made for voids ≤ 0.5m^2. Therefore, fixed light area (0.3 m^2) does not require deduction.*
	(door	
1.20		
1.20	(composite wind.	
	- End of internal finishes -	
	<u>Extl girth of wall</u>	
	intl girth 26000	
	4 corners 4/2/300 <u>2400</u>	
	<u>28400</u>	
	<u>Height</u>	
	2250	
	roof <u>300</u>	
	<u>2550</u>	
	<u>Finishes; Externally</u>	*HKSMM, XVI(a).1.1*
28.40	Spatterdash to all surf. of conc.	*SMM does not require spatterdash to external areas and internal areas to be measured separately. However, other finishing work including screeding, painting and tiling are deemed to be internal work unless otherwise described (XVI(c).1.D.1 & XXI(a).1.D.1). For convenience, all work including spatterdash is measured separately under "external" and "internal" headings.*
2.55	(wall	
2.00	<u>Ddt</u>	
2.00	Ditto	
	(door	
0.60		
0.50	(fixed light	
1.20		
1.20	(composite wind.	
28.40	20mm thk c/s (1:3) render to walls in 1 ct, wood floated finish, to receive ceramic wall tiles	*HKSMM, XVI(d).1.1*
2.55		

28.40 2.55	200 x 100 x 10mm thk light green ceramic wall tiles laid on render; bedding and jointing in c/s mortar (1:4); pointing in white cement	*HKSMM, XVI(e).5.1*
2.00 2.00	<u>Ddt</u> Last 2 items (door	*HKSMM, XVI(c).1.M.5*
1.20 1.20	(composite wind.	
32.40 1.00	25mm thk waterproof c/s (1:3) screed to floor in 1 ct; steel trowelled finish (walkway	*HKSMM, XVI(d).6.1*
36.40 0.03	(edges of walkway	*Narrow widths and small quantities are included in the associated item (XVI(c).C.1(b)).*
	- End of external finishes -	
	<u>Windows</u> <u>Designing, supplying and fixing colour anodised aluminum windows; including 6mm clear sheet glass; framing, water bars, fittings, fixing lugs, brackets, bolts and ironmongery; aluminum glazing beads; PVC weatherstrip; assembling, jointing, cutting and pinning lugs; painting back of frames with one coat of bituminous paint before fixing; bedding frames in waterproof cement mortar; as Drawings no. xxx.</u>	*Assuming detailed drawings for the aluminum windows are available, drawing reference should be put in the description (XV(n).2.1). The current drawing (* ***Figure 19.4w*** *) is only for illustration and is not detailed enough for incorporation into the tender / contract document.* *The measurement for metal windows is comparatively straight-forward as all the fixing, ironmongery, sealing of joints, bedding and pointing frames, sealant, water bars, glazing and glazing beads are included (XV(n).C.1).*
1	600 x 500mm overall; 1 fixed light	*HKSMM, XV(n).2.1.1* *Overall size has to be stated, the number and type of window (fixed light or opening light) has to be described.*
1	1200 x 1200mm overall; including 2 top hung opening lights and 2 side hung opening lights	*Any composite windows and doors with coupling transoms and mullions have to be so described clearly (XV(n).4.1.1). In this example, the windows are simple windows without coupling transoms / mullions.*

		Doors Fire rated galvanised mild steel doors and frames; all framed and welded together, complete with angle perimeter frames, horizontal and vertical angle rails and including all necessary painting, hardwares, assembling and fixing accessories; bedding frames in cement mortar; pointing joints with silicone sealant; ironmongery measured separately; as drawing nr. xxx	*HKSMM, XV(d).1.1* *Similar to windows, a steel door item is deemed to include door and door frame, fixing accessories, bedding and pointing frames (XV(d).M.1,C.1 and C.2). Ironmongeries can be included or alternatively measured separately (XV(d).1.1.1.1). Description should include dimensions, drawing reference (XV(d).1.1), kind and quality of materials, fixing method and fire rating (XV(d).S.1-S.7).*
	1	1-hour fire rated double leaf doors; overall door size 2000 x 2000mm high	
		End of Warehouse Measurement	

20 Processing measurements

Processing measurements

In the previous chapters, we have examined how to measure quantities of work in a standardised format (in accordance with HKSMM4 Rev 2018) from design information. The next step is to process the measurements to prepare bills of quantities or summaries of valuation of works. This chapter will focus on the working-up process – to process the measurements for the preparation of bills of quantities. If using the measurements for valuation of variations or assessment of payments, the processing method is similar, but much simpler, and therefore not elaborated here.

Conventional method of bill preparation

Traditionally, after the taking-off process, the quantities will go through squaring, abstracting and billing processes to produce the bills of quantities. These stages are collectively known as the 'working up' process.

Squaring

Squaring refers to calculating numbers, lengths, areas, volumes and masses and recording the results in the third column (also called the squaring column) on the dimension paper (see the shaded area in **Figure 20.1**). It is always a good practice to have the squaring independently checked by another QS to eliminate errors. All squared dimensions and waste calculations should be ticked in red if they are correct. Any corrections should be crossed out or marked in red.

For the items measured in weight, the bracketed conversion will be completed in the squaring stage (as shown in **Figure 20.1**).

Question 20.1

Complete the squaring for **Worked Example 19.2** in **Chapter 19**.

Abstracting

Abstracting is to transfer the squared dimensions to the abstract sheet where they are assembled in bill order under appropriate section headings. When taking-off

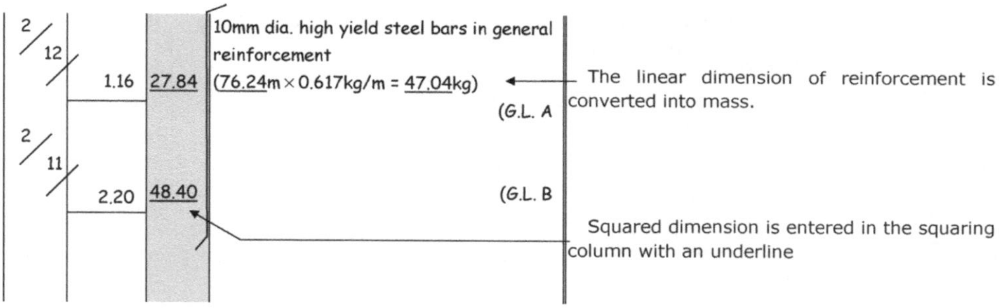

Figure 20.1 Squaring on dimension paper.

quantities for BQ preparation, a building is often split into different portions for measurement by several surveyors. Therefore, the same trade or even work item may appear in different sets of taking-off. Abstracting becomes a useful process to compile similar items following the HKSMM order and to put them in a sensible sequence, such as from smallest to largest, for bill preparation.

The abstract sheet is usually an A3-sized paper ruled in columns (as shown in **Figure 20.2**). Every sheet should be headed with the project title, the section and the subsection of work. The division and sequence of work sections in the abstract should follow the HKSMM. The same order will be applied in the bills of quantities.

To carry out abstracting, we start from the left side of the abstract paper. The item description will be written across two wide columns, with the measurement unit written against the item description. The associated dimensions will be entered below in the left-hand column and any deductions in the right-hand column (as shown in **Figure 20.2**). Against each dimension, the page number of the dimension paper from which the quantity has been abstracted is written (as shown in the shaded columns in **Figure 20.2** and **Figure 20.3**). When several similar items need to be transferred, such as those belonging to the same type of work but in different sizes, these should be grouped under a single heading with each size entered in a different column (**Figure 20.3**).

As each dimension is entered to the abstract sheet, the dimensions in the dimension paper should be crossed through with a line to indicate that the dimensions have

Project: Proposed residential development at 123, xx Road, H.K. EXCAVATION											**Sheet 1**
m³	Excavating Surf. exc.; av. 150mm dp			m³	Excavating Surf. tr. n.e. 1.5m dp commencing from red. lev.			m³	Disposal Disposal of exc. matl from site to a tip provided by Employer		
2	15.80		Ddt	3	28.70		Ddt	3	3.71		Ddt
2	25.50	2	3.37	4	14.84	3	5.21	4	2.64	3	0.82
3	11.93	3	6.58		43.54				6.35	4	0.50
4	21.35		9.95		−5.21				−1.32		1.32
	74.58				38.33				5.03		
	−9.95				= 38m³				= 5m³		
	64.63										
	= 65m³										

Figure 20.2 Squared dimensions and descriptions on an abstract paper.

Project: Proposed residential development at 123, xx Road, Hong Kong						**Sheet x**
m	UPVC pipes & fittings; BS 3505:1986 Class E; solvent welded joints; fixed to soffits					
	40mm dia.		50mm dia.		65mm dia.	
3	56.35			3	15.75	
4	27.83	2	2.30	4	28.13	
4	42.40	3	4.26	4	17.06	

Figure 20.3 Entry of similar items but of varying sizes in an abstract paper.

been transferred and to avoid duplicate entries. After all dimensions associated with an item description have been transferred, the figures are totalled and any deductions are subtracted to give a nett figure. The nett figure should then be rounded to the nearest whole number, which will be transferred to the billing sheet later.

All the entries and totals should be checked and ticked in red as described in the squaring process.

Billing

Billing is the final stage of bill preparation, where the description and quantity of each measured item are transferred to the standard bill paper. The bill paper is set in a format which is friendly to tendering contractors (**Figure 20.4**).

Some of the key features of the bill are listed below:

- All quantities in the bill should be rounded to the nearest whole number.
- Any abbreviations used in the item descriptions should be changed to full words in the bill except for those abbreviations used in HKSMM (see Section II of HKSMM4 Rev 2018).
- Each item should be indexed by letter and each page by number (see **Figure 20.4**).
- The order of billed items should follow the abstract and be grouped under suitable section headings. As in **Figure 20.4**, a work section heading, 'In-situ concrete', and a subsection heading, 'Concrete; grade 10/20', have been used for

					CONCRETE WORKS
Item	Description	Quantity	Unit	Rate	HK$
	IN-SITU CONCRETE				
	Concrete; grade 10/20				
	Blinding under foundations and beams				
A	50mm thick	30	m^3		
B	75mm thick	20	m^3		
				To Collection $	
		B.Q. 1/4			

Figure 20.4 Typical sample of a bill paper.

categorisation of items. The work section heading will be repeated at the start of each page wherever it applies. A further heading, 'Blinding under foundations and beams', is used to partially describe a group of items that follow. This can save repetition of item descriptions.

Question 20.2

If the total quantity of an item is $0.1m^2$, do we need to include this item in the bill? If yes, how should we state the quantity in the bill?

Question 20.3

Can 'ditto' be used in the bill? Is it considered to be an abbreviation?

- A total sum should be calculated on each page of the bill and carried to a collection at the end of each trade (or bill). A 'Collection' can be provided at the end of each bill showing the total (**Figure 20.5**). A 'Summary' is used to calculate the total cost, which constitutes the tender sum, at the end of the entire BQ (**Figure 20.6**).

Question 20.4

Transfer all the items in **Worked Example 19.2** of **Chapter 19** to a bill format.

Final bulk check

Independent checking of the calculations and transfer of items should be carried out in every stage of the working-up process. Having transferred all items from the abstract, the draft bill is ready for proofreading. As each item is transferred from the

		CONCRETE WORKS
		HK$
	COLLECTION	
A	Page No. B.Q.1/4	
B	Page No. B.Q.2/4	
C	Page No. B.Q.3/4	
	Bill No. B.Q.4 - CONCRETE WORKS	
	To Summary $	
	B.Q. 4/4	

Figure 20.5 Typical sample of a collection page.

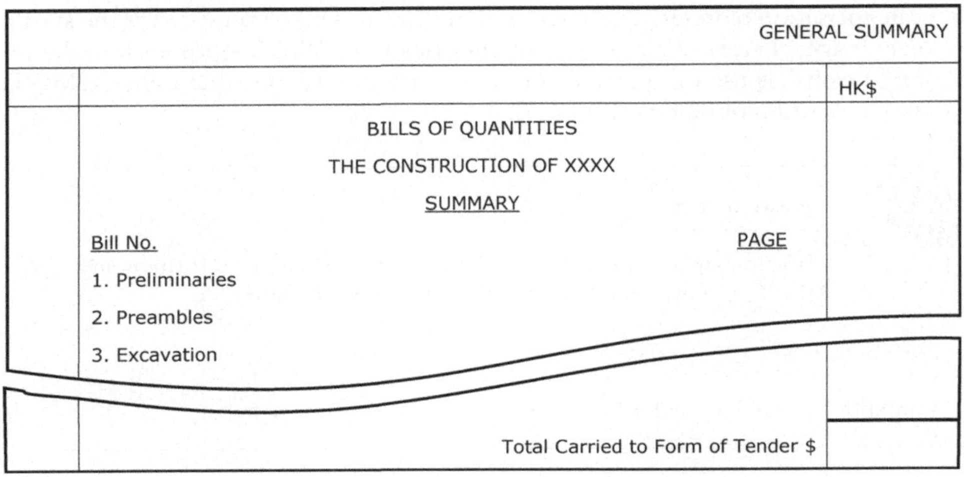
GENERAL SUMMARY

HK$

BILLS OF QUANTITIES

THE CONSTRUCTION OF XXXX

SUMMARY

Bill No. PAGE

1. Preliminaries

2. Preambles

3. Excavation

Total Carried to Form of Tender $

Figure 20.6 Typical sample of a summary page.

abstract to the draft bill the entry in the abstract should be lined through to avoid a duplicated entry.

Before printing and binding the bill, the entire draft bill has to be checked and ticked in red. All dimension papers and abstract papers involved should be reviewed again to ensure that all items have been ticked. A final bulk check of the major items is advisable by comparing the quantities of different items to see if any significant errors have been overlooked. Examples of bulk check items include:

- Check the total quantity of excavation to see whether it is equal to the total quantities of backfill, filling and disposal.
- Compare the total area of internal floor finishes in the bill and the total area of ceiling finishes in the bill.
- Compare the total wall area (without openings) with the total area of wall finishes in the bill.
- Count the total number of doors from the drawings and compare with the totals in the bill. Do the same for the windows.
- Count the number of toilets from the drawings and compare the total with the number of water closets in the bill.
- Count the number of toilets and/or kitchens from the drawings and compare the total with the number of sanitary fittings (basins, sinks, water taps etc.) in the bill.

These checks are not supposed to be highly accurate or thorough, but act as a final scan of the entire document. The bill is considered acceptable and ready for printing if there is no significant deviation between the bill quantities and the expected quantities.

Alternative methods of bill preparation

The traditional method of bill preparation is time-consuming, and many QSs have adopted simplified processes to save resources. There are two alternative methods available: cut and shuffle method and direct billing method.

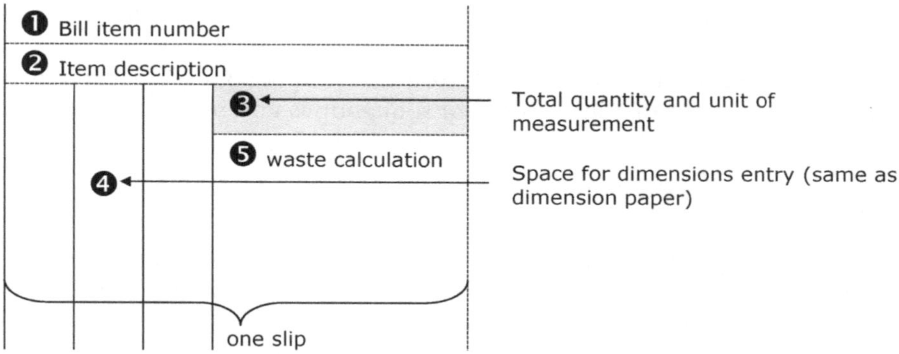

Figure 20.7 Format of a cut and shuffle slip.

Cut and shuffle method

The cut and shuffle method is a modified measurement and processing method that eliminates the abstracting process. This method starts with taking-off the dimensions using a 'cut and shuffle' paper. The paper is A4 or foolscap-sized, divided into three small dimension sheets (which are referred to as 'slips') by perforations (see **Figure 20.7**).

Taking-off and squaring

When taking-off, each item must be written on a separate slip, and only one side of the cut and shuffle paper is used. Each slip should be numbered, with the project reference and trade/section reference stated clearly at the top of each slip. If the measurement of a single item cannot be finished within one slip, the measurement will be continued on the second slip with cross-reference to the first slip. In principle, taking-off is carried out in the same manner as in the traditional method, except that full description has to be written for each item. After completion of the taking-off, squaring can be carried out and checked.

Cutting and shuffling

Before cutting the sheets into slips, they should be photocopied first. Alternatively, cut and shuffle sheets with carbonised papers can be used at the start. The carbon copy or photocopied sheets are retained for record purpose, and the original sheets will be cut along the perforations. The slips are shuffled or sorted according to the bill section order described in the HKSMM. Blank slips marked with group headings can be inserted where necessary. The unit of the item will be entered in the box provided on each slip (box 3 in **Figure 20.7**).

Totalling quantities

The total quantity for each item is calculated and inserted in the box (box 3 in **Figure 20.7**) provided on the slip (or on the first slip in case if more than one slip is used for an item). Collections and summary totals are inserted on new slips.

Independent checking should be done after the totalling to ensure that the calculations are error-free.

All slips are tagged in order and handed over for typing. Once the bill has been typed, check to ensure that all descriptions, quantities and units have been transferred correctly.

Direct billing method

In the direct billing method, items from the dimension sheets are transferred directly to the bill. Care must be taken to ensure that the order of items and sections follows the sequence prescribed in the HKSMM. Once an item in the dimension paper has been transferred, it should be lined through in red. An independent check on all calculations as well as all transfers should be carried out by another QS.

The direct billing method is normally used where the project is simple and the measurement is done in a trade-by-trade pattern.

Suggested answers

Question 20.1

			Steel Frame Structure
			Figures 19.2a & 19.2b
			Excavation
			Length of excavation
			1200
			w/s 2/600 1200
			2400
			Width of excavation
			1000
			w/s 2/600 1200
			2200
4/	2.40		Exc. for col. bases
	2.20		commencing at red. level, n.e. 1.50m deep
	0.90	19.01	(base
4/	1.20		
	1.00		(blinding
	0.10	0.48	&
			Backfilling with sel. exc. matl

			Concrete
4/			20P plain conc. in 100mm thk
	1.20		blinding under col. bases
	1.00		&
	0.10	0.48	Ddt
			Backfilling a.b.d.
			&
			Add
			Disposal of exc. matl from site to a tip provided by Contractor
4/			20D/20 R.C. in col. bases
	1.20		&
	1.00		
	0.90	4.32	Ddt
			Backfilling a.b.d.
			&
			Add
			Disposal of exc. matl from site to a tip provided by Contractor

			Formwork
			Girth of column base
			1.20
			1.00
			2/ 2.20
			4.40
4/			Sawn fwk to vert. surf. of
	4.40		col. bases
	0.90	15.84	
			Supply, fabricate and erect grade S355 steel struct. to BS EN 10210-2, approx. plan area 45m^2 ; single storey; 5.9m overall height above grd; incl. all nec. welding and bolting
			Columns
			Height of columns
			5500
			Less
			Cap plate 18
			Base plate 28
			5454
4/			356 x 406 x 340kg universal
	5.45	21.80	columns
			(21.80m x 340kg/m = 7412.00kg)
			Beams
			Length of beams
			7500
			proj 2/800 1600
			9100
2/			356 x 171 x 67kg universal
	9.10	18.20	beams
			(18.20m x 67kg/m = 1219.40kg)

2/			356 x 127 x 39kg universal
	6.00	12.00	beams
			(12.00m x 39kg/m = 468.00kg)
			Connections
			Mass per area (base plates)
			= 7843 x 0.028
			= 219.6kg/m^2
			Mass per area (cap plates)
			= 7843 x 0.018
			= 141.2kg/m^2
4/			Connections including cleats,
	0.70		plates and the like
	0.65	1.82	(base plates
			(1.82m^2 x 219.6kg/m^2 = 399.67kg)
4/			
	0.40		(cap plates
	0.41	0.66	(0.66m^2 x 141.2kg/m^2 =93.19kg)
4/			
2/			(equal cleats
	0.20	1.60	(1.60m x 14.3kg/m = 22.88kg)
4/			(unequal cleats
	0.13	0.52	(0.52m x 19kg/m = 9.88kg)
			(Connections Total = 525.62kg)
			Special bolts
4/			16mm dia. x 450mm long holding down bolts casting
	4	16	into conc. with 6mm thk washers and 75mm dia. expamet bolt boxes
			End of S.S. Frame Measure.

Question 20.2

According to HKSMM 4 Rev 2018, II.7.3, *'Where the application of this (Units of billing) clause would cause an entire item to be eliminated, such item shall be billed as one whole unit'.* Therefore, the item quantity should be stated as 1m^2 in the bill.

Question 20.3

'Ditto' is not an abbreviation. It can be used in the bill.

Question 20.4

	Item Description	Quantity	Unit	Rate (HK$)	Total (HK$)
	EXCAVATION				
	Excavating				
	Excavation for column bases commencing at reduced level				
1	not exceeding 1.50m deep	19	m^3		
	FILLING AND DISPOSAL				
	Filling				
	Excavated materials				
2	backfilling to excavation	15	m^3		
	Disposal				
	Surplus excavated materials				
3	removal from site to a tip provided by Contractor	5	m^3		
	CONCRETE WORKS				
	In-situ Concrete				
	Plain concrete; grade 20P				
	Blinding under column bases				
4	100mm thick	1	m^3		
	Reinforced concrete; grade 20D/20				
	Column bases				
5	generally	4	m^3		
	FORMWORK				
	Sawn formwork				
	Vertical surfaces				
6	column bases	16	m^2		
	STRUCTURAL STEEL WORK				
	Supply, fabricate and erect grade S355 steel structure to BS EN 10210-2, approximate plan area 45m^2; single storey; 5.9m overall height above ground; including all necessary welding and bolting				
	Columns				
7	356 x 406 x 340kg universal columns	7,412	kg		
	Beams				
8	356 x 171 x 67kg universal beams	1,219	kg		
9	356 x 127 x 39kg universal beams	468	kg		
	Connections				
10	cleats, plates and the like	526	kg		
	Holding down bolts with nuts and washers; casting into concrete				
11	16mm diameter x 450mm long; with 6mm thick washers and 75mm diameter expamet bolt boxes	16	nr		

Part 5

Estimating unit rates

21 Pricing in general

Factors of production

In **Part 4** we reviewed the basic method to measure the quantities of work for BQ preparation or for other contract administration functions such as procurement and valuation of payment. With the quantities of work ready, a contractor has to price each item to arrive at the total cost of the project. Pricing of work/project requires a thorough consideration of the resources used during the production process, which can be represented by **Figure 21.1**.

As shown in **Figure 21.1**, there are five categories of resource input required by a contractor to complete a project:

- Labour
  The direct labour employed by the contractor.
- Plant (or equipment)
  The mechanical plant, equipment and tools owned or hired by the contractor to carry out specific trade work, such as a roller or compactor for soil compaction.
- Materials
  The materials procured for the project.
- Subcontractors
  The subcontractors hired for the project.
- Preliminaries
  Also known as project overheads. The site-based facilities and staff provided to the project as a whole. Examples include site management staff, insurance, electricity and water supply for the site and so forth.

All the above resources contribute to the direct cost of a project to be spent by the contractor. When pricing the trade items, the first four resource categories will be ascertained accordingly. The last category, preliminaries, usually appears as a separate section in the bills of quantities as discussed in **Chapters 6 and 7.** Its likely cost is estimated in a holistic approach, not apportioned to each trade work item.

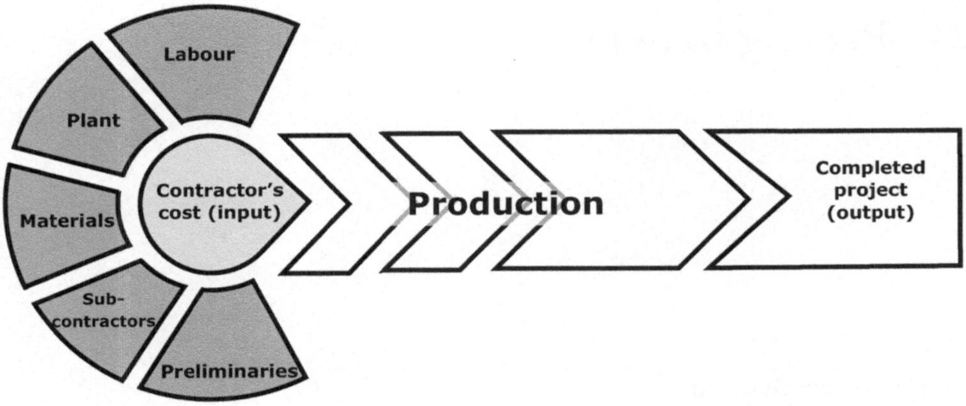

Figure 21.1 Production process – changing input (resources) into output (completed project).

Points to note

Besides the five categories of input required for the production process, a contractor also provides head office support to each project. This is regarded as the indirect cost of a project (as such expense is to keep the business running and will be expended even without the project). When we estimate the unit rate or price for each BQ item (see **Figure 4.1** for the illustration of unit rates), we will not consider the head office overheads or otherwise the estimating task will become very complicated. The head office overheads, together with the profit margin, will be considered by top management in the tender adjudication process (see **Chapter 6** for details).

Pricing methods

To draw up a tender bid, a unit rate for each item of work involved in the project has to be estimated. The estimator needs to predict the physical resources (which can be labour, plant, materials, subcontractors or a combination of them) required for each BQ item and estimate the likely cost contributory to each unit of work. The unit rates calculated are referred to as build-up rates.

Build-up rates are not only applied in tender preparation but also in valuation of variation orders during the construction period. When new work items are executed under variation orders, new rates (usually called star rates) are built up pending further negotiation between the contracting parties. Sometimes, the new work items in a variation order and the BQ items consist of a slight difference in the material quality or output. In that case, a pro-rata rate can be applied (subject to agreement between the parties) instead of building up a new rate. The pro-rata rate is calculated by making adjustment to the existing BQ rates on a pro-rata basis.

Build-up rates

Many publications, including the CIOB New Code of Estimating Practice (2018), give detailed illustrations on the calculation of unit rates. Although the principles of estimating are easy to understand, an accurate estimation of unit rates demands a thorough understanding of the construction method and sequence as well as the productivity of the resources to be hired or used. Take excavation as an example. The operation can be executed by hand digging or by machine digging. If machine-dug is used, the choice of equipment, such as large tractors, scrapers or draglines, can affect the cost significantly. The type of equipment, model, maintenance condition, skill level of operator etc. all have an impact on the productivity of work. That is why an estimating team comprising estimator, project planner and project manager is required to prepare a bid. Readers should therefore appreciate the diversity of construction operations and site conditions and make relevant adjustment when tackling estimating problems.

Having established the physical resources required for a work item, a unit rate (for the work item) will be estimated, which can be represented by:

$$U_x = (C_{xl} + C_{xp} + C_{xm} + C_{xs} + C_{xo})(1 + p\%)$$

Where

U_x = unit rate of work item x
C_{xl} = labour cost per unit of item x
C_{xp} = plant cost per unit of item x
C_{xm} = material cost per unit of item x
C_{xs} = subcontractor cost per unit of item x
C_{xo} = other sundry cost per unit of item x
p = mark-up percentage for head office overheads and profit

Reminder

Head office overhead is normally allowed in the overall mark-up, whereas the project overheads items are priced in the Preliminaries Bill.

As shown in **Figure 21.2**, the cost of each required resource will be calculated to establish the unit rate. The all-in hourly rates for labour and plant will be converted to the labour and plant rates (per unit of the BQ item) based on their respective productivities. The all-in rate for material (if applicable) will also be calculated for feeding into the unit rate calculation. If input from a subcontractor is required, a subcontractor price will be included. Other necessary sundry cost that is not allowed in the Preliminaries Section of the BQ but required for the execution of the work item will be converted to a cost per unit of work. Head office overheads and profit are added to the aggregated unit rate at a later stage during tender adjudication.

Points to note

All-in rate refers to the total cost (per unit) of an item. For instance, the all-in rate for an operative is the hourly cost of employing the operative including the basic wage, together with costs of providing provident fund, holidays, allowances and so on.

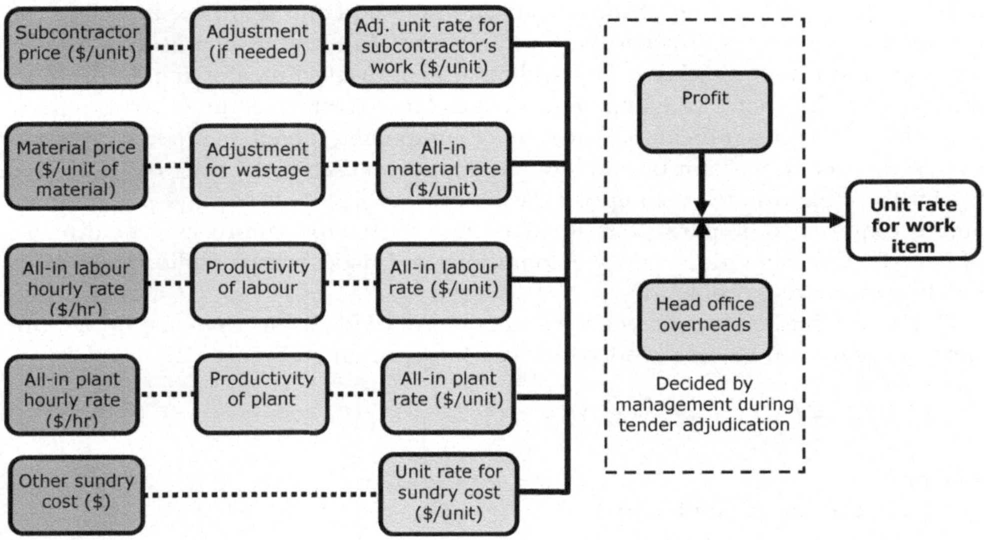

Figure 21.2 Building up of a unit rate in the bills of quantities.

Subcontractor rate

There are two main types of subcontractors: domestic subcontractors and nominated subcontractors. The nature of the subcontract work can be a labour-and-materials subcontract or a labour-only subcontract.

Comparatively, the subcontractor unit rate is the easiest to estimate. Nominated subcontractor costs will not be used to calculate unit rates as they are included in the BQ separately in the form of a P.C. sum. The estimator only needs to allow a profit and attendance percentage to the P.C. sum (more details can be referred to in **Chapter 7**). For domestic subcontractor costs, normally material and labour components are covered. The estimator needs to choose the most competitive quotation from the various received quotations and apply a profit and overheads mark-up to price the corresponding BQ item. Although the risk of underestimation lies with the subcontractor, the estimator should make sure that the subcontractors have full understanding of the scope of work to avoid future disputes.

As shown in **Figure 21.2**, in case there are any non-priced items or qualifications in a subcontractor's quotation, the appropriate cost adjustment should be made before finalising the unit cost for tender adjudication. If the subcontract work covers the labour component only, the relevant material rate and/or plant rate will be added with the subcontractor rate to build up the unit rate.

Points to note

Sometimes, a subcontractor's or supplier's quotation may include pricing for testing, shop drawings production and the like. Depending on the drafting of the BQ, these costs can be incorporated in the all-in rate if the preambles of that trade item have made such provision. Otherwise, these costs are often removed from the unit rate and priced in the Preliminaries Bill.

All-in material rate

If material cost has to be estimated for a unit rate, the cost information can be collected from suppliers in the form of quotations or from company records. When using the latest quotations to build up rates, estimators should watch out for several aspects:

- Discounts: Discounts can be in the form of bulk purchase discounts, discount for prompt payment or cash payment, which means the cost per unit can be reduced if specific condition is fulfilled.
- Delivery arrangement: Normally, each supplier has its own preferred delivery arrangement. These arrangements include ex works; free on board (FOB); cost, insurance and freight (CIF); and delivered duty paid (DDP) (details can be referred to in **Chapter 6**).
- Surcharge on special conditions such as partial shipment/delivery, quantity less than minimum order.

The above are usually stated in the supplier's quotations and relevant allowance should be made if any of them exists.

Besides the supplier's quoted price, material wastage should be included in the all-in material rate. Since the quantities in the BQ are measured nett in accordance with the HKSMM, estimators have to estimate the likely wastage (usually in the form of a percentage) based on their experience. For instance, 1 m^3 concrete to wall will require more than 1 m^3 ready-mixed concrete as there will be spillage and wastage. Besides, many trade work items like ceramic tiles require materials to be sold in specific packaging unit (in this case, boxes or pieces) which is different from the unit of measurement in the SMM. Therefore, conversion with due consideration of wastage should be made when calculating the all-in rates for materials.

Material wastage is affected by a lot of risk factors. Some of the prominent ones include:

- Nature of materials
  Some materials are having a higher risk of damage by their nature. For instance, ceramic tiles, glass and the like are fragile. Cement is moisture-sensitive and must be protected from dampness to prevent hardening.
- Design
  Many materials are manufactured in modular sizes to enable easy handling and waste minimisation. However, design may involve irregular shapes or odd sizes which can produce more cutting waste than usual.
- Skill level of operatives
  If the operatives employed are skilful and experienced, waste due to the wrong use of material, poor handling, wrong mixing or installation can be much reduced.
- Site management
  The quality of frontline supervision and management on site can also affect the wastage level. Prompt instructions from supervisors to operatives on the proper use and storage of materials can minimise abortive work and production waste. A tidy working environment can also enhance working efficiency and waste control.

- Condition of site store
  The likelihood of theft and vandalism is difficult to predict but site management can try their best to reduce the loss by proper storage and locking of materials as well as by employing sufficient watchmen. Besides the security concern, if the site store is too small or without proper weather protection, materials stored inside may be subject to a higher level of damage.

Other associated material costs such as the mechanical plant cost for unloading and distribution of materials, setting up cost for storage huts within the site etc. is usually allowed in the Preliminaries.

Labour and plant rates

Labour and plant rates can be dealt with in the same manner as both of them are depended on two factors:

- productivity of the resource (labour/plant)
- hourly 'cost' of the resource (i.e. the all-in hourly rate)

With the all-in hourly rate for labour or plant established, the labour or plant cost per unit can be calculated as follows:

$$C_{xl} \text{ or } C_{xp} = R_x \cdot p_x$$

Where

C_{xl} or C_{xp} = labour or plant cost per unit of item x
R_x = all-in hourly rate for labour or plant for item x
p_x = productivity of labour or plant (i.e. productive time required per unit of item x)

All-in labour rate

Labour may be paid on a daily, piecework or monthly basis, depending on the employment contract between the operative and the contractor. In Hong Kong, many large contractors maintain a certain number of direct labour (instead of employing labour-only subcontractors) to accommodate the fluctuating labour market. Although the Hong Kong Construction Industry Employees General Union reviews and suggests wages for different tradesmen on a regular basis, the published wages are not mandatory. Subject to the recruitment policy of individual companies and the market force, the labour wages vary across different contractors. If workers are employed as full-time employees of the contractor, benefits such as leaves, holidays, insurance, mandatory provident fund, overtime payment, bonus, training cost, and other allowances have to be considered when calculating the all-in rate. In other words, the all-in rate for labour not only includes the explicit wage received by an operative, but also other fringe benefits. **Table 21.1** illustrates the all-in labour rate calculation.

In simple terms, the all-in rate for an operative is:

$$\text{All-in hourly rate for operative} = \frac{\text{Total cost of employing an operative in the wage period}}{\text{Productive hours in the wage period}}$$

Table 21.1 Calculation of the All-in Hourly Rate for Labour.

Total Productive Hours (wage period = 1 year)		
No. of hours per day[i]	7.5	
No. of weeks per annum	52	
No. of calendar days per annum	365	365
Less:		
Rest days (once per week)	52	
Statutory holidays	12	
Annual leave (assumed 3 years of service)	8	
		(72)
Sickness (say 5 days)		(5)
Allow bad weather in summer (3 days)		(3)
Working days per year		285
Total productive hours for payment		2,137.5
Total Cost of Operative (wage period = 1 year)		
12-month wages	12 × 20,000	240,000
1-month bonus		20,000
Mandatory provident fund (5%)[ii]	20,000 × 5% × 11 + 1,500	12,500
Training (safety)[iii]		300
Training (excl. safety)		700
Employees' Compensation	Incl. in Prelim.	
Total cost of operative[iv]		273,500
Operative cost per hour (HK$):		**128.0**

Note: [i] Allowance has been made for rest and meal breaks required by the operative. Operatives for different trades/operations and operatives working in different site conditions may require different allowance for rest and meal breaks.

[ii] The labour receives 1-month wage plus bonus in one of the months. The MPF payable by the employer in this month will reach the maximum contribution and therefore only $1,500 is charged for MPF.

[iii] Assumed zero allowance in the Preliminaries Bill.

[iv] Other possible costs which are not included in the table such as overtime allowance, meal allowance, travel allowance and medical insurance can be allowed where appropriate.

Points to note

Mandatory provident fund (MPF) in Hong Kong is governed by the Mandatory Provident Fund Schemes Ordinance (Chapter 485). Wages, salary, leave pay, bonus, commission, gratuity, cash allowance and benefits (except for reimbursement for the expenses which are necessary in the performance of an employment duty incurred by that employee and the severance payment) are relevant income items subject to the MPF contribution by an employer. The current maximum contribution (by employer) per month is $1,500 (MPFA, 2020).

Also, note that profit and overheads mark-up is not allowed here as the all-in rate is assumed to be used for calculating the unit rates of trade work items. However, if the hourly rate is for pricing a daywork labour item, profit and overheads is often allowed. Final checking with the preambles for the daywork pricing should be made to avoid duplication or omission of allowance.

Reminder

When allowing the fringe benefits and the like in all-in labour rate calculation, care should be taken not to duplicate with the preliminaries allowance. For instance, safety training has to be provided to all workers. The estimator may allow the training cost in the all-in rate for labour but more often in the safety management item of the Preliminaries Bill. Another example is insurance expenses. Employees' Compensation (EC) is a statutory insurance policy required to be secured by all contractors for their staff and operatives. Normally, this item is priced in the Preliminaries Bill but not in the all-in labour rate.

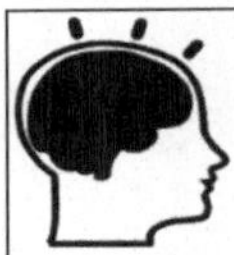

Question 21.1

Calculate the all-in rates for the following staff/labour based on the employment terms provided below:

i Hourly rate of a scaffolder who is employed on a daily basis
ii Daily rate of a building services engineer employed on a 1-year contract (contract will end in Dec)

	Scaffolder	*BS Engineer*
Salaries / wages	$2,000/day	$28,000/month
Productive hours/day	6.5 hours	7.5 hours
Double pay	No	Yes, paid in Dec
Bonus	No	10% of 1-month basic salary, paid in Dec
Training allowance	No	Max. $3,000/year
Sick leave, annual leave per year	Not applicable	Sick leave: max. 10 days Annual leave: 10 days
Contract-end gratuity	Not applicable	10% of the 12-month basic salary (excluding any double pay, bonus, and employer's contribution to the MPF)
MPF	Yes	Yes

All-in plant rate

There are two main types of mechanical plant with different pricing implications: plant for general use across a wide range of operations and plant only required for a specific operation. Plant and equipment for general use, such as tower crane, material hoist, water pump, folk lift and the like, are normally priced in the Preliminaries Bill, which will be discussed later in **Chapter 23**. Plant and equipment for a specific operation, such as rollers and backhoes, are usually priced with the trade item and is discussed below.

Normally, plant and equipment are either owned by the contractor or hired from a plant supplier. In general, when estimating the all-in plant rate, the following items must be priced:

- cost of machine per hour
- fuel cost
- sundry consumables such as replacement parts

If the machine is hired, the cost of machine per hour is equal to the hiring cost divided by the hiring period. Care should be taken to check if there are any minimum hire charges. If the machine is owned by the contractor, the cost of machine per hour should include the depreciation cost, financing charge (cost of using the money on plant purchase), maintenance cost, license fee, tax and so forth. **Table 21.2** illustrates the calculation of typical all-in rates for plant.

Points to note

The straight line depreciation method is the simplest way to calculate the annual fixed cost of a machine.

$$m = (M - S) / n$$

Financing cost is the cost of using the money to purchase the machine. It can be the interest paid to the bank or the opportunity cost that the money will earn if invested elsewhere. Some companies may not consider this cost, assuming it to be covered by the profit. In Table 21.2, a simple method is used to calculate the financing cost, by multiplying the average annual investment (AAI) by the interest rate.

$$AAI = \frac{M(n+1) + S(n-1)}{2n}$$

(Sundberg and Silversides, 1988)

Some machinery involves transportation charges, which should be included in the total plant cost calculation. In addition, plant rental may include the hiring of an operator in the rental agreement. In that case, the operator cost can be included in the all-in plant rate. If the operator is hired separately, the operator cost can be priced in the labour cost of the work item.

Other sundry costs

Besides labour, material, plant and subcontractors, a contractor may have other sundry expenses such as an extra insurance cover for a material delivery or a tailored product design to meet the architect's requirement. These sundry costs, if not specified as a separate BQ item, will be allowed in the unit rate of the trade work item

Table 21.2 Calculation of the All-in Rate for Plant.

Productive Hours per Year		
Estimated no. of hours operated per year	1,500	
Cost of Plant per Year		
Expected life of machine (n)	8 years	
Straight line depreciation method is assumed*		
Cost of machine (M), say	100,000	
Less:		
Scrap value (S)	(2,000)	
	98,000	
Annual machine cost (m)	12,250	
Machine cost (per hour)	8.2	8.2
Other Costs		
Annual maintenance cost (20%) = 12,250 × 20%	2,450	
Financing cost (interest rate i = 4%)		
= average annual investment × interest rate		
4% × (100,000 × (8 + 1) + 2,000 × (8 – 1))/(2 × 8)	2,285	
Annual fuel cost (for 1,500 hours operation)	39,000	
(2 litre/hour @ $13/litre)		
Consumables (grease, lubrication, cables etc.)	4,000	
Other cost (per year)	47,735	
Other cost (per hour)	31.8	31.8
Plant cost per hour (HK$):		**40.0**

Note: * Other depreciation methods such as declining balance method and units of production method can be used if more applicable.

concerned. Since these costs are usually lump sum in nature, the amount will be apportioned to each unit of the work item:

$$C_x = \frac{V_x}{Q}$$

Where

C_x = sundry cost per unit of item x
V_x = total sundry cost for item x
Q = quantity of item x

Pro-rata rates

There are several ways to establish pro-rata rates and much practical experience is required to decide which method is appropriate to the work item in question. The following examples illustrate how a pro-rata rate can be established.

Example 21.1 – By simple proportion

The BQ rate for supply and installation of 500mm L × 600mm W × 1,000mm H timber shelf is HK$800/no. A shelf of the same specification but in a different size:

500mm L × 600mm W × 1,500mm H is required in a variation order. Assuming that the labour and materials required for a shelf is more or less proportional to the furniture height, the simplest way to find the timber shelf rate for the variation is to adjust the BQ rate for the height difference as follows:

$$\text{HK\$}800 \times 1{,}500 / 1{,}000 = \text{HK\$}1{,}200 \ (/\text{no.})$$

Example 21.2 – By derivation

The following items and rates appear in the BQ of a project:
<u>Screeds; cement and sand (1:3); wood floated finish</u>
20mm thick to floor; to receive ceramic tile: HK\$45/m^2
30mm thick to floor; to receive granite tiling: HK\$55/m^2
40mm thick to floor; to receive homogeneous tile: HK\$65/m^2

Now, a new item of 25mm thick screed in the same mix laid to a similar specification is required. Although the price of floor screed is related to its thickness, it is obvious that the two are not in direct proportion. Just referring back to the BQ rates above, 40mm thick screed costs more than 20mm thick screed but the cost is not doubled. This can be explained by the fact that the materials required in 40mm screed is twice as much as that in 20mm screed, but the labour required is not doubled (at least the surface area that requires trowelling remains unchanged).

By inspection, we can see that the rate for floor screed increases by HK\$10/m^2 for every increase of 10mm thickness. It is therefore reasonable to assume that the unit rate for 25mm thick screed can be derived from the BQ rates in the same way as follow:

$$\text{HK\$}45 + (55 - 45)(5/10) = \text{HK\$}50 \ (/\text{m}^2)$$

Example 21.3 – By cost analysis

A BQ item is found as follows:
Vitrified ceramic tiles; 300 × 300 × 10mm to floor; bedding and jointing in cement and sand (1:3); pointing in white cement: HK\$300/m^2

Now, a new item of 300 × 300 × 10mm polished tile bedded and jointed in the same specification is required. Since the labour cost required for both items should be quite similar, the major cost difference should come from the materials. By breaking down the unit rate of the BQ item (vitrified ceramic tiling) into labour, materials and mark-up, the unit rate for polished tiling can be established.

Supplementary information:

- Cost of vitrified ceramic tiles: HK\$170/m^2
- Cost of polished tiles: HK\$150/m^2
- Wastage of tiles: 10%
- Profit mark-up: 15%

		HK$ / m²
Vitrified ceramic floor tiling:		300.0
Deduct 15% mark-up (HK$300 / 1.15):		
	Nett rate:	260.9
Deduct materials cost:		
Vitrified ceramic tiles	HK$ 170 / m²	
Wastage 10%	HK$ 17 / m²	(187.0)
	Labour cost:	73.9
Add new materials cost:		
Polished tiles	HK$150 / m²	
Wastage 10%	HK$ 15 / m²	165.0
		238.9
Add 15% profit mark-up:		35.8
	Pro-rata rate (HK$):	274.7

Points to note

In Example 21.3, note that the equation we used to arrive at the nett rate is: Nett rate = BQ rate/(1 + p%); not Nett rate = BQ rate x (1 – p%). This is based on how we apply the profit mark-up during the tender adjudication process:

$$\text{Nett rate} \times (1 + p\%) = \text{BQ rate}$$

Therefore, if we need to work back a nett rate from a BQ rate,

$$\text{Nett rate} = \text{BQ rate} / (1 + p\%)$$

As shown in examples 21.1 to 21.3, pricing a unit rate on a pro-rata basis may not be an accurate estimation of the cost required for the work item. The potential error depends on how similar the referenced item and the new item are. Due to the time-saving benefit which is often crucial in valuation of variations and final account settlement, this method is often used by Quantity Surveyors. However, if complicated or substantial adjustments have to be made on the BQ rates (referenced rates) in order to establish the pro-rata rate, a new rate should be built instead.

Readers should be reminded that pro-rata rates are normally used for pricing additional work items in a variation order where no contract rates are applicable. When preparing BQ rates for tender, estimators seldom use the pro-rata method due to the concern of accuracy.

Question 21.2

Calculate a pro-rata rate for each of the following variation items:

i BQ items: Teak wood boarding to walls
- 15mm thick HK$310/m²
- 20mm thick HK$380/m²
- 30mm thick HK$520/m²

Variation item: 25mm thick teak wood boarding to walls. (per m²)

ii BQ items: Deposit surplus excavated material on site in heaps
wheeling not exceeding 100m HK$45/m³
wheeling 100–150m HK$50/m³
Backfill excavation with selected spoil materials, wheeling not exceeding 100m. $60/m³

Variation item: Backfill excavation with selected spoil materials, wheeling 150–200m. (per m³)

Suggested answers

Question 21.1

i All-in hourly rate of a scaffolder:

Productive hours/day	6.5
Wage period = 1 day	
Daily wage	2,000.0
Employees' Compensation	Incl. in Prelim.
MPF (Industry Scheme)*	50.0
Other benefits	-
All-in hourly rate of scaffolder (HK$):	**315.4**

ii All-in monthly rate of building services engineer:

wage period = 1 year			
No. of days/year		365	
Less: Rest days	52		
Statutory holidays	12		
Annual leave	10		
Sick leave	10		
	84	(84)	
No. of productive days		281	
12-month salary ($28,000 × 12)		336,000.0	
Double pay		28,000.0	
Bonus		2,800.0	
		366,800.0	366,800.0
Gratuity ($28,000 × 12 × 10%)		33,600.0	
Jan to Nov MPF (5%)		15,400.0	
Dec MPF (upper limit)*		1,500.0	
Training allowance		3,000.0	
Medical insurance		Incl. in Prelim.	
Employees' Compensation		Incl. in Prelim.	
		53,500.0	53,500.0
Total staffing cost of engineer for 12 months			420,300.0
All-in daily rate of engineer (HK$):			**1,495.7**

Note: * Upper limit of employer's contribution to the MPF Industry Scheme is HK$50 per day as at Feb 2020. For Master Trust Scheme, the upper limit of employer's contribution is HK$1,500 per month as at Feb 2020. Such limits will be subject to revision by the Mandatory Provident Fund Scheme Authority, H.K.

Question 21.2

i Pro-rata rate for 25mm thick teak wood boarding to walls
By derivation method:
= HK$380 + (380 – 310)(5/5)
= HK$450 (/$m^2$)

ii Pro-rata rate for backfill excavation with selected spoil materials, wheeling 150–200m
By derivation method:
From the two deposit items in the BQ: To wheel an extra 50m = HK$50 – 45 = HK$5 (/m^3)
Therefore, the pro-rata rate for the variation item:
= BQ rate for backfill + wheeling extra 100m
= HK$60 + 5 × 2
= HK$70 (/$m^3$)

22 Pricing trade work

Introduction

Today, with the vast development in information technology and the World Wide Web, contractors can obtain quotations from suppliers or subcontractors and standard rates from institutions more effectively than before. Assuming an estimator knows the construction method to be employed very well and thus the required resources for each work item, there are still two sets of information required for building the unit rates:

- the coverage of each BQ item
- the productivity of the required resource

Coverage of BQ items

In principle, the scope and details of project work is described in drawings and BQ. From these documents, the method of work and resources required can be formulated. However, before pricing each BQ item, the estimator needs to know the extent of work covered in each BQ item. By referring to the standard method of measurement in BQ preparation, estimators can tell what is included/excluded in each item. While the SMM helps to eliminate ambiguities and misunderstanding, estimators must have a good knowledge of the SMM measurement rules and coverage rules. At the same time, attention must be drawn to the deviations from the SMM rules that are highlighted in the preambles.

Factors affecting productivity

As mentioned in **Chapter 21**, labour costs and plant costs are heavily dependent on a dynamic factor – productivity. Unlike cost data that can be obtained from many organisation or company websites, realistic productivity data is difficult to acquire. One of the main reasons is due to the sensitivity of productivity estimates, which makes suppliers and service providers unwilling to provide them. Factors impacting productivity include:

1 Weather conditions

Poor weather conditions such as heavy rainstorms and typhoons affect the productivity adversely, especially if the operation is executed outdoors (e.g. excavation, concreting, roofing etc.). Estimators have to check the tentative programme to see which operations will take place in the summer when typhoons happen frequently. Knowing that poor weather may affect those operations adversely, estimators can make adjustment to the productivity of the concerned activities.

2 Project complexity

It is reasonable to foresee that productivity will decrease if the project is complex in nature, such as implementing new technology or new material. However, the extent of such impact to productivity is quite difficult to assess. The judgement relies on the experience and knowledge of the estimator.

3 Motivation of operatives

Motivation of operatives directly affects productivity. For labour and operatives, motivation can be best accomplished by financial incentives such as bonuses. However, remuneration package design is a corporate decision and estimators can only assess the level of motivation from past projects. Any existence of demotivating factors such as payment delays, lack of proper transportation support, lack of training sessions etc. (DeCenzo and Holoviak, 1990) should scale the productivity down appropriately.

4 Experience and skill level of operatives

It is obvious that if an operative is experienced and skilful, productivity of work will be higher. Evidence suggests that proper training can improve teamwork and efficiency of site activities (Tabassi et al., 2012). Estimators can check the licensure and training record of operatives when assessing their productivity.

5 Site restrictions

Each project may inherit different restrictions due to design and site conditions. A congested site may result in having too many workers working in a limited area or multiple trades assigned to work in the same area. Under these circumstances, the probability of interference rises and productivity may be reduced. Moreover, if the site layout is undesirable, requiring workers to walk a long way to lunch, rest areas, washrooms, entrances and exits, overall productivity will also be hampered (AACE International, 2004).

6 Quality of tools and equipment

Inefficient equipment or old-fashioned tools often decreases the overall productivity. Even a small accident resulting from scrap timber with protruding nails can decrease productivity on site (Sanders and Thomas, 1993). Machinery which is lacking proper maintenance can lead to frequent breakdowns and serious injuries to operatives. Estimators should review the quality of equipment owned and rented by the company, as well as the maintenance policy of the self-owned equipment when considering the productivity level of the equipment used in a project.

Although there are clear measurement and coverage rules stated in the HKSMM for trade work, the database and methodology adopted by contractors to calculate unit rates may differ. The following worked examples on unit rate pricing cover excavation, concrete works, brickwork, wood work, plastering work, painting work and P.C. rate items. These examples are based on a variety of productivity data, which may not be applicable in all circumstances. The main objective of the worked examples is to illustrate the general principles and procedures in estimating unit rates. Readers should appreciate the diversity of available resources and project conditions. There is no substitute for a good database developed by the contractor that is based on its own resources and past projects.

Question 22.1

Productivity and cost data can be recorded in different formats, which may not be directly applicable to the estimating task concerned. Convert the following productivity or cost data to suit the task.

i A painter can finish one coat of paint on $100m^2$ of wall surface within 12 hours. What is his average productivity (hr/m^2)?

ii A labourer requires 0.5 hour to compact $1m^2$ of the bottom of an excavation. A 20m(L) × 3m(W) × 2m(D) foundation excavation is carrying out. What is the productivity of compaction (hr/m^3 of excavation) if this labourer is employed?

iii The material all-in rate for cement sand mortar is $\$900/m^3$. The mortar is used to for a 40mm screed. What is the cost of mortar material per m^2 of screed?

In the worked examples, direct labour and materials are used for each trade work item. While such an approach is mostly applied at the subcontractor level, the principles and techniques required in pricing can be demonstrated in a more pertinent way. In practice, general contractors normally employ subcontractors to carry out the trade work, and therefore, the unit rates can be calculated easily by adding profit and overheads to the subcontractors' (and suppliers') unit rates. The cost and productivity data required for each example is summarised in the 'database snapshot' boxes for easy reference.

Pricing excavation

Excavation is one of the risky trades in construction because the subsoil information including underground water and soil type is not fully accessible in most cases. The uncertainty affects the decision on:

- the use of excavation method such as hand-dug or machine-dug;
- the use of equipment such as large tractors, scrapers or draglines; and
- the necessary soil supports.

Since the cost of excavation should cover excavating in any ground encountered (HKSMM4 Rev 2018 VI(a).C.3), estimators have to predict the soil type based on all tender information available including observations in the site visit.

To build up the unit rate for excavation, the following should be considered at the outset:

- Type of excavation such as surface excavation, trench excavation etc.
- Existing site profile and typography.
- Volume and size of excavation including the overall size of excavation and the total volume to be excavated.
- Type of soil to be excavated such as rock, clay, sand and so forth.
- Distance of tipping the disposed soil.

The above considerations help us to make decisions on the method of excavation and soil support, the productivity of work and the actual volume of soil to be dealt with.

Method of excavation and productivity

Excavation can be carried out by two basic methods: by machine or by hand. If choosing mechanical excavation, there are many equipment types available providing different productivities. The choice of method normally depends on the nature and scale of excavation. For instance, extensive surface excavation can be carried out effectively by a bulldozer, whereas trench excavation should be done by a backhoe. However, if the scale of excavation is very small, hand excavation may be a better choice to eliminate the cost of mobilising the heavy equipment. Further, some site conditions like sloping sites, sites with limited headroom or soft soil may also favour hand excavation.

The nature of soil also influences the use of excavation method. To deal with different soil types including rock, deep excavation for tunnels often employs a tunnel boring machine (TBM) and for bored piles often uses a reverse circulation drilling rig (RCD). When rock is encountered in open excavation, an excavator-mounted hydraulic jackhammer will be a common option.

Obviously, productivity of excavation varies with the machine and method used as well as the soil type encountered. Examples of output rates in **Table 22.1** illustrate significant variations in excavation performance when different methods are used under different soil conditions.

Actual volume of soil

According to HKSMM, excavation should be measured nett in volume (except for surface excavation) (SMM VI(a).3). No allowance is made for subsequent variations to bulk (SMM VI(a).3.M.4). The principal of 'measured nett' is critical in pricing excavation works because the volume of soil is subject to change. We describe the volume change as bulking or compaction.

The soil extracted from an excavation will occupy more space than it is in the ground. This is described as bulking. For instance, if 10m^3 of soil is excavated, the actual volume of soil to be disposed may be 13 or 14m^3. Since the quantity measured for the BQ is the nett quantity of soil to be excavated, estimators have to make allowance for the bulking effect when pricing the disposal item. On the contrary, when backfilling or filling the site, the actual amount of soil required will be more than the volume of the hole to be filled because the soil will be compacted after filling.

The extent of bulking and compaction varies in different soil types. **Table 22.2** summarises the bulking factor for some soil types. When pricing soil disposal items,

Table 22.1 Excavation Output Rates.

Method / Equipment	*Output Rates (m^3 / hr)*		
	Sandy Loam	*Hard Clay*	*Wet Clay*
Hand (with pick and shovel)	1.5	0.7	0.4
Power shovel	53.5	34.4	19.1
Dragline	50.0	30.6	15.3
Backhoe	42.1	26.8	19.1

Adapted from: U.S. Department of the Army, 1999

Table 22.2 Bulking Factors for Different Soils.

Soil	*Bulking Factor*	*Soil*	*Bulking Factor*
Granular	10–15%	Igneous	50–80%
Cohesive	20–40%	Metamorphic	30–65%
Peat	25–45%	Sedimentary	40–75%
Topsoil	25–45%	Chalk	30–40%

Source: Trenter, 2001

a bulking factor has to be applied to the soil volume. When we measure backfilling or filling items, an appropriate adjustment for compaction will be required.

Worked example – unit rate build-up for excavation

Example 22.1 Trench excavation by hand

In this example, cost for the total quantity of excavation is calculated first in order to derive the cost per m³ unit rate.

	Labour $	Matl $	Total $	Supplementary notes
<u>Excavate trench for foundation commencing at reduced level not exceeding 1.5m deep</u> Refer to **Figure 22.1** Length of trench: 10m Width of trench: 0.6m Depth of trench: 1.2m Volume of excavation: $=10 \times 0.6 \times 1.2\ m^3 : 7.2m^3$				When excavation is done by hand, we have to consider the maximum height that a man is capable of throwing the excavated soil. Normally, a maximum height of 1.5m is assumed. Therefore, if hand excavation has to be carried out down to a depth of more than 1.5m (but less than 3.0m deep), a second lift (1.5 to 3.0m lift) will be required.
<u>Excavation</u> Method: by hand Soil type: clay First throw stage 1.5 m deep: = volume of excavn x P_e x R_e = 7.2 m^3 x 0.7 hr/m^3 @ $130 /hr	655.2			**Database Snapshot** • Hand excavation < 1.5m deep: 0.7 hr/m^3 • Spreading trench bottom: 0.2 hr/ m^2 • Labourer all-in rate: $130/hr
Spread and level at the trench bottom = area of excavn x P_s x R_s = 6.0 m^2 x 0.2 hr/m^2 @ $130 /hr	156.0			
<u>Planking and strutting</u> Materials Size of poling boards: 200 x 38 x 1200 mm; 1m c/c				Assumed planking and strutting details are prepared by engineer as per **Figure 22.1**.
Size of struts: 100 x 100 x 524mm; 1m c/c No. of polling boards required: = 2 sides x [(10m / 1m) + 1] = 22 no. No. of struts (2 layers) = 2 layers x [(10m / 1m) +1] = 22 no.				Length of strut: 600 − 2 × 38 = 524(mm)

Quantity of timber required: Polling boards: = 22 no. x 0.2 x 0.038 x 1.2 = 0.20m^3 Struts: = 22 no. x 0.1 x 0.1 x 0.524 = 0.12m^3 Timber required (0.20 + 0.12) x (1+5% wastage)				It is common to reuse the timbering to save cost. Note that no. of reuse can have a significant impact to the material cost. Wastage should be added to the materials but not the labour content.
= 0.336m^3 Material cost per use = $6,300 x 0.336/15		141.1		**Database Snapshot** Timber material: • Timber all-in rate: $6,300/m^3 • No. of reuse: 15 times • Wastage: 5%
Plant: Nil				
Labour = 0.32m^3 x 7hr/m^3 @ $130	291.2			Labour productivity for timbering: • Fixing of timbering: 6 hr/m^3 • Removal of timbering: 1 hr/m^3 • So, output rate is 7 hr/m^3 of timbering
Sundry item	0	0		
Contingencies	0	0		
Sub-total:	1,102.4	141.1		
Total cost:			1,243.5	
Cost per m^3 excavation:			172.7	In case if pricing BQ item, profit and overheads will be added at a later stage (during tender adjudication) by the management. If building up a rate for a variation item, profit and overheads allowance will be applied here.
Profit and overheads: 15%			25.9	
Unit rate (per m^3):			**198.6**	

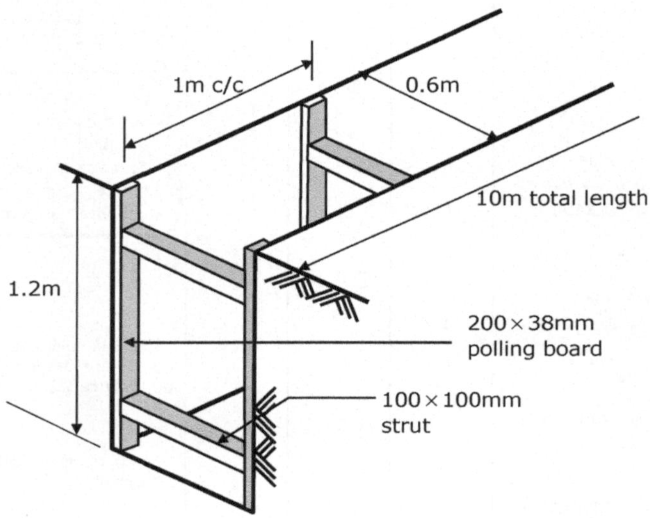

Figure 22.1 Planking and strutting details.

Example 22.2 Open excavation by machine

In this example, the oversite excavation covers an area of $300m^2$ to an average depth of 0.8m. Note that idling of the operative and the machine reduces the overall productivity and eventually increases the unit rate. Besides, the machine still consumes fuel while it is left idling, though many estimators ignore this in their estimates.

Oversite excavation to reduced level commencing from existing ground level not exceeding 1.50m deep	Labour $	Plant $	Total $	Supplementary notes
Method: by backhoe Soil type: hard clay				Assumed no soil support is required to the sides of excavation in view of the depth and soil type.
Plant Normal output/day (8am to 6pm): 9 hrs Adjusted output/day: 9hrs x 70% = 6.3hrs				**Database Snapshot** Backhoe cost: • Rental charge: $4,800/day • Fuel consumption: 14 litre/hr (operating); 2 litre/hr (idling) • Fuel cost: $13/litre • Transportation to site (round-trip): $3,000 Output rate: • Normal soil: 15 m^3/hr • Hard clay: 75% normal output • Ramming bottom of excavation is included Plant operator and labour: • Operator all-in rate: $175/hr • Banksman all-in rate: $130/hr • Overall idling: 30% (exclude lunch hour)
Rental/hr = $4,800 / 6.3 hrs		761.9		
Fuel: Operating: 14 litre/hr × 0.7 = 9.8 litre Idling: 2 litre/hr x 0.3 = 0.6 litre				
Total fuel/hr: (9.8+0.6) x $13/litre		135.2		
Maintenance:		incl.		
Plant operator and labour				
Operator (1 no.) = $175/0.7	250.0			
Banksman (1 no.) = $130/0.7	185.7			
Sub-total:	435.7			
Sub-total cost (per hour):	435.7	897.1		
Normal output of backhoe: 15 m^3/hr Adjust for hard clay: 75% = 15 m^3/hr x 0.75 = 11.25 m^3/hr = (1/11.25) hr/m^3 = 0.09 hr/m^3				
Sub-total cost (per m^3):	39.2	80.7		Banksman is required to direct the operation of the mechanical plant.
Transport backhoe to and from site = $3,000 / (300 x 0.8)		12.5		Transportation cost for backhoe is spread evenly in 240 m^3 excavation.
Sundry item	0	0		
Contingencies	0	0		
Sub-total (cost per m^3):	39.2	93.2		
Total (cost per m^3):			132.4	Contingencies are additional allowances for risk factors such as inflation, poor weather and the like.
Profit and overheads: 15%			19.9	
Unit rate (per m^3):			**152.3**	

Example 22.3 Soil disposal by wheelbarrow

Unlike the earlier example, costs per m^3 for the labour and plant components are used to build up the unit rate here. The impact of soil bulking is also illustrated.

Cart away excavated material and deposited to a dumping area 100m away from site	Labour $	Plant $	Total $	Supplementary notes
Output rate per m^3 = 1/5 + 1/4 +1/5 = 0.65 hr				**Database Snapshot** Labour productivity: • Load soil: 5 m^3/hr • Wheel 100m: 4 m^3/hr • Return 100m: 5 m^3/hr • Labourer all-in rate: $130/hr • Bulking of soil: 30%
Bulking for soil 30%				
Adjusted output rate after bulking = 0.65 × 1.3 = 0.85 hr/m^3				
Labour = 0.85 hr/m^3 @ $130	110.5			
Equipment (wheelbarrow) – incl. in preliminaries		0		
Contingencies	0	0		Bulking will lead to an increase in soil volume: 1m^3 excavated soil becomes 1.3 m^3 disposed soil. Therefore, the output rate is adjusted as shown.
Sub-total:	110.5	0		
Total:			110.5	
Profit and overheads: 15%			16.6	
Unit rate (per m^3):			**127.1**	
Output rate per m^3				

Example 22.4 Filling by machine

In this example, 800m^3 imported soil is used for filling to make up levels. The effect of soil compaction is considered.

Suitable materials obtained from specified off-site area; filling to make up levels; depositing and compacting by vibratory roller	Labour $	Plant $	Matl $	Total $	Supplementary notes
Total filling volume: 800m^3					**Database Snapshot** Backhoe cost: • Rental charge: $4,800/day • Fuel consumption: 14 litre/hr (operating); 2 litre/hr (idling) • Fuel cost: $13/litre • Transportation to site (round-trip): $3,000 • Output rate: 25 m^3/hr (no adjustment required) Plant operator and labour: • Operator all-in rate: $175/hr • Banksman all-in rate: $130/hr • Overall idling: 30%
Method: by backhoe and roller					
Plant					
Backhoe rental per hour (8am to 6pm): = $4,800 / (9 x 0.7) hrs		761.9			
Fuel:					
Operating: 14 litre/hr × 0.7 = 9.8 litre					
Idling: 2 litre/hr x 0.3 = 0.6 litre					
Total fuel/hr: (9.8+0.6) @ $13/litre		135.2			
Maintenance:		incl.			
Backhoe operator and labour					
Operator (1 no.) = $175/0.7	250.0				
Banksman (1 no.) = $130/0.7	185.7				
Sub-total:	435.7				
Sub-total cost (per hour):	435.7	897.1			
Normal output of backhoe: 25 m^3/hr = (1/25) hr/m^3 = 0.04 hr/m^3					
Sub-total cost (per m^3):	**17.4**	**35.9**			

<u>Roller</u>				
Roller rental per hour (8am to 6pm):				
= \$3,000 / (9 x 0.9) hrs		370.4		
Fuel:				
Operating: 10 litre/hr x 0.9 = 9 litre				
Idling: 1 litre/hr x 0.1 = 0.1 litre				
Total fuel/hr: (9+0.1) @ \$13/litre		118.3		
Maintenance:		incl.		
<u>Roller operator</u>				
Operator (1 no.) = \$140/0.9	155.6			
Sub-total cost (per hour):	155.6	488.7		
Normal output of roller: 65 m³/hr				
= (1/65) hr/m³ = 0.015 hr/m³				
Sub-total cost (per m³):	**2.3**	**7.3**		
Transport machines to and from site				
= \$3,000 x 2 / 800		7.5		
<u>Material</u>				
Filling material delivered to site (\$/m³)			250.0	
Add 15% compaction			37.5	
Sundry item	0	0		
Contingencies	0	0		
Sub-total:	19.7	50.7	287.5	
Total:				357.9
Profit and overheads: 15%				53.7
Unit rate (per m³):				**411.6**

Database Snapshot
Roller cost:
- Rental charge: \$3,000/day
- Fuel consumption: 10 litre/hr (operating); 1 litre/hr (idling)
- Fuel cost: \$13/litre
- Transportation to site (round-trip): \$3,000
- Output rate: 65 m³/hr (no adjustment required)

Plant operator and labour:
- Operator all-in rate: \$140/hr

- Overall idling: 10%

Transportation cost is spread evenly in 800m³ filling.

Database Snapshot
Imported filling material:
- All-in rate: \$250/m³
- Compaction factor: 15%

The soil must be compacted after filling. The actual volume of soil required has to be increased by the compaction factor.

Pricing concrete works

Concrete works involve three main groups of activities:

- Supplying and placing concrete
- Fixing and removal of formwork
- Supplying and fixing steel reinforcement

Referring to SMM VII(a).C.1, the price of concrete should include vibrating, curing, and protecting concrete, and complying with temperature control requirements. Also, the price is deemed to include forming grooves, chases, mortices and the like, forming holes for pipes ≤ 150mm diameter, forming openings ≤ 0.50m² sectional area, and forming construction joints and waterstops (except those that are designed by the engineer). Cost for test cubes and testing certificates should be included as well.

Quantity of concrete/ingredients required

To price the concrete work accurately, cost of materials must be calculated. Concrete can be either mixed on site or ordered from a supplier in the form of ready-mixed concrete. If concrete is to be site-mixed, the estimator has to calculate the quantity and cost of concrete ingredients together with the cost of the batch plant to mix the concrete. The amount of ingredients required to produce a cubic metre of concrete

Table 22.3 Amount of Ingredients Required to Produce 1 m^3 Concrete.

Concrete Mix	*Materials Required to Produce 1 m^3 Concrete*		
	Cement (kg)	*Sand (m^3)*	*Gravel (m^3)*
1:4:8	150	0.48	0.96
1:3:6	200	0.48	0.96
1:2:4	275	0.44	0.88
1:2:3	325	0.52	0.78

Source: Johannessen, 2008

with a specific strength has to be established from manuals and not from the mixing ratio because shrinkage will occur during the mixing process. As shown in **Table 22.3** below, a cubic metre of concrete requires more than a cubic metre of materials to produce. The figures shown in the table are some indicative guidelines only. The moisture content of sand and size of gravel will also affect the final amount of materials required for each batch.

To ensure high consistency in mixing quality, ready-mixed concrete is mostly used in local projects. From the estimating point of view, pricing ready-mixed concrete is much easier as quotations that include supply and delivery costs of the material can be obtained from the ready-mixed concrete suppliers. Nevertheless, no matter if site-mixed or ready-mixed concrete is used, sufficient allowance should be made for the shrinkage, consolidation, wastage and spillage of the material during distribution and compaction. Normally, a factor that lies between 5% and 10% will be applied to the fresh concrete.

Concreting productivity

Depending on the site conditions, there are several methods available for concreting:

- by wheelbarrow
- by crane and skip
- by concrete pump

The wheelbarrow method is usually used in small concrete jobs. The crane and skip method is popular in Hong Kong as tower cranes are often employed in building projects. If tower cranes are not available on site, a concrete pump provides an efficient alternative.

Points to note

When estimating the cost for placing concrete, the plant cost such as tower crane and concrete pump is normally included in the Preliminaries Bill. Further details will be illustrated in **Chapter 23**.

Besides concreting method, there are other factors affecting the productivity of concrete placing. For instance, the mix of concrete can affect the workability that in turn influences the productivity of concreting. In general, higher slump concrete is more workable, easy to place and consolidate. Extreme sizes and odd shapes of concrete structure, limited working space and congested rebar design can hammer the efficiency of concreting.

Worked example – unit rate build-up for concrete

Example 22.5 Concreting to beds by wheelbarrow

The following illustrates the pricing of ready-mixed concrete work, which involves distributing, pouring, vibrating and curing of the concrete. The plant and equipment cost is allowed in the Preliminaries.

Reinforced concrete (1:2:4) in beds; 250mm thick	Labour $	Matl $	Total $	Supplementary notes
Distributing concrete (per m³)				**Database Snapshot** Materials: • Ready-mixed concrete all-in rate: $650/m³ • Shrinkage and spillage allowance for concrete: 10% • Curing matt all-in rate (including wastage): $7/m² Labour: • A gang of 1 vibrator operator, 1 concreter, 1 labourer and 1 supervisor is required for placing concrete (gang productivity: 4m³/hr) • Distributing concrete (2m³/hr) & curing concrete (4m³/hr) require 1 labourer in each task • All-in rates for labour: Vibrator operator $260/hr Concreter $240/hr Labourer $130/hr Supervisor $270/hr
Ready-mixed concrete delivered to site				
Cost per m³		650.0		
Add: shrinkage and wastage 10%		65.0		
Distribution labour (per m³)				
= 1/2 @ $130/hr	65.0			
Placing concrete (per m³)				
Vibrator operator = 1/4 @ 260/hr	65.0			
Concreter = 1/4 @ 240/hr	60.0			
Labourer = 1/4 @ 130/hr	32.5			
Supervisor = 1/4 @ 270/hr	67.5			
Equipment (incl. in prelim.)				
Curing concrete (per m³)				
Matt for curing				
= 1/0.25 @ $7/m²		28.0		
Curing concrete labour (per m³)				
= 1/4 @ $130/hr	32.5			
Sundry item - testing (incl. in prelim.)	0	0		
	322.5	743.0		
Contingencies: 2%	6.5	14.9		
Sub-total:	329.0	757.9		
Total:			1,086.9	The estimator may include a contingency allowance if there is potential risk involved (e.g. lower productivity or work suspension due to bad weather).
Profit and overheads: 15%			163.0	
Unit rate (per m³):			**1,249.9**	

Materials allowed for reinforcement fixing

Typically, the reinforcement in concrete is either steel bar or fabric reinforcement. Reinforcement bars can be either 'site cut and bent' in accordance with the bending schedules or to be 'cut, bent and tagged to the site'. For reinforcement bars that are site cut, a minimum of 3–5% wastage (Poon et al., 2004) should be allowed. In addition to the supply and installation cost of rebar, reinforcement prices should include all tie wires and spacers. Usually around 7kg of tie wires will be allowed for every 1,000kg of reinforcement bars. While fabric reinforcement should be measured nett (SMM VII(c).M.3), extra allowance must made for the side and end laps when pricing fabric reinforcement.

> **Reminder**
>
> The lap lengths in rebar are specified by the engineers and measured in the quantities. Therefore, no allowance will be made for the laps in rebar when pricing.

The productivity of rebar fixing can be affected by factors such as the size and lengths of bars, bending shapes and complexity of the structures. In Hong Kong, most general contractors prefer to employ subcontractors, rather than direct labour, to bend and fix the rebars. In some cases, some general contractors even transfer the risk of waste estimation to the subcontractor by using a supply and fix contract.

Worked example – unit rate build-up for reinforcement

In the following worked example, the scenario of employing direct labour to cut, bent and fix the rebar is assumed.

Example 22.6 Reinforcement (cut, bend & fix by direct labour)

12mm diameter high yield steel reinforcing bars in general reinforcement	Labour $	Matl $	Total $	Supplementary notes
Materials (per tonne)				
1000 kg of 12mm bars in stock lengths		4,650.0		
Add: waste 3%		139.5		
Tie wire (per tonne): 7kg @ $12/kg		84.0		
Chairs and spacers (per tonne), say		150.0		
Labour (per tonne)				
Unload and stack				
= 1/2 @ $130 /hr	65.0			
Cut, bend and fix by bar bender and fixer				
= 25 @ $285 /tonne	7,125.0			
Sundry item - testing	Incl.			
Contingencies	-			
Sub-total (per tonne):	7,190.0	5,023.5		
Total (per tonne):			12,213.5	
Total (per kg):			12.2	
Profit and overheads: 15%			1.8	
Unit rate (per kg):			**14.0**	

Database Snapshot

Materials:

- 12mm high yield steel bars all-in rate: $4,650/tonne
- Cutting waste: 3%
- Tie wires: 7kg required per tonne, all-in rate (including wastage): $12/kg
- Chairs and spacers: allow a lump sum of $150/tonne of rebar

Labour:

- All-in rates for labour:
  Labourer $130/hr
  Bar bender $285/hr

Labour productivity:

- Loading: 2 tonne/hr
- Cut, bend & fix: 25 hr/tonne

Example 22.7 Fabric reinforcement (cut & fix by direct labour)

Square steel fabric reinforcement; BS4483; ref. A252; weighing 3.95 kg/m²; 200mm minimum side and end laps to concrete	Labour $	Matl $	Total $	Supplementary notes
Materials (per m²)				
Fabric reinforcement				
= $400 / (2.4×4.8)		34.7		
Add: 17% for laps + waste		5.9		
Tie wire				
0.2 kg @ $12 /kg		2.4		
Spacers and chairs, say		3.0		
Labour (per m²)				
Cut and fix				
Bar fixer = 1 / 60 @ 285/hr	4.8			
Labourer = 1 / 60 @ $130 /hr	2.2			
Sundry item	0			
Contingencies	0			
Sub-total:	7.0	46.0		
Total:			53.0	
Profit and overheads: 15%			8.0	
Unit rate (per m²):			**61.0**	

> **Database Snapshot**
>
> Materials:
> - Fabric reinforcement all-in rate: $400/sheet (Sheet area: 2.4×4.8m²)
> - Laps and waste: 17%
> - Tie wires: 0.2kg required per m², all-in rate (including wastage): $12/kg
> - Spacers and chairs: allow a lump sum of $3/m² of fabric reinf.
>
> Labour:
> - All-in rates and productivity of labour:
>
> | | | |
> |---|---|---|
> | Bar fixer | $285/hr | 60 m²/hr |
> | Labourer | $130/hr | 60 m²/hr |

Formwork cost coverage

Generally, formwork to the concrete surfaces which require support during casting (SMM VII(d).M.1) is measured in m² (SMM VII(d).M.2). Although the measurement rule looks simple, a complete set of formwork does not include the boardings in direct contact with the concrete surface only. For instance, a complete set of formwork for columns includes boardings, yolks, ties and wedges. Besides materials, formwork items are deemed to include labour and material costs for erection, support, propping, striking and removal (SMM VII(d).C.1). Furthermore, as stipulated in SMM VII(d).C.2 and C.4, application of release agents, forming of fillets, chamfered edges, splayed edges, boxing for openings ≤0.5m², forming holes for pipes and the like should all be included in the formwork items.

Formwork type and amount of reuse

Formwork pricing requires a fundamental understanding of the system to be employed. The most critical aspects which determine the material cost of formwork items are the type of formwork applied and amount of reuse. There are many types of formwork with respect to materials (such as timber, steel, aluminium and plastic) and design (including non-mechanised type, table form, climb form, jump form and the like). Complicated mechanised formwork systems are more costly involving substantial equipment cost. However, those systems can be reused many times and a shorter floor cycle can be achieved. These benefits often justify the use of mechanised formwork in a high-rise building construction. For traditional timber formwork, although it can be fit for up to 12 times of reuses theoretically (Lu et al., 2011), it is often reused for several times only in practice (Gammon Construction Limited, 2020).

Besides the amount of reuse, complexity of the concrete component also affects the material cost as well as the labour productivity of formwork. Non-typical structure, odd size components and members in complex shapes require custom-made formwork which results in higher material cost.

Worked example – unit rate build-up for formwork

Similar to reinforcement fixing, direct labour is used in the worked example to illustrate a more complicated scenario. In typical building projects, subcontractors are normally employed for the formwork package.

Example 22.8 Timber formwork to wall (fix by direct labour)

Sawn formwork to vertical surfaces of walls	Labour $	Matl $	Total $
Refer to **Figure 22.2**			
Overall Size: 7.6×2.85m^2 = 21.66m^2			
Materials			
19mm thick plywood board			
= 7.6×2.85m^2×2 faces @ $80/$m^2$		3,465.6	
Studs (50×75×3000mm at 400mm c/c)			
No. of studs on one face = (7.6 / 0.4)+1 = 20 no.			
Timber for studs = (0.05×0.075×3.0×20 no.×2 faces) @ $6,650 /$m^3$		2,992.5	
Waling (2 nos. of 25×100mm)			
No. of waling on one face: 3			
Timber for waling = (0.025×0.1×7.6×2 pc×3 no.×2 faces) @ $6,650 /$m^3$		1,516.2	
Struts (50×75×2200mm at 600mm c/c)			
No. of struts = (7.6 / 0.6) + 1 = 14 no.			
Timber for struts = (0.05×0.075×2.2×14 no.) @ $6,650/$m^3$		768.1	
Wedges and sill, say		200.0	
No. of tie rods			
Horiz.: 7.6m / 1m = 8 no.			
Vert.: 2.85m / 1m = 3 no.			
Tie rods (including washers, nuts) = 8×3 @ $8/no.		192.0	
Sub-total (4 uses):		9,134.4	
Sub-total (1 use):		2,283.6	
Add: 10% waste		228.4	
Consumables			
Nails			
=1 kg @ $60 /kg		60.0	
PVC sleeve (same no. as tie rods) = 8×3 @ $2/no.		48.0	
Shutter oil to formwork, say		30.0	
Material cost (per 21.66m^2 both faces)		2,650.0	
Material cost (per m^2 both faces)		**122.3**	
Labour			
Unloading = 0.5 hr/m^2 @ $(2×130)/hr	130.0		
Cutting & erecting = 0.25 hr/m^2 @ $(270+2×130)/hr	132.5		
Stripping = 0.25 hr/m^2 @ $(270+2×130)/hr	132.5		
Cleaning = 0.25 hr/m^2 @ $(2×130)/hr	65.0		
Sundry item	0		
Contingencies	0		
Sub-total (per m^2 both faces):	460.0	122.3	
Sub-total (per m^2 one face):	230.0	61.2	
Total:			291.2
Profit and overheads: 15%			43.7
Unit rate (per m^2):			**334.9**

Supplementary notes

Plywood boards are sold in pieces and therefore priced in m^2. For other parts of the timber formwork, sawn timber is sold in m^3.

Database Snapshot

Materials:

4 uses for the following:

- 19mm thick plywood board all-in rate: $80/$m^2$
- Timber all-in rate: $6,650/$m^3$
- Wedges and sill: allow a lump sum $200 (for the whole set)
- Tie rods all-in rate: $8/no.
- Wastage allowance for reusing material: 10%

Consumables:

- Nails: 1kg required for the whole set; all-in rate (including wastage): $60/kg
- PVC sleeve all-in rate: $2/no.
- Shutter oil: allow a lump sum $30 (for the whole set)

Timber wedges and sill can be calculated according to the volume of timber required but a sum of money is allocated here for simplicity.
As mentioned before, the number of uses is an important factor affecting the overall cost of material. If 5 uses, the material cost for timber per use will be reduced to $1,826.9.

Database Snapshot

Labour:

- 2 labourers are required for unloading and cleaning materials
- A gang of 1 carpenter and 2 labourers is required for erecting and stripping formwork
- All-in rates for labour:

Carpenter	$270/hr
Labourer	$130/hr

Labour productivity:

- For each m^2 of wall formwork (2 faces):

Unloading	0.5 gang hour
Cut & erect	0.25 gang hour
Stripping	0.25 gang hour
Cleaning	0.25 gang hour

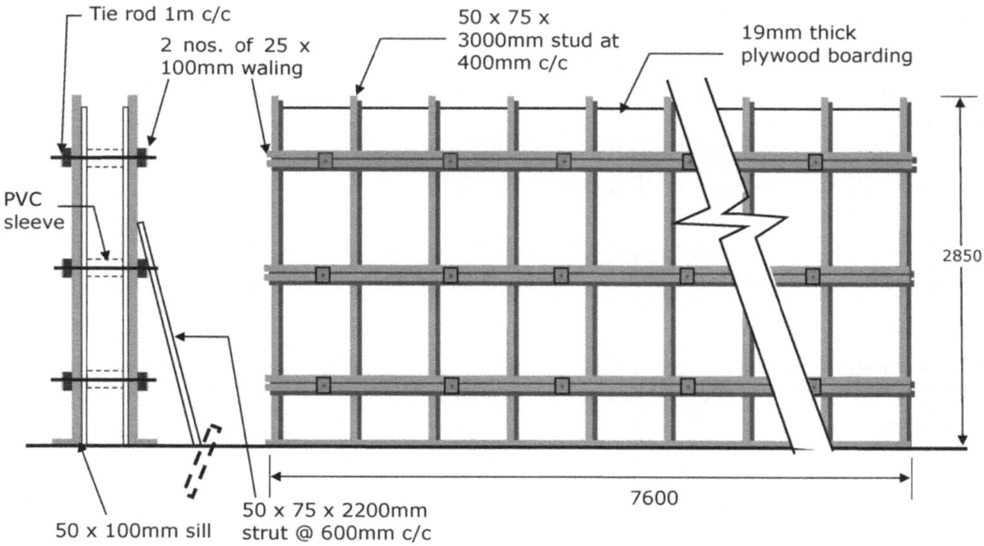

Figure 22.2 Wall formwork details.

Pricing brickwork and blockwork

Brick walls and block walls are generally measured in m². According to SMM VIII(a).C.1, the items should include all costs associated with cutting, raking out joints and pointing. Although brick/block walls can be built by a wide range of bricks in different bonds and joints, the pricing method for each of them is similar.

To price brick or block wall, details about the brick size, type and mortar joint must be known. All this information has a direct impact on material cost and productivity. In general, if common bricks of 215 × 102.5 × 65mm are used, 60 bricks will be required for each metre square of half-brick wall. One-brick walls will need 120 bricks; and so on. The type of bond (such as English bond and Flemish bond) used will only have a slight impact on the number of bricks required. However, if a fair face brick wall is to be built, the number of facing bricks required in different bonds will be different (see **Table 22.4**). The amount of mortar required varies with the type of bricks (e.g. $0.03m^3$ is required for $1m^2$ half-brick wall in commons), but in all cases, wastage must be allowed. Average bricklayer productivity is between 55 to 80 bricks per hour in most situations. For curved work, the productivity may be reduced by 40% to 50%.

Table 22.4 Number of Bricks Required in Each Type of Bond.

Bond	*Wall Thickness*	*No. of Facing Bricks / m²*	*No. of Common Bricks / m²*	*Total no. of Bricks /m²*
Stretcher	½ brick thick	60	0	60
English	1 brick thick	90	30	120
Flemish	1 brick thick	80	40	120

Source: KSSP and HMP Liverpool, 2013

Worked example – unit rate build-up for brick wall

Although brick laying is often done by a gang of two to three bricklayers, the labour productivity is based on the output of one skilful bricklayer in the following example.

Example 22.9 One-brick wall in commons

<u>One brick wall in common bricks in cement mortar (1:3)</u>	Labour $	Matl $	Total $
<u>Materials</u>			
1:3 Mortar (per 4 m^3)			
1 m^3 cement = 1.281 tonne @ $730/ tonne		935.1	
3 m^3 sand = 3×1.922 tonne @ $280/ tonne		1,614.5	
		2,549.6	
Add: shrinkage & wastage (20%)		509.9	
Mortar (per 4m^3)		3,059.5	
Mortar (per m^3)		764.9	
Mortar (per 1 m^2 brick wall) = 0.06 m^3		45.9	
Bricks (per 1 m^2 brick wall) 120 no. @ $1/no.		120.0	
Add: 5% wastage for bricks		6.0	
<u>Labour</u>			
Bricklaying per m^2			
Unloading	incl.		
Mixing mortar = 1.5×0.06 @ $140/hr	12.6		
Laying bricks = 120 / 90 @ $140	186.7		
Sundry item	0.0	0.0	
Contingencies	0.0	0.0	
Sub-total:	199.3	171.9	
Total:			371.2
Profit and overheads: 15%			55.7
Unit rate (per m^2):			**426.9**

Supplementary notes

Noticed that the cement and sand is priced by weight (not volume). For the sake of simplicity, mixing of mortar is based on the ratio of ingredients measured by volume.

Database Snapshot

Material:

- Cement all-in rate: $730/tonne
- Sand all-in rate: $280/tonne
- Weight of:
  Cement 1.281 tonne/m^3
  Sand 1.922 tonne/m^3
- Shrinkage and wastage of mortar: 20%
- 1m^2 1B wall required: 0.06m^3 mortar, 120 bricks
- Common bricks: $1.0/no.
- Wastage of bricks: 5%

The volume of mortar will shrink after mixing and 20% has been added here for shrinkage and wastage. It is easier to encounter the shrinkage in this way instead of calculating the reduced volume of mortar. Having applied the shrinkage factor, the mortar mix will remain as 4m^3.

Database Snapshot

Labour:

- All-in rate for bricklayer: $140/hr

Labour productivity:

- Unloading: included
- Mixing mortar: 1.5 hr/m^3 mortar
- Bricklaying: 90 no./hr

Pricing wood works and steel and metal works

Both wood work and steel and metal work cover a wide variety of work ranging from structural components to furniture and fittings. Material cost of these items is mainly dependent on the raw materials used, i.e. the species of timber or type of metal. While the measured quantities in the BQ are the 'finished dimensions', allowance for cutting in timber and rolling margins in steel should be made in pricing. The work items should also include the costs for all fixing accessories such as fasteners, nails, screws, bolts, nuts, washers, welding and the like. Preservative treatment against fungal and insect attack to timber and galvanisation to metal surfaces should be included in the pricing where necessary.

> **Reminder**
>
> Rolling margin is the allowable unit weight difference between the actual delivered steel unit weight and the nominal steel unit weight. In other words, it is the tolerance in nominal mass of the steel sections.

Any bedding and adhesive (for instance, the timber backing in case of timber flooring) shall be included in the item cost. For proprietary items such as demountable partitions, all accessories which form an integral part of the components are included.

Woodworks and metal works are trade works that demand a high level of craftsmanship and are normally carried out by specialist contractors. In practice, the productivity of the craftsmen depends on the nature and complexity of the item.

Again, direct labour is assumed in the following worked examples.

Worked examples – unit rate build-up for wood works and steel and metal works

Example 22.10 Timber stud partition

	Labour $	Plant $	Total $
115mm thick timber partition comprising with timber studs, timber head and sole plates and plywood board fixed on both surfaces			
Refer to **Figure 22.3**			
Materials			
Size: 5.4 mL x 2.4 mH			
19mm Plywood boards			
= 5.4×2.4×2 sides @ $80/m^2		2,073.6	
Sole plate and head plate (50×75mm)			
= (0.05 x 0.075 x 5.4 x 2 no.) @ $6,650 /m^3		269.3	
Studs (50×75×2300mm at 450mm c/c)			
No. of studs required			
= (5.4 / 0.45)+1 = 13 no.			
Timber for studs			
= (0.05×0.075×2.3×13 no.) @ $6,650 /m^3		745.6	
Noggins (50 x 75mm)			
Timber for noggins:			
= (0.05×0.075×4.75) @ $6,650 /m^3		118.5	
Nails:			
= 1 kg @ $60 /kg		60.0	
Sub-total:		3,267.0	
Add: 5% for waste		163.4	
Material cost (per 12.96 m^2)		3,430.4	
Material cost (per m^2)		264.7	
Labour (per m^2)			
Unloading = incl.			
Erecting partition = 0.5hr @ $180/hr	90.0		
Sundry item	0	0	
Contingencies	0	0	
Sub-total:	90.0	264.7	
Total:			354.7
Profit and overheads: 15%			53.2
Unit rate (per m^2):			**407.9**

Supplementary notes

Database Snapshot

Material:

- Plywood board all-in rate: $80/m^2
- Timber all-in rate: $6,650/m^3
- Nails: 1kg required for the whole task; all-in rate: $60/kg
- Wastage for all materials: 5%

Length of stud: thickness of head plate and sole plate should be deducted.
= 2.4m - 2(0.05m) = 2.3m

Total length of noggin: thickness of studs should be deducted.
= 5.4m - 13(0.05m) = 4.75m

Database Snapshot

Labour:

- All-in rate for joiner: $180/hr

Labour productivity:

- Unloading: included
- Cut & erect partition: 0.5 hr/m^2 partition

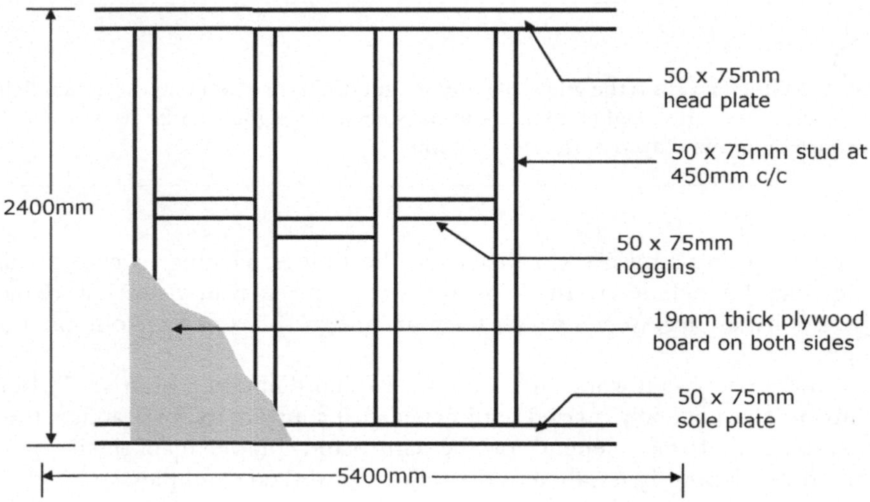

Figure 22.3 Timber stud partition details.

Example 22.11 Steel joists to floor

Hot-rolled steel to BS EN 10025; grade S355; galvanised; welded and bolted fabrication	Labour $	Matl $	Total $	Supplementary notes
203 x 102 x 25.33 kg I-joists				
Materials				
I-joists delivered per tonne		18,500.0		
Waste and rolling margin 5%		925.0		
Fixing bolts, etc.		200.0		
Labour				
Unloading				
3 hrs @ $130 /hr	390.0			
Fixing gang: 1 structural steel erector and 4 labourers				
4 hr @ ($210+$130×4)	2,920.0			
Sub-total (per tonne):	3,310.0	19,625.0		
Sub-total (per kg):	3.3	19.6		
Sundry item	0.0	0.0		
Contingencies	0.0	0.0		
Sub-total:	3.3	19.6		
Total:			22.9	
Profit and overheads: 15%			3.4	
Unit rate (per kg):			**26.3**	

Database Snapshot

Material:
- I-joists all-in rate: $18,500/tonne
- Rolling margin: 5%
- Fixing bolts, etc.: allow a lump sum $200/tonne

Labour:
- A gang of 1 structural steel erector and 4 labourers is used
- All-in rates for labour:
  Labourer $130/hr
  Structural steel erector $210/hr

Labour productivity:
Unloading 3 labourer hr/tonne
Erect & fix 4 gang hr/tonne

Pricing plastering and painting works

Finishing work including preparation of surfaces such as screeding, tiling, painting and so forth is included in the plastering section and the painting section of the SMM. Estimation of wastage and productivity in these work items is challenging. For instance, application of different paints or painting on different surfaces can produce varying output rates. Patterned wallpaper and marble can result in unexpected wastage and reduced productivity. To cope with the large variety of finishing materials and designs, an estimator has to maintain a large and updated database in order to prepare accurate price estimation.

Worked examples – unit rate build-up for screeds, tiling and painting works

Example 22.12 Floor screed

The calculation of the mortar cost is similar to the that in the bricklaying item. However, the wastage level is usually different.

<u>50mm thick cement and sand (1:3) screed laid to floors to receive ceramic tiles</u>	Labour $	Matl $	Total $	Supplementary notes
<u>Materials</u>				**Database Snapshot**
1:3 Mortar (per 4 m^3)				Material:
1 m^3 cement = 1.281 tonne @ $730/ tonne		935.1		• Cement all-in rate: $730/tonne
3 m^3 sand = 3 x 1.922 tonne @ $280/ tonne		1,614.5		• Sand all-in rate: $280/tonne
		2,549.6		• Weight of: Cement 1.281 tonne/m^3 Sand 1.922 tonne/m^3
Add: shrinkage & wastage (25%)		637.4		• Shrinkage and wastage of mortar: 25%
Mortar (per 4m^3)		3,187.0		
Mortar (per m^2; 50mm thk)		39.8		The mortar required for 1m^2 screed should be adjusted by: $(3,187 /4$m^3$)(0.05m)
Labour				
Gang: 2 plasterers				
Mix mortar, lay, spread and float 1/10 hr/m^2 @ (2 @ $170/hr)	34.0			**Database Snapshot** Labour:
Sundry item: small tools	Incl.	0		• A gang of 2 plasterers is required
Contingencies	0	0		• Plasterer all-in rate: $170/hr
Sub-total:	34.0	39.8		
Total:			73.8	
Profit and overheads: 15%			11.1	Labour productivity:
Unit rate (per m^2):			**84.9**	• 10m^2 /gang hour

Example 22.13 Ceramic tiles to floors

When calculating the material cost for tiling work, we often estimate the cost for a 100 tiled area for convenience. The material cost will be reduced to 1m^2 tiled area afterwards.

<u>150 x 150 x 6mm thick ceramic tiles; including 10mm bed of cement and sand mortar (1:3), jointing and pointing with cement and sand mortar (1:3) to floor</u>	Labour $	Matl $	Total $	Supplementary notes
				Database Snapshot Material:
<u>Materials</u> (per 100 tiled area)				• Ceramic tiles all-in rate: $80/box (36 pcs/box)
100 tiles = 100 / 36 @ $80		222.2		• Wastage of tiles: 5%
Add: 5% wastage		11.1		• Joint between tiles: 4mm
Area covered by 100 tiles (4mm joints)				• Grout for joint: Allow 0.005m^3 mortar for 100 tiles
100 x 0.154 x 0.154 = 2.37 m^2				
Mortar bed (per m^3) = $764.9				Mortar bed = $764.9/$m^3$ has been calculated for the brickwork **example 22.9**, which included a shrinkage and wastage factor of 20%. The allowance is applicable in this case.
Mortar bed (per 100 tiled area)				
2.37 x 0.01 @ $764.9		18.1		

Mortar grout (per 100 tiles)				
0.005m^3 @ $764.9		3.8		
Sub-total (mortar only):		21.9		Grout has to be allowed to fill the space between the tiles after the mortar has been cured.
Sub-total (mortar and tiles):		255.2		
Sub-total (per m^2):		107.7		
Labour				
Unloading				
0.25 / 2.37 @ $130/hr	13.7			
Gang: 2 plasterers				
0.6 hr / 2.37 @ (2 x $170/hr)	86.1			
Sundry item: small tools	Incl.			
Contingencies	0	0		
Sub-total:	99.8	107.7		
Total:			207.5	
Profit and overheads: 15%			31.1	
Unit rate (per m^2):			**238.6**	

Database Snapshot

Labour:
- A gang of 2 plasterers is required
- Plasterer all-in rate: $170/hr
- Labourer all-in rate: $130/hr

Labour productivity:
- Unloading: 0.25 labourer hr /100 tiles
- Preparing mortar and laying tiles: 0.6 gang hr /100 tiles

Example 22.14 Emulsion painting to walls

Normally, painting work involves more than one coat and its productivity data is recorded in hour/100m^2 per coat.

One coat of water-based lime resistant primer & two full coats of emulsion paint on plastered walls over 300mm girth	Labour $	Matl $	Total $	Supplementary notes
Materials (per 100m^2)				
1st coat primer				
5.5 @ $55 /litre		302.5		
1st coat emulsion paint				
7 @ $60 /litre		420.0		
2nd coat emulsion paint				
7 @ $60 /litre		420.0		
Sub-total:		1,142.5		
Add: 5% wastage		57.1		
Labour (per 100m^2)				
Unloading paints	Incl.			
1st coat primer				
8 hr @ $160 /hr	1,280.0			
1st coat emulsion paint				
10 hr @ $160 /hr	1,600.0			
2nd coat emulsion paint				
9 hr @ $160 /hr	1,440.0			
Sundry item: small tools		50.0		
Contingencies	0	0		
Sub-total:	4,320.0	1,249.6		
Total (per 100 m^2):			5,569.6	Small tools such as brushes wear off quickly and a lump sum per 100m^2 is allowed here. However, scaffolding, if required, will be allowed in the Preliminaries.
Total (per m^2):			55.7	
Profit and overheads: 15%			8.4	
Unit rate (per m^2):			**64.1**	

Database Snapshot

Material:
- Primer all-in rate: $55/litre
- Emulsion all-in rate: $60/litre
- Primer: 5.5 litre/100m^2
- Emulsion paint: 7 litre/100m^2 per coat
- Wastage of primer & paints: 5%
- Sundry item e.g. brushes: Allow $50 for 100m^2 (in 3 coats) of work

Labour:
- Unloading paints: included
- Painter all-in rate: $160/hr

Productivity of labour:

Primer	8hr /100m^2
1st emulsion paint ct	10hr /100m^2
2nd emulsion paint ct	9hr /100m^2

Pricing items with P.C. rate

As mentioned in **Chapter 11**, a P.C. rate that represents the material cost allowed may be applied to a trade work item. Normally, a P.C. rate is allowed for the material cost of the item delivered to site. When pricing items with a P.C. rate for material, the material cost will be replaced by the P.C. rate. Note that wastage and ancillary materials for fixing must be allowed in the estimate. For instance, if a P.C. rate of HK$150/m^2 is allowed in the ceramic tiling item in **Example 22.13**, the pricing of the item will be modified as below.

Example 22.15 Ceramic tiles to floor (with P.C. rate)

150 x 150 x 6mm thick ceramic tiles (P.C. Rate HK$150 per m^2); including 10mm bed of cement and sand mortar (1:3), jointing and pointing with cement and sand mortar (1:3) to floor	Labour $	Matl $	Total $
Materials			
Area covered by 100 tiles (4mm joints)			
100 x 0.154 x 0.154 = 2.37 m^2			
Mortar bed (per m^3) = $764.9			
Mortar bed (per 100 tiled area)			
2.37 x 0.01 @ $764.9		18.1	
Mortar grout (per 100 tiles)			
0.005m^3 @ $764.9		3.8	
Sub-total (per 100 tiles):		21.9	
Sub-total (per m^2):		9.2	
Tiles (P.C. rate: $150/m^2)		150.0	
Add: 5 % wastage		7.5	
Labour			
Unloading			
0.25 / 2.37 @ $130/hr	13.7		
Gang: 2 plasterers			
0.6 hr / 2.37 @ (2 x $170/hr)	86.1		
Sundry item: small tools	Incl.		
Contingencies	0	0	
Sub-total:	99.8	166.7	
Total:			266.5
Profit and overheads: 15%			40.0
Unit rate (per m^2):			**306.5**

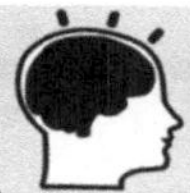

Question 22.2

The preambles of the BQ state that P.C. rates in the items included only the material cost of an item delivered to site. The contract rate will be adjusted by the nett difference between the prime cost rate and the actual price, and will be applied to the nett quantity of the item measured as fixed in position.

If the nett quantity of the ceramic tiling item (in **Example 22.15**) is 50m^2 and the confirmed ceramic tiles are HK$160/m^2, what is the adjustment to the contract sum?

Suggested answers

Question 22.1

i Given: 12 hours/100m^2 by painter
Productivity in m^2 = 12/100 = 0.12 hr/m^2

ii Given: 0.5 hour/m^2 compaction
Excavation area on plan = 20m × 3m = 60m^2; Excavation volume = 20m × 3m × 2m = 120m^3
Productivity of compaction (hr/m^3 of excavation) = 0.5 × 60/120 = 0.25 hr/m^3

iii Given: all-in rate for mortar = \$900/m^3
Cost of mortar per m^2 of screed = \$900 × 0.04 = \$36/m^2

Question 22.2

Adjustment to the contract sum:
= HK\$(160 – 150) × 50
= HK\$500 (to be added to the contract sum)

23 Pricing preliminaries

What are preliminaries?

Preliminaries are also known as project or site overheads. Preliminaries items cover the provision of site facilities, administration services and general attendance provided by the contractor. Examples of preliminaries include site staffing, site office, power and water supply, mechanical plant, testing and so forth.

General practice

The pricing method of preliminaries varies from company to company. Some small contractors simply apply a fixed percentage to the work items for preliminaries. Large contractors usually price preliminaries by analytical estimating, to estimate the resources required for each item in the Preliminaries Bill.

Difficulties in pricing preliminaries

Unlike trade work, we can hardly expect detailed drawings and specifications to be provided for preliminaries items. This is because the arrangement of project overheads depends on how the contractor plans and organises its work, which is impractical for an architect to design. In reality, even the contractor has difficulty predicting how a project will turn out, not to say allocating suitable allowance for all the problems that the contractor may incur. For instance, site accommodation is a requirement in most projects. Provision of site accommodation can range from a well-furnished temporary office to a few containers with some basic office furniture. If the site is too congested, the site office may need to be relocated within the site or in between floors. Office space from nearby commercial buildings may have to be rented for a project office. The actual expense on different arrangements can vary considerably but the details of accommodation requirement may not be found in the Preliminaries Bill. Very often, the Preliminaries Bill only provides a brief description on the site accommodation item emphasising the provision to be 'up to the satisfaction of the architect'. Some examples of briefly described site office item in local contracts are quoted below for illustration:

> The Main Contractor shall provide for his own use all necessary workshops, mess rooms, offices and sheds of suitable construction for the storage of materials, maintain them in good order to the satisfaction of the Architect, remove on completion of the Works and make good the Site.

In another contract document, the item is written as:

> The Main Contractor is to provide and erect sheds, workshops and offices (including an adequate office and necessary furniture for the use of the Site Agent) of suitable construction for the storage of such materials as he may require and maintain them and keep them in good order to the satisfaction of the Architect. The sheds, workshops and offices are to be removed on completion of the works and the sites made good.

From the above examples, the lack of clear and detailed description on the preliminaries requirements is the key problem. In addition, there are a number of uncontrollable factors that may affect the estimating accuracy of preliminaries, such as: project complexity, project duration, site layout, stakeholders' interest, regional economic uncertainty, financial and insurance charges and procurement arrangement (Chan, 2012).

Basic approaches to preliminaries pricing

To price preliminaries, contractors must thoroughly analyse the project and site conditions and derive all the site overhead costs required for the project. In general, preliminaries can be classified into two main cost areas:

- Fixed costs:
  These include fixed costs for initial set up, installation and removal of site facilities or a fixed cost for service. For instance, fixing hoardings on site involves materials and labour costs. The same applies to any intermediate alterations and final removal of hoardings after project completion. By estimating the materials and labour costs incurred in every installation and removal, the fixed costs can be calculated.
- Variable costs:
  These are time-related or works-related costs. Expenses of this nature will rise if the duration of project or the value of contract increases. For examples, management salaries and plant rental charges are depended on the hiring period. The amount of levies charged on a project is based on the value of the contract sum. When pricing preliminaries costs that fall into this category, availability of an accurate resources programme or value of work estimate will be highly important.

Worked examples of preliminaries pricing

HKSMM4 Rev 2018 Section III portrays a typical list of preliminaries items, which are subdivided into three categories: Preliminary Particulars, Conditions of Contract and General Matters. Some items that do not have direct cost impact (such as Employer and Consultants) or items that can be measured with the trade work (such as working hours) are not discussed here. To illustrate the estimating methods in a more precise manner, common preliminaries items are classified into eight groups, with those that share a similar nature grouped together for further explanation (see **Table 23.1**). Although the costs of all items within a group are added into a total sum, readers should appreciate that a sub-total of items can be extracted or totalled according to the terms in the Preliminaries Bill.

Table 23.1 Grouping of Preliminaries Items for Estimation.

Site Management	*Site Accommodation and Facilities*	*Mechanical Plant*	*Temporary Works*	*Miscellaneous Services*	*General Attendance*	*Insurances and Levies*	*Sundries*
• Site management staffing • Watchman and security • Safety management and equipment • Environmental management and equipment • Ground water control instruments	• Site offices • Furniture • Office equipment • Stationery • Latrines • Stores • Transport • Power and lighting • Water • Telephone and internet service	• Tower cranes • Material hoists • Passenger hoists • Mobile crane • Concrete mixers • Concrete pumps	• Hoardings • Gantries • Sign boards • Shoring and temporary support • Working platforms • Access roads • Scaffolding • Catch fans	• Setting out • Protection • Testing • Drawings	• Unloading and distribution • Cleaning	• Contractor All Risk Insurance • Employees' Compensation • Performance bonds • Levies	• Entertainment • Mock-ups and samples • Photographs

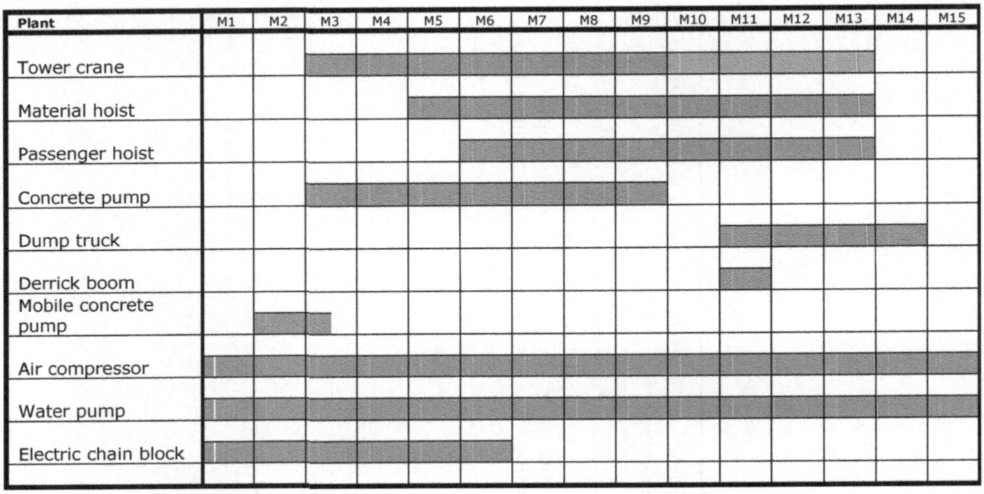

Figure 23.1 Plant schedule of a hypothetical project.

A hypothetical commercial project is used to illustrate the pricing of preliminaries. The project is located in city centre, comprises 25 storeys that will take 15 months to complete. The estimated contract sum is HK$500 million. It is assumed that the estimating team has studied the tender documents thoroughly and all the key factors relating to the project have been identified. A plant schedule has been prepared as shown in **Figure 23.1**. The costs shown in the worked examples are the likely preliminaries costs to be incurred by the contractor.

Site management

A contractor needs to assign a team of site management staff to supervise the project work. These personnel include project manager, site agent, foremen, project coordinators, engineers, QSs, storekeeper and safety officer. At the same time, support staff including clerical staff, watchmen and technicians such as rigger and mechanic are required.

Site management staff cost is the largest part of the preliminaries cost (Chan, 2006; Chartered Institute of Building, 2009). It is a time-related cost and the duration of employment should be properly estimated. Any staff employment during the defects liability period should be included in the estimate as well. An all-in rate that includes basic salary, bonus, MPF and other fringe benefits for each staff (see **Table 23.2**) should be used to estimate the total staffing cost.

In **Table 23.2**, ground water control, safety management and environmental management has been allowed in the site management cost. To maintain consistency, the costs for all necessary equipment and instruments to fulfil these management requirements are also included here.

Site accommodation and facilities

Site accommodation refers to the site offices as specified in the contract conditions. In addition to the contractor's site office, a separate office for employer's representatives

Table 23.2 Worked Example – Site Management Cost Estimation.

	Rate	*Months*	*No.*	*Sub-total*	*Total (HK$)*
1 Site Management					
Staff Salary including MPF, bonus & benefits					
During construction period					
• Project Manager	75,000	15	1	1,125,000	
• Site Agent	50,000	15	1	750,000	
• General Foreman	40,000	15	1	600,000	
• Foreman	25,000	15	3	1,125,000	
• Assistant Foreman	19,000	15	3	855,000	
• Building Services Coordinator	35,000	15	1	525,000	
• Project Coordinator	30,000	15	2	900,000	
• Project QS	45,000	15	1	675,000	
• QS	33,000	15	1	495,000	
• Assistant QS	22,000	15	1	330,000	
• Site Engineer	35,000	6	1	210,000	
• Storekeeper	15,000	15	1	225,000	
• Site Clerk	12,000	15	2	360,000	
• Site Secretary	18,000	15	0	0	
• Safety Officer	28,000	15	1	420,000	
During DLP					
• Site Agent	52,500	12	1	630,000	
• Foreman	26,250	12	3	945,000	
• Project Coordinator	31,500	12	1	378,000	
• Project QS	47,250	12	1	567,000	
• QS	34,650	12	1	415,800	

	Rate	*Months / Storeys*	*No.*	*Sub-total*	*Total (HK$)*
Watchman and Security					
• Watchman	23,000	15	2	690,000	
• Guard house				15,000	
• Access control system	10,000		1	10,000	
Safety Management					
Safety Precautions					
• Safety handrail	5,000	25		125,000	
• Safety barrier (lift shafts)	1,000	25	4	100,000	
Fire Protection					
• Extinguishers	700	25	2	35,000	
• Sand bucket	200	25	4	20,000	
• Fire pumps				30,000	
Environmental Management					
• Dust control				30,000	
• Sound level meter				40,000	
• Noise barrier				200,000	
• Mosquito prevention				30,000	
• Vehicle washing bay	70,000		1	70,000	
Ground Water Control					
• Dewatering system				50,000	
					12,975,800

including resident architect, resident engineer and clerk of works is usually required. Furniture, stationery and basic office equipment such as computers, printers, photocopiers, fax machines, telephones and the like should be provided for both offices. As far as practicable, storage areas should also be provided. In very large and remote projects (such as construction of a power plant), cafeteria or catering facilities may be provided on site. On the other hand, if the site is congested, storage space may be provided by building temporary platforms or using offsite storage. Site accommodation cost and the associated equipment costs are normally fixed costs (including initial set up and removal costs). In case when nearby office or storage space is rented for the project, provision of site accommodation will incur a time-related cost.

Provision of separate toilets or latrines for employees of each gender is a legislative requirement under the Occupational Safety and Health Regulation Cap. 509. Contractors may install temporary toilets that are connected to the sewer system or employ portable chemical toilets. For temporary toilets, only fixed costs are involved. If chemical toilets are used, rental charges as well as delivery, installation and cleaning costs will be incurred.

To support site accommodation and construction operations, a number of utility services including telephone, internet connection, electricity, water supply and drainage are required. All these items involve an installation cost and a cost for the service consumed during the period. For telephone and internet connection services, a fixed rate can be applied to each month of the construction period. For the electricity and water bills, historical data from past projects can be used to predict the consumption level and thus the likely expenses.

Transportation cost may include provision of transportation facilities such as car, shuttle bus or ferry for consultants and site staff (and even workers if the site is very remote). Running cost including fuel cost, insurance and license fee must be included.

The worked example in **Table 23.3** illustrates the cost estimation for a typical site accommodation provision such as relocation of site offices to completed floors, connection of utility services etc.

Mechanical plant

Large mechanical plant items that are shared by a number of activities are normally priced in the Preliminaries Bill. These include cranes, hoists, trucks, compressors, water pumps etc. As discussed in **Chapter 22**, when calculating the all-in-rates for plant, rental charges (or depreciation cost and financing cost if the equipment is purchased by the contractor), fuel cost, maintenance charges and other sundry consumables such as grease and cables have to be included. The same concept applies to the estimation of mechanical plant cost in the Preliminaries Bill. Large items such as hoists involve mobilisation, fixing, testing and dismantling expenses, which should not be overlooked.

As shown in **Table 23.4**, the fuel charges for plant are included in the plant costs, whereas the electricity charges used by the electrical machines remain in the Light & Power sub-section under the Site Accommodation and Facilities Category. Operator cost is also priced here and should match with the plant schedule. In addition, it is a common practice for the main contractors to provide concreting plant and equipment for the subcontractors. As more than one concreting method may be adopted

Table 23.3 Worked Example – Site Accommodation and Facilities Cost Estimation.

	Rate	*Months*	*No. / trip*	*m²*	*Sub-total*	*Total (HK$)*
2 Site Accommodation and Facilities						
Contractor Office						
• Container office						
• 5 no., rental 3 months	1,800	3	5		27,000	
• delivery and removal	1,500		2 × 5		15,000	
• Relocate site office					50,000	
• Office inside building						
• Set up site office (180 m²)	1,800			180	324,000	
• Removal					70,000	
• Office equipment						
• Air conditioner	4,000		5		20,000	
• Photocopier	8,000		1		8,000	
• Computer & software licenses	7,000		12		84,000	
• Printers	2,500		3		7,500	
• Fax machine	2,000		1		2,000	
• Office furniture					60,000	
• Stationery ($2,000 / month)	2,000	15			30,000	
Office for Employer's Representatives						
• Container office						
• 3 no., rental 3 months	1,800	3	3		16,200	
• delivery and removal	1,500		2 × 3		9,000	
• Relocate site office					50,000	
• Office inside building						
• Set up site office (120 m²)	1,800			120	216,000	
• Removal			inclu. in contractor office removal			
• Office equipment						
• Air conditioner	4,000		4		16,000	
• Photocopier	8,000		1		8,000	
• Computer & software licenses	7,000		3		21,000	
• Printers	2,500		1		2,500	
• Fax machine	2,000		1		2,000	
• Office furniture					30,000	
• Stationery ($800 / month)	800	15			12,000	
Latrines						
• Chemical toilet	2,000	15	8		240,000	
• Delivery & removal	1,200		2 × 8		19,200	
• Cleaning					incl.	
• Connection / disconnection					10,000	
• Urinal, WC down pipes					5,000	
• Sanitary fittings					8,000	
Store						
• Set up and removal					12,000	

Table 23.3 (Continued)

	Rate	*Months*	*No. / trip*	*m*^{2}	*Sub-total*	*Total (HK$)*
Telephone						
• Telephone line and internet	150	15	18		40,500	
• Fax line					incl.	
• Mobile phone service	500	15	3		22,500	
Radio communication						
• Inverter					10,000	
• Base station and antenna					10,000	
• Walkie talkie					25,000	
Light & Power						
• Electrician (from labour schedule)	21,000	15	1		315,000	
• Distribution to structure						
• 3 phase distribution board and switch board					25,000	
• TPN isolator					10,000	
• Cables					150,000	
• Connection to site offices					5,000	
• Temporary lighting, switches, sockets	15,000		25		375,000	
• Cable to tower crane					40,000	
• Electricity bills						
• Site office	5,000	15			75,000	
• General site operations	34,000	15			510,000	
• Tower crane	23,000	11	1		253,000	
• Material hoist	3,500	9	1		31,500	
• Passenger hoist	5,000	8	1		40,000	
• Water pump (2 no.)	19,000	15	2		570,000	
• Derrick boom	6,500	1	1		6,500	
• Electric chain block	34,000	6	1		204,000	
• Main distribution board						
• Connection					200,000	
• Temporary substation / transformer room					0	
• Main distribution board and panel					200,000	
• Removal and make good					20,000	
Water						
• Fitter (from labour schedule)	21,000	15	1		315,000	
• Connection					15,000	
• Distribution						

Table 23.3 (Continued)

	Rate	*Months*	*No. / trip*	*m²*	*Sub-total*	*Total (HK$)*
• Temporary pumps, water tanks	19,000		2		38,000	
• Pipes, hose and fitting					30,000	
• Materials for maintenance					10,000	
• Water bills						
• Site office	1,200	15			18,000	
• General site operations	10,000	15			150,000	
• Connection to site office					5,000	
• Removal and make good					15,000	
Transportation						
• Contract car					N/A	
						5,108,400

in a project, it is rather difficult to tell how much plant cost should be allocated to specific parts of the concrete structure. As a result, concreting plant costs are priced in this part of Preliminaries Bill. In the hypothetical case, concreting will be completed by two methods: placing by concrete pump and by crane and skip method.

Temporary works

This group covers a wide variety of temporary enclosures and temporary works which are possibly required in a project:

- Hoardings and fences to enclose the site so as to restrict access and to provide protection to the public.
- Gantries to provide overhead and side protection at the site entrances.
- Covered walkways to provide protection to pedestrians against falling objects.
- Signboards to give information about the project, project owner, consultants and contractor.
- Shoring to support the sides of excavation to prevent collapse.
- Access roads to ensure adequate access is available for vehicles and site personnel.
- Temporary working platforms to provide working areas for plant or operatives.
- Scaffolding to support workers to carry out operations at height.
- Catch fans at the scaffold to trap fallen objects.

In each case, the total cost should include the cost of labour, materials and any necessary equipment for fixing the temporary structure, as well as removing and making good the area. For shoring work and other temporary support structures, the cost for employing a qualified engineer to prepare the structural design has to be added.

Among the various items grouped under the temporary work category, pricing on site access demands careful attention. Many construction sites in Hong Kong have

Table 23.4 Worked Example – Mechanical Plant Cost Estimation.

	Rate	*Months*	*No.*	*Sub-total*	*Total (HK$)*
3 Mechanical Plant					
Plant hire					
• Tower crane	65,000	11		715,000	
• Material hoist	35,000	9		315,000	
• Passenger hoist	55,000	8		440,000	
• Concrete pump	54,000	7		378,000	
• Dump truck	9,000	4		36,000	
• Derrick boom	35,000	1		35,000	
• Mobile concrete pump	54,000	1.5		81,000	
• Air compressor	25,000	15		375,000	
• Water pump	35,000	15	2	1,050,000	
• Electric chain block	17,000	6		102,000	
Operator					
• Rigger A	22,000	10		220,000	
• Rigger B	22,000	7		154,000	
• Tower crane	28,000	11		308,000	
• Material hoist	15,000	9		135,000	
• Passenger hoist	15,000	8		120,000	
• Concrete pump	18,000	7		126,000	
• Dump truck	20,000	4		80,000	
• Derrick boom	15,000	1		15,000	
• Mobile concrete pump	18,000	1.5		27,000	
Fuel					
• Concrete pump	16,000	7		112,000	
• Dump truck	1,000	4		4,000	
• Mobile concrete pump	16,000	1.5		24,000	
• Air compressor	38,000	15		570,000	
Miscellaneous plant					
• Vibrator - motor & poker				80,000	
• Power tools				18,000	
• Concrete pump pipe line				50,000	
• Concrete skip				7,000	
• Day marking & warning lights				12,000	
• Temporary landing platform for material hoist				60,000	
• Temporary landing platform for passenger hoist				40,000	
• Hiring of permanent lift	61,000	2	2	244,000	
Assembly, climbing, dismantle and transportation					
• Tower crane				460,000	
• Material Hoist				65,000	
• Passenger hoist				120,000	
• Concrete pumps, derrick boom, air compressor, water pump				7,500	
Others					
• Base and temporary materials for crane, hoist, boom				400,000	
• Mechanical hardware	9,000	15		135,000	
• Testing for crane, hoist, derrick				60,000	
					7,180,500

Table 23.5 Worked Example – Temporary Works Cost Estimation.

	Rate	*No. / Months*	*m² / m*	*Sub-total*	*Total (HK$)*
4 Temporary Works					
Hoarding					
• Renew existing hoarding				80,000	
• Maintenance				incl.	
• Site gantry	30,000	1		30,000	
• Removal and make good				110,000	
Signboard and maintenance				25,000	
Temporary platforms and support					
• Temporary loading platform				150,000	
• Temporary support to voids				105,000	
• Temporary barriers				44,000	
Access					
• Traffic control lights				7,000	
• Traffic controller	15,000	15		225,000	
Scaffolding					
• External					
• Bamboo scaffolding - double line	55		45,000	2,475,000	
• Nylon mesh	25		45,000	1,125,000	
• Catch fan (in 1800m)	190		1,800	342,000	
• Internal					
• Vertical bamboo scaffolding - single line	45		4,500	202,500	
• Vertical bamboo scaffolding - double line	60		1,000	60,000	
• Horizontal ceiling platform	140		2,000	280,000	
• Mobile platform				80,000	
					5,340,500

problems related to access, such as limited width of access roads, lack of proper access to sloping sites, lack of parking space nearby the site, high traffic in major access and so forth. Contractors may need to provide temporary roads, hardstandings or paving including regular maintenance. A detailed site visit can provide practical assistance to establish the exact requirement for access provision. Sometimes, access problem can occur within the site if it is congested. Temporary ramps are required in many local projects to provide vehicular access and underground construction simultaneously. In the worked example, although temporary access roads are not required, expense on traffic control is allowed (see **Table 23.5**).

Miscellaneous services

The items categorised into this group are the general services possibly provided by the main contractor, including setting out, protection to finished work, testing,

Table 23.6 Worked Example – Miscellaneous Services Cost Estimation.

	Rate	*Months*	*No.*	*Sub-total*	*Total (HK$)*
5 Miscellaneous Services					
Setting out					
• Senior leveller	21,000	12	2	504,000	
• Leveller	17,000	12	4	816,000	
• Survey equipment and consumables				25,000	
Protection to finished work					
• Granite floor				60,000	
• Carpet tile				30,000	
• Lift cars				30,000	
• Sanitary fittings			by subcontractor		
• Others				24,000	
Testing					
• Concrete cube sampling, curing	40		960	38,400	
• Concrete cube operator (from labour schedule)	17,000	8		136,000	
• Hammer test	20		600	12,000	
• Rebar bending & tensile test	600		800	480,000	
• Concrete trial mixes				by supplier	
• Plumbing & Drainage			by subcontractor		
• Other building services			by subcontractor		
• Window, roofing			by subcontractor		
Drawings					
• CAD technician (allow 80%)	26,000	5	1	130,000	
• CAD technician (allow 30%)	10,000	10	1	100,000	
• CAD engineer (allow 20%)	16,000	15	1	240,000	
• Printing charges: CAD drawings	40		200	8,000	
• Printing charges: As-built and submission	40		600	24,000	
					2,657,400

drawings and so forth. Traditionally, these services are provided by the main contractor. With the increasing complexity in some specialist work, main contractors may not be capable or willing to provide such service to every subcontractor. For example, many trade-specific tests and shop drawings can be more efficiently provided by the subcontractors as they are experts in their areas. The protection item in **Table 23.6**, for instance, does not include any protection to sanitary fittings. Many main contractors prefer the protection to these items to be provided by subcontractors to minimise the disputes on damaged items. Therefore, the extent of service provided by the main contractor has to be negotiated with subcontractors.

Similar to other preliminaries items, the required level of miscellaneous services can vary substantially in different projects. For example, in **Table 23.6**, expenses on standard tests are allowed. If there are any special fire resisting installations or structural features in the project design, the main contractor has to arrange extra tests, which will incur additional expenses.

Table 23.7 Worked Example – General Attendance Cost Estimation.

	Rate	*Months*	*No.*	*Sub-total*	*Total (HK$)*
6 General Attendance					
Site Cleaning					
• Daily cleaning					
• Labour (from labour schedule)	15,000	15	15	3,375,000	
• Hire of skip	900		15	13,500	
• Rubber refuse chute				84,000	
• Final clean prior to handover					
• Labour (from labour schedule)	15,000	1	15	225,000	
• Floor hacking (31,000m^2)				430,000	
• Rubbish truck					
• Waste collection	1,000		520	520,000	
• Dumping charge	1,000		520	520,000	
Unloading and distribution					
• Labour (from labour schedule)	15,000	15	4	900,000	
					6,067,500

General attendance

General attendance covers the main contractor attendance to all subcontractors, such as keeping the site clean, unloading the materials and distributing the materials to various working levels. As shown in **Table 23.7**, most of the cost in general attendance belongs to labour cost.

Site cleaning cost is generally divided into two types: daily cleaning and final cleaning. Usually, fixing and removal of refuse chutes and dumping of refuse is included in the daily cleaning cost. Some contractors may sublet the final cleaning work to subcontractors but direct labour is allowed in the worked example here.

Insurances and levies

As shown in **Table 23.8**, pricing of insurances and levies is relatively straight-forward. Major insurance policies that a contractor has to arrange are the Contractors All Risks (CAR) and Employees' Compensation (EC).
A standard CAR policy provides two main coverages:

- Material damage due to unforeseen physical loss of or damage to contract works (such as damages due to fire).
- Third party bodily injury or damage to third party's physical property (described as third party liability) that arises out of the execution of the contract works.

EC is a legislative requirement under the Employees' Compensation Ordinance (Cap. 282). As stipulated in the Ordinance, all employers in Hong Kong have to compensate their employees if they suffer from bodily injury or death in an accident arising out of and in the course of the employees' employment.

Table 23.8 Worked Example – Insurance and Levies Cost Estimation.

	Contract Value	*%*	*Years*	*Sub-total*	*Total (HK$)*
7 Insurances and Levies					
Insurances					
• Contractor All Risk Insurance Allow 0.3% of contract sum	500,000,000	0.30%		1,500,000	
• Employees' Compensation Allow 1.5% of contract sum	500,000,000	1.50%		7,500,000	
Performance Bonds					
Amount of bond: Annual HK$5M for 15 months Allow 1.3% per annum	5,000,000	1.30%	1.25	81,250	
• No allowance during DLP				0	
Levies					
• Construction Industry Levy	500,000,000	0.53%		2,650,000	
• Pneumoconiosis and Mesothelioma Levy	500,000,000	0.15%		750,000	
					12,481,250

The amount of premiums for insurance policies depends on the terms and agreement between the contractor and the insurance companies. When pricing, estimators usually apply a percentage to the contract value for such allowance. For example, the premium charge for EC normally ranges from 1.5% to 2.5% of the contract value.

Regarding the levies, there are two levies payable by the contractor:

- Construction Industry Levy is required under the Construction Industry Council Ordinance (Cap. 587) and the Construction Workers Registration Ordinance (Cap. 583). The levy amount is 0.53% of the contract sum, including 0.5% to support the Construction Industry Council and 0.03% as the Construction Workers Registration Authority Levy. All contracts with a contract sum (including M&E works) higher than HK$3 million are required to pay such levy.
- Pneumoconiosis and Mesothelioma Levy is required under the Pneumoconiosis and Mesothelioma (Compensation) Ordinance (Cap. 360). The levy amount is 0.15% of the contract sum. Projects with a contract value $3 million are exempted from the payment of levy.

Reminder

The threshold for these levies has been raised from HK$1 million to HK$3 million since 30 July 2018.

Table 23.9 Worked Example – Sundry Cost Estimation.

	Rate	*Months*	*No.*	*Sub-total*	*Total (HK$)*
8 Sundries					
Entertainment					
• Ceremonies (commencement, completion)	20,000		2	40,000	
• Meals (commencement, Lunar New Year, Lo Pan Festival, completion)	8,000		4	32,000	
• Petty cash	6,000	15		90,000	
Progress photos				5,000	
Mock-up				120,000	
					287,000

Performance bond is often found in contract conditions requiring the contractor to arrange with an insurance company (the surety) to guarantee the employer that the contractor (the principal) will complete the work satisfactorily according to the terms in the contract. Estimation of the performance bond cost is similar to that for the insurance policies.

As shown in **Table 23.8**, an assumed percentage is applied to the contract sum of HK$500 million to calculate the amount of premium for the various insurance plans. Similarly, the amount of levy to be paid is estimated by applying the levy percentage to the contract sum.

Sundries

Besides the seven categories of costs described above, there are some general expenses that may be incurred in a project but may not be always encountered. Some of the popular items are listed in the worked example in **Table 23.9**. An interesting item is the entertainment cost which is allowed by many large contractors in Hong Kong. This is somehow related to the customs of the local industry. For progress photos, although an electronic archive of project information is popular, a certain extent of photo printing is necessary for submissions.

If mock-ups or show flats have to be completed by the main contractor, they are usually priced separately so that adjustment can be made easier in case if there is any variation. Normally, sample material submission is allowed in the trade work items unless the sample involves substantial amount of preparatory work or material.

Points to note

When prepare estimates for preliminaries items, estimators often break down the estimate for each item into five categories: labour, material, plant, subcontractor, and others (for costs such as insurance premiums, utility bills and the like). This can provide a source of checking on the cost allocated to each resource category, which is particularly useful for cost control in the contract stage. However, such breakdown is not illustrated in the worked examples above for simplicity.

24 Computer applications in measurement and estimating

Quantity take-off using computer applications

It is true that manual take-off using dimension paper is a tedious and error-prone process. Over the years, information technology has been changing rapidly and providing new solutions to building measurement. Along with the popularity of computer-aided drafting, take-off tools have evolved from pen and paper to spreadsheets, and then to digitiser tablets. In the last decade, on-screen take-off based on electronic 2D drawings has been widely adopted in many QS firms. Today, building information modelling (BIM) is the mainstream in the construction industry and 3D model-based take-off will surely become a prevalent take-off and estimating tool.

It is not the intention here to elaborate the steps of writing a Visual Basic for Applications (VBA) program for taking-off or the method of using a specific take-off software. While construction projects have increased in size and complexity, more efficient and reliable solutions for quantity take-off and cost estimation tasks are necessary. Therefore, an overview of the computer applications in measurement and estimating is discussed so that readers can decide on the best method to adopt.

Designing spreadsheet templates for taking-off

Today, every business goes paperless. Many QSs have shifted to use spreadsheets instead of papers to measure quantities. Without any sophisticated software required, a spreadsheet such as Microsoft® Excel provides a strong computational tool that is simple to use for most people. By using spreadsheets to take-off, not only is the paper saved, but the working-up of measurements can be performed more effectively.

The take-off schedule introduced in **Chapter 9** (**Figure 9.6**) can be transformed into an electronic template ready for measurement, such as the one shown in **Figure 24.1**. Many QSs like to design templates with pre-set formats such as company logo and built-in equations to extend the quantities automatically. Since entering and modifying data in a spreadsheet is extremely easy, entering errors into a spreadsheet is just as easy. All equations set in the templates must be thoroughly tested. Unintentional changes to the equations or content can be restricted by protecting the cells.

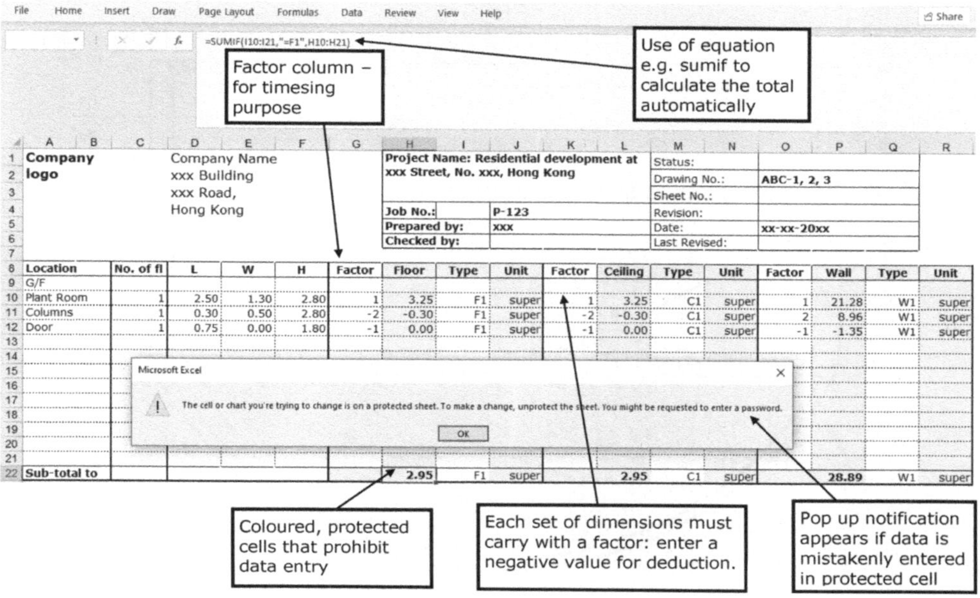

Figure 24.1 An example of spreadsheet template for taking-off finishes.

Designing a suitable template may take some time, but once it is created, all take-off for the same trade can be done quickly. Below are some tips for the design of take-off templates:

- Provide clear details of the project, drawing no. and the name of taker-off.
- Record complete description of the items or set a cross-reference to link the abbreviated description to the detailed list of item descriptions.
- Follow the sequence of length, width and height (or depth).
- Remember to incorporate the factor column for timesing.
- Protect the cells from changes by mistake.
- Make good use of the 'link' function to cross-reference the cell contents so that changes made in a cell can be automatically updated in other related cells/ spreadsheets.

On-screen take-off from 2D AutoCAD drawings

Today, most drawings, including the tender drawings, are prepared in AutoCAD file format. Consequently, a taker-off can measure the quantities directly on the screen using the AutoCAD 'measure' function.

As shown in **Figure 24.2**, the mean girth of the brick wall in **Worked Example 19.1** of **Chapter 19** can be measured easily by entering the end points of the mean girth. Alternatively, the value can be obtained from the 'properties' of a mean girth constructed by polyline. The 'measure' function is particularly useful when dealing with irregular shapes or layouts. For example, in **Figure 24.3**, the shaded brick paving area can be obtained accurately.

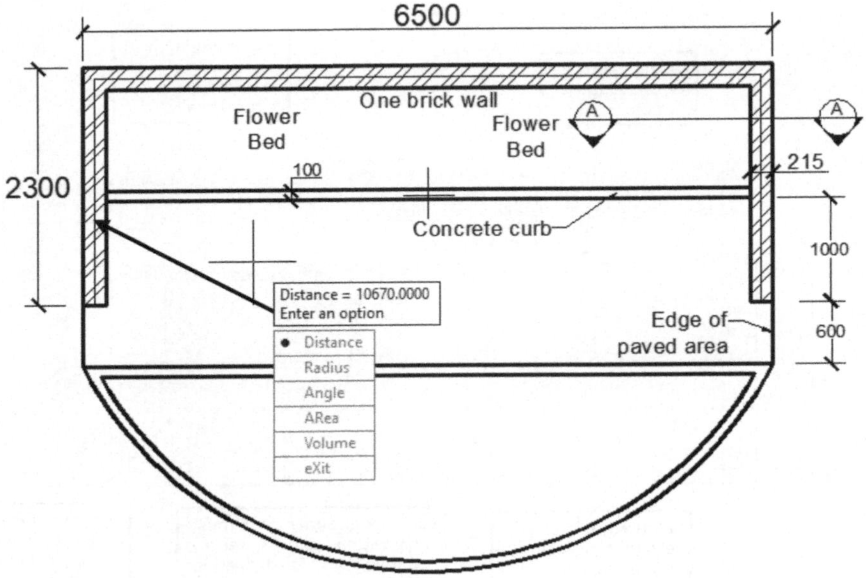

Figure 24.2 Measuring the mean girth of the brick wall.

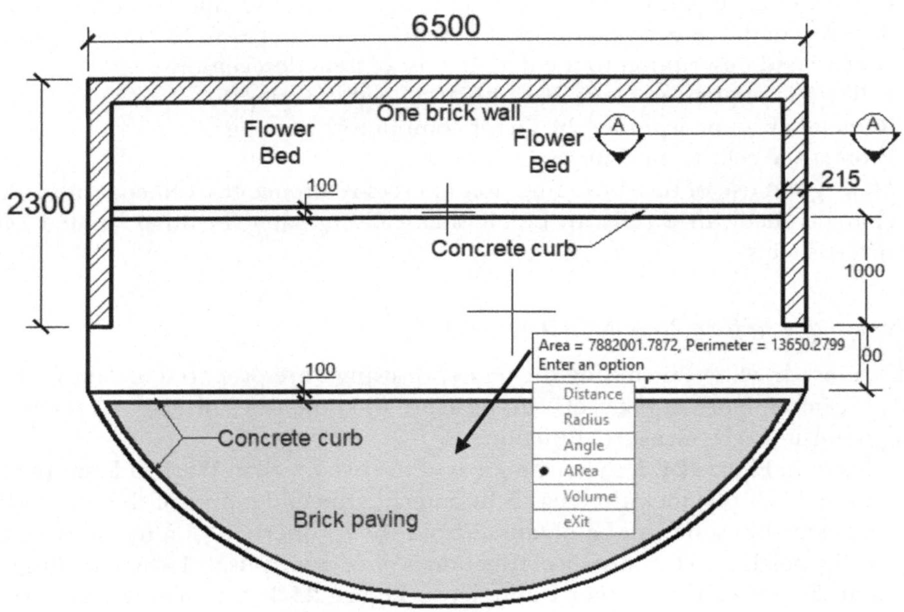

Figure 24.3 Measuring the shaded brick paving area.

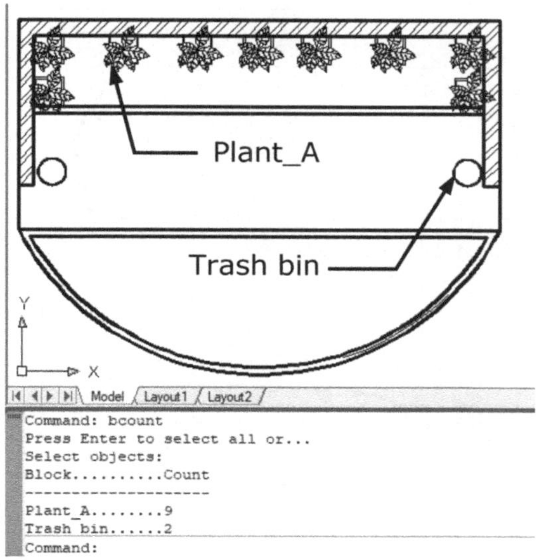

Figure 24.4 Counting of blocks on a drawing.

AutoCAD can also perform counts on blocks. As shown in **Figure 24.4**, two types of blocks, namely 'Plant_A' and 'Trash bin' have been added to the drawing. By using the 'bcount' function, the number of blocks on the drawing can be shown in the command window.

Some surveyors who are familiar with AutoCAD and Excel macros can extract the required quantities in an automated manner. Using Excel VBA, one can write macros to automate the take-off tasks such as extracting block counts or polyline lengths/enclosed areas from the 2D AutoCAD drawings and import the data to an Excel template without manual input. The imported data can then be computed into quantities of work.

When using 2D drawings to take-off, quantity adjustments due to windows, doors, columns, beams and the like have to be done manually. For instance, in **Worked Example 19.4**, the external girth of wall can be found easily on the screen, which is 28.4m as shown in **Figure 24.5**. However, the area of external wall finish is not 71.42m^2 (28.4m × 2.55m). The door and window openings (shown in the clouds in **Figure 24.5**) have to be deducted. These deductions should be done manually but are easily missed.

Although on-screen take-off eliminates a lot of waste calculations and gives us the totals directly, the ease and accuracy of data extraction depends heavily on the quality of the CAD drawings. For instance, a draftsman may have entered an overridden dimension text. As shown in the cloud in **Figure 24.6**, the displayed length (dimension text) of brick wall has been changed from 6500mm to 7000mm, which may not be conspicuous. The AutoCAD measuring tool will not capture the overridden dimension text but will still show 6500mm as its length. If using a manual method to take-off, such a 'mistake' will not happen because only the figured dimensions are used. Measurers should take steps to verify that there are no overridden dimensions and the CAD drawing is in full scale before extracting any dimensions.

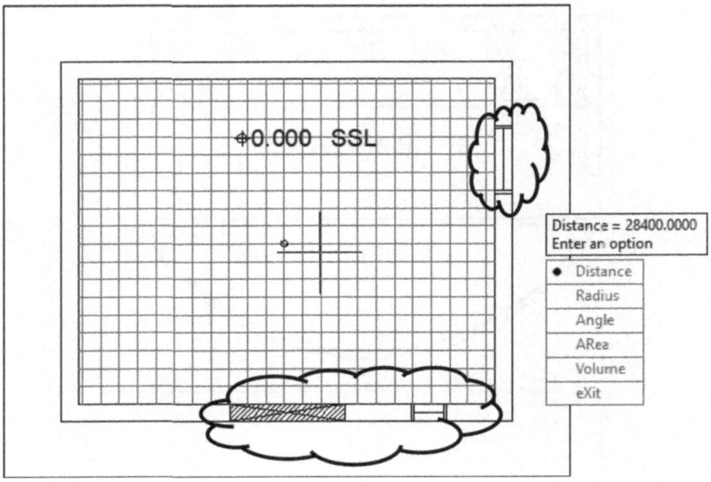

Figure 24.5 Adjustment for door and window areas.

Take-off by commercial software packages from 2D drawings

Many specialised quantity take-off software packages have been launched over the years, including CostX, Bluebeam and On-screen Takeoff. Since the packages are developed for quantity surveying but not for design purpose, the data can be extracted more precisely to suit the SMM needs. All these software packages emphasise accuracy, efficiency and user-friendliness. Most of them are characterised with the following features:

- Provide an all-in-one estimating platform with take-off, billing and estimating capabilities.
- Allow on-screen take-off from CAD file formats (such as .dwg), vector files (such as .eps) as well as raster files (such as .jpeg) using the mouse.
- Enable users to select and fill colours to lines and areas to differentiate take-off items.
- Enable users to hide the items that have been measured from the screen.

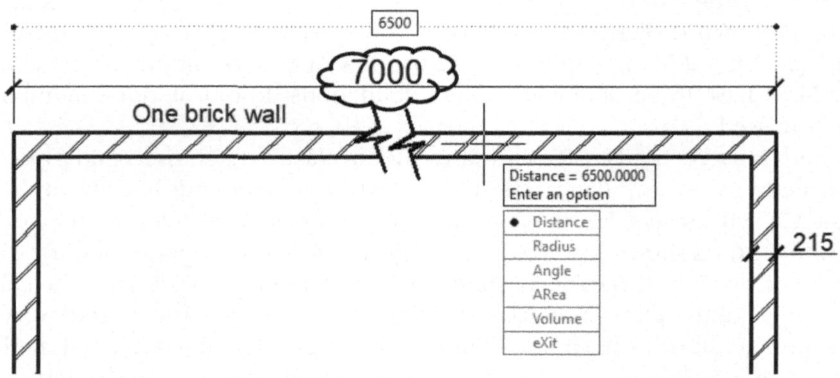

Figure 24.6 Original dimension (6500mm) is measured by AutoCAD instead of the overridden dimension text (7000mm).

- Automatically calculate areas and volumes from lengths by setting default height and/or thickness, enhancing efficiency and accuracy of calculations. Other similar default settings are available, for example conversion of steel weight.
- Count fittings and the like with the image recognition technique; this can be done automatically even though the fitting is not saved as an object in AutoCAD.
- Enable an audit trail since the quantities measured are saved with the electronic drawings.
- Integrate with spreadsheet software such as Microsoft® Excel to produce customised templates to sum up the quantities for each measuring task.
- Enable 3D model take-off to suit the BIM trend.

Disregarding the implementation and training costs, using take-off software is inevitably more productive and accurate when compared with the traditional manual method. More important, the working-up process to prepare bills of quantities from take-offs can be done automatically. Furthermore, project cost estimation is well-integrated in the system, making cost analysis and pricing more efficient than ever.

Automated quantities generated from 3D BIM

BIM is

> a process for creating and managing information on a construction project across the project lifecycle. One of the key outputs of this process is the Building Information Model, the digital description of every aspect of the built asset.
>
> (NBS, 2019)

BIM tools such as Revit and ArchiCAD are capable of compiling quantities of work automatically from the models, which saves a lot of tedious measurements.

To exploit the strengths of BIM in measurement and estimating, the best way is to implement it as soon as the design stage begins. However, the high implementation cost and the lack of local BIM experts impedes the use of BIM significantly. Most local designers are still preparing 2D digital drawings for tendering, which makes automated BQ production unfeasible. Nevertheless, many clients, especially the Hong Kong Government, realise the benefits of BIM and require contractors to prepare BIM as one of the contractual requirements. Once the BIM models are available, we can export the schedule of quantities to a spreadsheet for contract and cost management. Any revisions made in the model will be updated in all schedules instantaneously. This is extremely useful when dealing with a large amount of variations during the construction period.

While BIM sounds attractive to many people, it is not a magic wand for the measurement task. Since the BIM tools are developed overseas primarily for designers, the way that the drawing data are labelled, filtered and analysed have not considered any HKSMM rules. Users should anticipate that the quantities extracted may not fit our measurement requirements entirely. More importantly, the 'quality' of taking-off generated by BIM depends heavily on the level of detail of the model. To explain some of the deficiencies in Revit (the most popular BIM software used in Hong Kong), a 3D model has been built using Revit 2019 for **Worked Example 19.1** in **Chapter 19**. The Revit model produced an automated material take-off schedule, as shown in **Figure 24.7**.

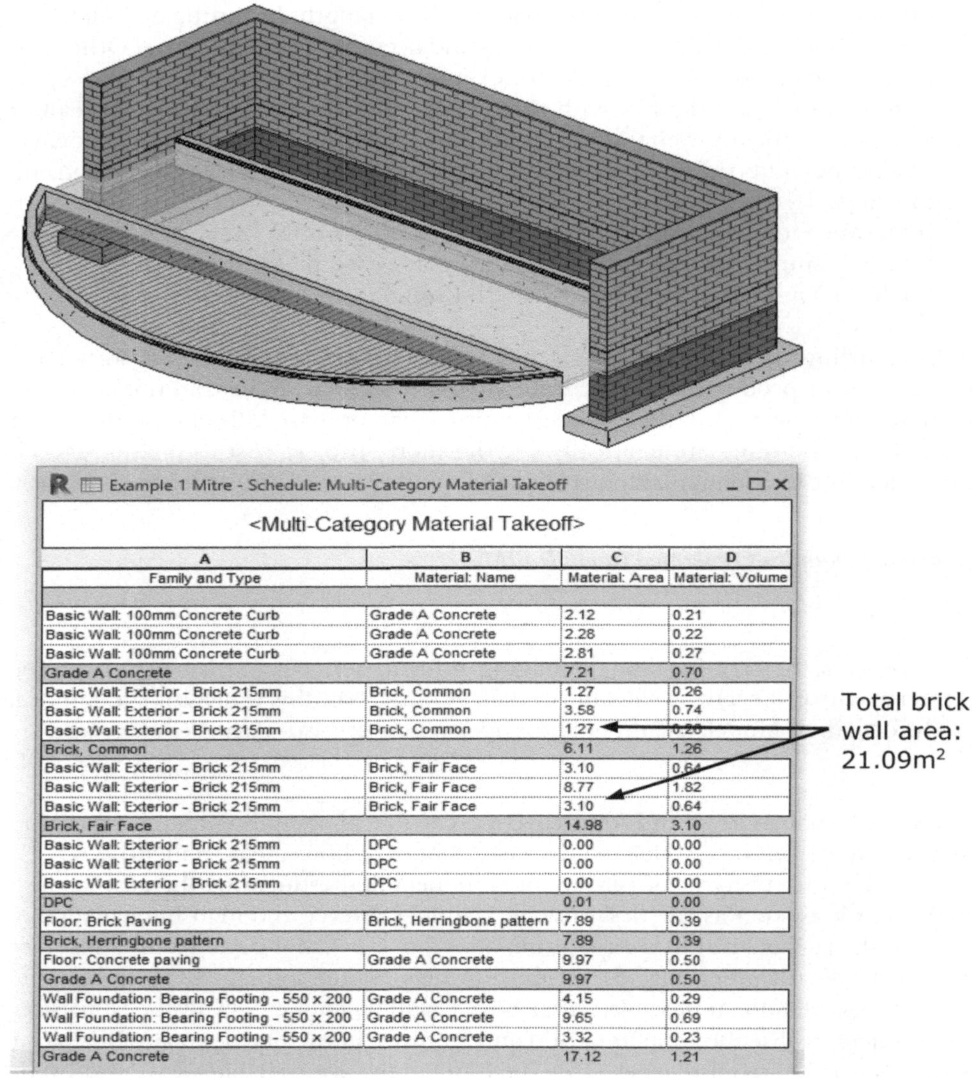

<Multi-Category Material Takeoff>

A	B	C	D
Family and Type	Material: Name	Material: Area	Material: Volume
Basic Wall: 100mm Concrete Curb	Grade A Concrete	2.12	0.21
Basic Wall: 100mm Concrete Curb	Grade A Concrete	2.28	0.22
Basic Wall: 100mm Concrete Curb	Grade A Concrete	2.81	0.27
Grade A Concrete		7.21	0.70
Basic Wall: Exterior - Brick 215mm	Brick, Common	1.27	0.26
Basic Wall: Exterior - Brick 215mm	Brick, Common	3.58	0.74
Basic Wall: Exterior - Brick 215mm	Brick, Common	1.27	0.26
Brick, Common		6.11	1.26
Basic Wall: Exterior - Brick 215mm	Brick, Fair Face	3.10	0.64
Basic Wall: Exterior - Brick 215mm	Brick, Fair Face	8.77	1.82
Basic Wall: Exterior - Brick 215mm	Brick, Fair Face	3.10	0.64
Brick, Fair Face		14.98	3.10
Basic Wall: Exterior - Brick 215mm	DPC	0.00	0.00
Basic Wall: Exterior - Brick 215mm	DPC	0.00	0.00
Basic Wall: Exterior - Brick 215mm	DPC	0.00	0.00
DPC		0.01	0.00
Floor: Brick Paving	Brick, Herringbone pattern	7.89	0.39
Brick, Herringbone pattern		7.89	0.39
Floor: Concrete paving	Grade A Concrete	9.97	0.50
Grade A Concrete		9.97	0.50
Wall Foundation: Bearing Footing - 550 x 200	Grade A Concrete	4.15	0.29
Wall Foundation: Bearing Footing - 550 x 200	Grade A Concrete	9.65	0.69
Wall Foundation: Bearing Footing - 550 x 200	Grade A Concrete	3.32	0.23
Grade A Concrete		17.12	1.21

Figure 24.7 Material take-off extracted by Revit for Worked Example 19.1.

Based on the manual taking-off of **Worked Example 19.1**, a schedule of quantities is compiled as shown in **Table 24.1**.

Comparing the quantities in **Figure 24.7** with the those in **Table 24.1**, several discrepancies are observed.

1 In the Revit schedule, item descriptions are based on the element and material descriptions entered by the modellers who focus on the design aspect rather than the HKSMM rules.
2 The elements in Revit are not grouped or totalled in a way that complies with the HKSMM. For example, we need to measure the DPC in length if it is 225mm wide

Table 24.1 Schedule of Quantities Calculated Manually Based on Measurement in Worked Example 19.1.

	Item Description	*Total quantity*	
1	Trench excavation for foundation commencing at existing ground level, not exceeding 1.50m deep	10.84	m^3
2	Trench excavation for 100 × 350mm curbs without beds commencing at existing ground level, not exceeding 0.25m average depth	19.39	m
3	Backfilling with selected excavated material	7.97	m^3
4	Disposal of excavated material from site to a tip provided by contractor	2.87	m^3
5	Reinforced concrete Grade A in foundation	1.21	m^3
6	Grade A plain concrete in curbs, with 13mm tooled recess and 13mm chamfered edge	0.71	m^3
7	Sawn formwork to vertical sides of foundation	4.61	m^2
8	Sawn formwork to vertical sides of curbs, straight on plan, exceeding 300mm wide	8.73	m^2
9	Ditto, circular on plan, exceeding 300mm wide	5.43	m^2
10	215mm thick vertical brick walls in commons in English bond in cement mortar (1:3) with flush pointing	20.27	m^2
11	Extra over ditto for fair face in English bond in cement mortar (1:3) with flush pointing	14.99	m^2
12	215mm wide and 0.5mm thick polythene sheeting DPC, laid horizontal	10.67	m
13	50mm thick grade A plain concrete paving, laid on existing ground	9.96	m^2
14	Paving in bricks (P.C. rate $180 per m^2); bedding and jointing with cement and pour sand on compacted existing ground; laid on face in herringbone pattern	7.76	m^2

but Revit provides elevation area and volume only in the 'multi-category material takeoff'. Although we can find the linear dimensions of the DPC from another material schedule, such as the 'wall material takeoff' (as shown in **Figure 24.8**), many required quantities are not extractable in any Revit schedules.

3 All items related to the excavation trade are missing in the Revit take-off, as designers seldom model excavation and backfilling activities in their design. Takers-off may have to modify the model themselves to generate the data.

4 Likewise, formwork quantities are missing in the Revit take-off. It is a typical example of an SMM measurement item that is not a model object in the BIM. Although the formwork areas/lengths can be extracted from quantities of other schedules such as the wall material take-off, the dimensions provided may not be the SMM required dimensions. This is due to the predefined way Revit calculate the material quantities within an element (see further explanation in the 'Points to Note' below).

5 Since Revit does not have the capability to handle or measure 'extra over' components, the fair face brickwork here was input as a different material to enable a separate fair face quantity to be extracted. The fair face brickwork area provided by Revit matches with the manual schedule quantity. However, the total brick wall area in Revit (21.09m^2) is different from that of the manual schedule (20.27m^2).

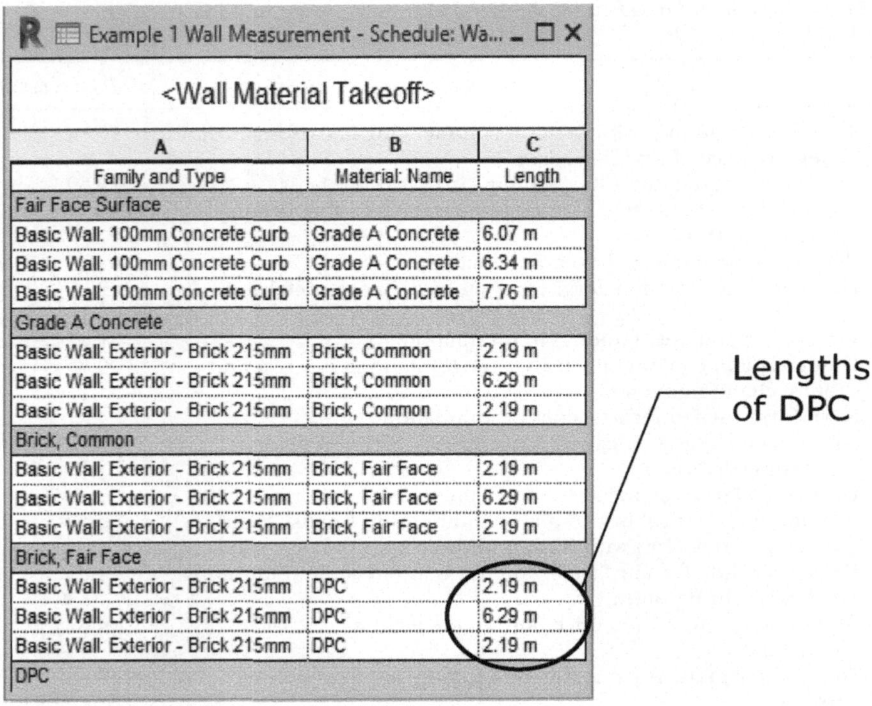

Example 1 Wall Measurement - Schedule: Wa...

<Wall Material Takeoff>

A	B	C
Family and Type	Material: Name	Length
Fair Face Surface		
Basic Wall: 100mm Concrete Curb	Grade A Concrete	6.07 m
Basic Wall: 100mm Concrete Curb	Grade A Concrete	6.34 m
Basic Wall: 100mm Concrete Curb	Grade A Concrete	7.76 m
Grade A Concrete		
Basic Wall: Exterior - Brick 215mm	Brick, Common	2.19 m
Basic Wall: Exterior - Brick 215mm	Brick, Common	6.29 m
Basic Wall: Exterior - Brick 215mm	Brick, Common	2.19 m
Brick, Common		
Basic Wall: Exterior - Brick 215mm	Brick, Fair Face	2.19 m
Basic Wall: Exterior - Brick 215mm	Brick, Fair Face	6.29 m
Basic Wall: Exterior - Brick 215mm	Brick, Fair Face	2.19 m
Brick, Fair Face		
Basic Wall: Exterior - Brick 215mm	DPC	2.19 m
Basic Wall: Exterior - Brick 215mm	DPC	6.29 m
Basic Wall: Exterior - Brick 215mm	DPC	2.19 m
DPC		

Figure 24.8 Wall material take-off by Revit for Worked Example 19.1.

Points to note

There are three basic types of wall join in Revit 2019, namely butt join, mitre join and squared join (as shown in **Figure 24.9**). If the walls are joined at 90°, butt join has the same effect as the squared join.

For simplicity, we only focus on the butt join and mitre joint. Assuming two brick wall enclosures of the same size (1m long × 1m wide × 5m high) are modelled with different join types, as shown in **Figure 24.10**.

According to the SMM rules, we should measure the brick wall area by its mean girth, which should be 16m² for the above cases. As shown in **Figure 24.10**, the total butt-joined wall area (A1 to A4) calculated by Revit is 16m², which is the correct value

Figure 24.9 Plan of basic wall joins at an angle in Revit.

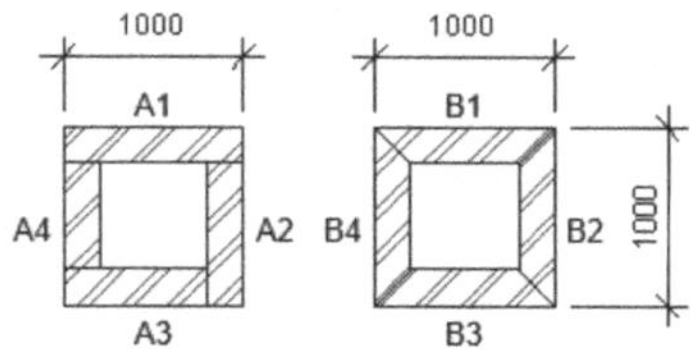

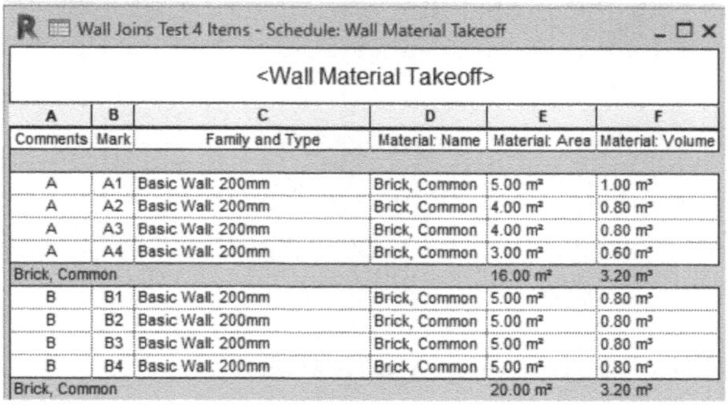
Wall Joins Test 4 Items - Schedule: Wall Material Takeoff

<Wall Material Takeoff>

A	B	C	D	E	F
Comments	Mark	Family and Type	Material: Name	Material: Area	Material: Volume
A	A1	Basic Wall: 200mm	Brick, Common	5.00 m²	1.00 m³
A	A2	Basic Wall: 200mm	Brick, Common	4.00 m²	0.80 m³
A	A3	Basic Wall: 200mm	Brick, Common	4.00 m²	0.80 m³
A	A4	Basic Wall: 200mm	Brick, Common	3.00 m²	0.60 m³
Brick, Common				16.00 m²	3.20 m³
B	B1	Basic Wall: 200mm	Brick, Common	5.00 m²	0.80 m³
B	B2	Basic Wall: 200mm	Brick, Common	5.00 m²	0.80 m³
B	B3	Basic Wall: 200mm	Brick, Common	5.00 m²	0.80 m³
B	B4	Basic Wall: 200mm	Brick, Common	5.00 m²	0.80 m³
Brick, Common				20.00 m²	3.20 m³

Figure 24.10 Screenshot of Revit for butt-joined and mitre-joined wall areas.

that we need. However, in Revit the total mitre-joined wall area (B1 to B4) is computed based on the longest wall length (i.e. the external girth), which amounts to 20m².

This simple example illustrates that the quantities given by Revit may not be the values we expect.

As highlighted in the 'Points to note', the way that the wall join is input in a Revit model will affect the length and area of the walls calculated. In the Revit model for **Worked Example 19.1**, the brick walls are connected by a mitre join. As a result, the wall areas extracted are larger than the actual value but the quantity for the fair face work area applied on the external side is correct. To deal with this, takers-off have to make modifications manually or build their own macros to obtain the correct quantities. Likewise, to extract the formwork quantities from Revit wall material take-offs, one must check the joins of each concrete element and make necessary adjustments. However, this modification/adjustment process is tedious and error-prone. Further, any real-time design changes implemented by the designers will not be captured in the QS modified models.

Regarding the lack of excavation quantity, this problem can be solved by adding a building pad using 'toposurface'. The total cut and fill volumes can be generated in the typography schedule by Revit.

For the HKSMM compliance problem, some large organisations such as the Housing Authority have developed their own library of labels to codify the elements and materials. Although designers may not use the same codes when they prepare the models, QS can put the standard code to each element using an add-on module. Other large contractors and design consultants have developed standard libraries of design objects that matches with the HKSMM rules so that SMM-based quantities can be generated by the BIM models.

The above example illustrates that the 'automated' material take-off schedules from Revit may not provide the required quantities to meet our measurement needs. Before using the Revit automated take-off, surveyors must understand that the accuracy of quantities measured by Revit depends on the level of detail and precision of the model, as well as the method of model input. Therefore, responsible surveyors must know the BIM software and measurement rules very well and use the automated material take-offs with care and understanding. However, considering the probability of human errors as well as the time and cost required in manual take-off (especially for complex designs and site conditions), BIM is a convenient tool to provide a reliable reference for quantity bulk check and project estimate.

While the industry is awaiting a unified BIM standard for modellers and designers, the local take-off rules and methods will evolve to cater for the technological changes. For today and the near future, manual taking-off remains an indispensable skill to supplement computer-aided take-off.

Billing by computer applications

Billing from spreadsheet take-off

Calculating the totals manually from traditional dimension papers is time-consuming and troublesome. However, with a spreadsheet, computation can be completed instantaneously and effortlessly using equations.

As shown in **Figure 24.11**, quantities can be automatically extended and totalled by using 'sumif' or 'countif' functions. Then, the bills can be compiled electronically by setting 'links' between spreadsheets and files. Furthermore, pivot tables can be used to analyse quantities of specific categories. For example, in **Figure 24.11**, floor finish quantities for the shopping areas can be extracted easily by pivot table. However, care must be taken when setting links or pivot tables to arrive at the correct summaries.

Billing by estimating software

Computerised estimating software such as CostX and CubiCost are undoubtedly helpful in mathematical computation and data processing. These commercial software packages usually provide automated bills of quantities after taking-off has been completed. Responsible takers-off must ensure that the quantities as well as the descriptions are input to the system correctly. Bulk checking must be done by an independent, experienced QS.

	A	B	C	D	E	F	G	H	I	J	K	L	M	N	O	P
9	**Location**	**Floors**	**Factor**	**L**	**W**	**Girth**	**H**	**F1**	**F2**	**F3**	**S1**	**S2**	**S3**	**W1**	**W2**	**W3**
10	G/F															
11	Management office	1	1	10.00	13.00	46.00	2.80	130.00			46.00			128.80		
12	Columns (isolated)	1	2	0.30	0.50	3.20	2.80	0.30			3.20			8.96		
13	Door	1	-1	0.75		4.35	1.80				-0.75			-1.35		
14	Window	1	-1	2.30		5.80	0.60							-1.38		
15	Shop 1	1	1	12.00	2.20	28.40	2.80		26.40			28.40			79.52	
16	S_Columns (isolated)	1	1	0.30	0.50	1.60	2.80		0.15			1.60			4.48	
17	Shop 2	1	1	10.00	3.00	26.00	2.80			30.00			26.00			72.80
18	**Sub-total to Summary:**							**130.30**	**26.55**	**30.00**	**48.45**	**30.00**	**26.00**	**135.03**	**84.00**	**72.80**
19																
20	**Floor finish of shop areas**															

21	**Shops**	Sum of F1	Sum of F2	Sum of F3	Sum of S1	Sum of S2	Sum of S3
22	Shop 1		26.40			28.40	
23	Shop 2			30.00			26.00
24	S_Columns (isolated)		0.15			1.60	
25	**Grand Total**		26.55	30.00		30.00	26.00
26							

Pivot table created from the take-off quantities

Figure 24.11 A pivot table created for the interior finishes take-off.

Estimating by computer applications

Estimating using a spreadsheet

In Hong Kong, most estimators make use of spreadsheets to save time when completing the tedious calculations. While the decisions on prices and resource requirements are made by the estimators, spreadsheets are used to analyse the sensitivity of the overall price to changes in parameters such as profit mark-up or contingency allowance. The speed of recalculation is mostly useful when management adjudicates the tender. Under the trend of electronic tendering, the use of spreadsheet to prepare estimates is further encouraged.

As in **Figure 24.12**, with the use of Microsoft Excel functions including Naming a List, Data Validation and VLookup, a simple cost database can be developed for cross-referencing by project administrators. Similar to taking-off using spreadsheet, care should be taken when designing templates to make sure that the equations used or copied are correct. Independent bulk checks should always be conducted.

Estimating by commercial software

In addition to estimating using Excel spreadsheets, some taking-off software packages are available with a cost estimating function. Packages such as CostX provide hierarchical workbooks and linkage to cost libraries. Estimators can draw cost data from online databases to their own libraries or make cost comparisons. Customised reports can be produced according to user-defined formats. For companies that apply computerisation in taking-off, these software packages undoubtedly provide a good integrative solution for quantity measurement and cost estimating.

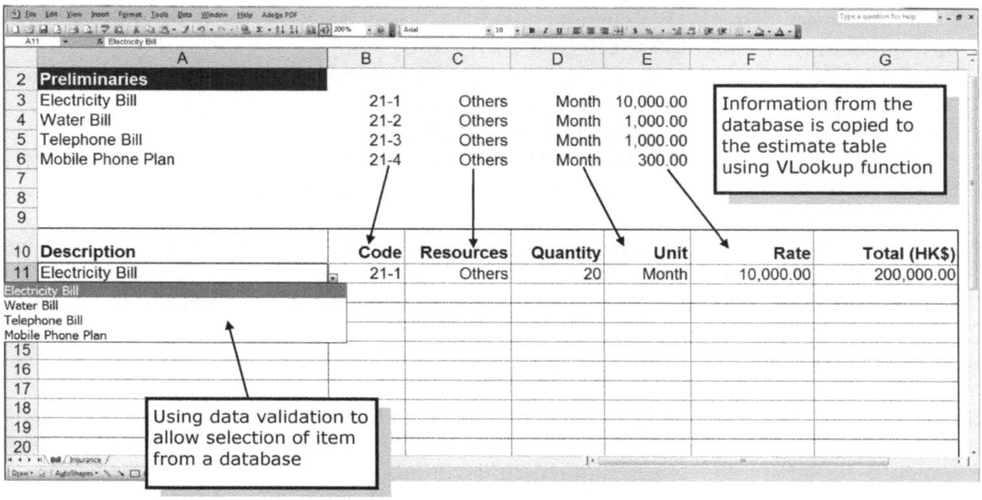

Figure 24.12 Developing estimating template using spreadsheet.

References

AACE International, 2004. AACE International Recommended Practice No. 25R-03 Estimating Lost Labor Productivity in Construction Claims.

Aqua Group, 2007. *The Aqua Group Guide to Procurement, Tendering and Contract Administration*. Blackwell Pub./Davis Langdon, Oxford ; Malden, MA.

Aqua Group, 2003. *Pre-Contract Practice and Contract Administration for the Building Team*. Blackwell Science, Oxford.

British Standards Institution, 2010. *PD 6697:2010 - Recommendations for the design of masonry structures to BS EN 1996-1-1 and BS EN 1996-2.*

Brook, M., 2017. *Estimating and Tendering for Construction Work, 5th Edition*. Routledge, New York.

Buildings Department, HKSAR, 2019. *Codes and References, Modular Integrated Construction* https://www.bd.gov.hk/en/resources/codes-and-references/modular-integrated-construction/index.html.

Buildings Department, HKSAR, 2011. *Practice Note for Authorized Persons, Registered Structural Engineers and Registered Geotechnical Engineers, APP-2, Calculation of Gross Floor Area and Non-accountable Gross Floor Area, Building (Planning) Regulation 23(3)(a) and (b).*

Chakra, H., Ashi, A., 2019. Comparative Analysis of Design/Build and Design/Bid/Build Project Delivery Systems in Lebanon. *Journal of Industrial Engineering International* 1–6. https://doi.org/10.1007/s40092-019-00323-1

Chan, C.T.W., 2012. The Principal Factors Affecting Construction Project Overhead Expenses: An Exploratory Factor Analysis Approach. *Construction Management and Economics 30*, 903–914. https://doi.org/10.1080/01446193.2012.717706

Chan, C.T.W., 2006. A Decision-Making Matrix Model for Estimating Construction Project Overhead, in: *Proceedings of PMI Research Conference 2006*. PMI Research Conference 2006. Montreal, Canada.

Chan, C.T.W., Sin, H.C., 2009. *Construction Project Management: From Theory to Practice*. Prentice Hall, Singapore.

Chartered Institute of Building, 2009. *Code of Estimating Practice, 7th Edition*. John Wiley & Sons, Chichester, UK.

China Trend Building Press Ltd., 2008. Tamar Development. *Building Journal Hongkong China* 24–36.

Civil Engineering and Development Department, HKSAR, 2018. *Project Administration Handbook for Civil Engineering Works, 2018 Edition*. https://www.cedd.gov.hk/eng/publications/standards-spec-handbooks-cost/stan-pah/index.html.

Civil Engineering and Development Department, HKSAR, 2014. *General Specification for Civil Engineering Works*, Volumes 1 and 2.

Commission for Architecture and the Built Environment, U.K., 2013. *Creating Excellent Buildings - A Guide for Clients*. http://webarchive.nationalarchives.gov.uk/20110118095356/http://www.cabe.org.uk/files/client-guides/buildings-client-guide.pdf.

Competition Commission (HK), 2019. *Bid-rigging.* https://www.compcomm.hk/en/media/reports_publications/bid_rigging_1.html.

Construction Industry Board, U.K., 1997. *Code of Practice for the Selection of Main Contractors.* Thomas Telford Pub. for the Construction Industry Board, London.

Construction Industry Review Committee, H.K., 2001. *Construct for Excellence, Report of the Construction Industry Review Committee.*

DeCenzo, D.A., Holoviak, S.J., 1990. *Employee Benefits.* Prentice Hall, Englewood Cliffs, N.J.

Dragages Hong Kong Ltd., 2013. *Latest News.* http://www.dragageshk.com/news/.

Drainage Services Department, HKSAR, 2019. *All Projects.* https://www.dsd.gov.hk/EN/Our_Projects/All_Projects/index.html.

Duggan, T., Patel, D., 2014. *Design-Build Project Delivery Market Share and Market Size Report.* Reed Construction Data/RSMeans Consulting.

Efficiency Unit, HKSAR, 2008. *An Introductory Guide to Public Private Partnerships (PPPs), 2nd Edition.* http://centaur.reading.ac.uk/4307/1/Roles_in_construction_projects_v6c.pdf.

Environment, Transport and Works Bureau, HKSAR, 2004a. Environment, Transport and Works Bureau Technical Circular (Works) No. 32/2004, *Reference Guide on Selection of Procurement Approach and Project Delivery Techniques.*

Environment, Transport and Works Bureau, HKSAR, 2004b. Environment, Transport and Works Bureau Technical Circular (Works) No. 35/2004, *Prequalification of Tenderers for Public Works Contracts.*

Environment, Transport and Works Bureau, HKSAR, 2009. Environment, Transport and Works Bureau Technical Circular (Works) No. 8/2004, *Tender Evaluation of Works Contracts with Amendments.*

Environmental Protection Department, HKSAR, 2013. *Forecast of Works Tender.* http://www.epd.gov.hk/epd/english/business_job/business_opp/forecast_tender.html.

Finch, R., 2011. *NBS Guide to Tendering: For Construction Projects.* RIBA Publishing, London.

Flanagan, R., Jewell, C., 2018. *The CIOB New Code of Estimating Practice.* Hoboken, New Jersey Wiley-Blackwell.

Friedman, A.L., Miles, S., 2006. *Stakeholders: Theory and Practice.* Oxford University Press, New York.

Gammon Construction Limited, 2020. *Avoiding Timber Waste at One Taikoo Place.* https://www.gammonconstruction.com/en/sustainability-case-studies-details.php?sustainability_highlights_id=2

Gammon Construction Limited, 2019. *Projects.* Projects. https://www.gammonconstruction.com/en/project-listing.php?category_id=2.

Gammon Construction Limited, 2013. *Ocean Shores Phase 2, Hong Kong.* Projects. http://www.gammonconstruction.com/en/html/projects/projects-db8348311728410e916e55b50bdb14f2.html.

Georgia State Financing and Investment Commission, 2003. *Selecting the Appropriate Project Delivery Option - Recommended Guidelines, Project Delivery Options,* Volume 2 of 2.

Greenhalgh, B., 2013. *Introduction to Estimating for Construction.* Routledge, Abingdon.

Highways Department, HKSAR, 2019. *About HZMB.* http://www.hzmb.hk/eng/contact.html.

HKIS, ACQS, HKCA, 2012. *Practice Notes for Quantity Surveyors - Tendering.*

Home Affairs Bureau, HKSAR, 2019. LC Paper No. CB(2)1266/18-19(01) - *Legislative Council Panel on Home Affairs Progress Report on Kai Tak Sports Park.*

Hong Kong Institute of Surveyors, 2016. *Practice Notes for Quantity Surveyors - Pre-Contract Estimates and Cost Plans.*

Hong Kong Institute of Surveyors, 2018. *Hong Kong Standard Method of Measurement of Building Works, 4th Edition Revised 2018.*

Hudson, A.A., 1994. *Hudson's Building and Engineering Contracts: Including the Duties and Liabilities of Architects, Engineers and Surveyors, 11th Edition.* Sweet & Maxwell, London.

JCT, 2002. *JCT Practice Note 6: Main Contract Tendering.* Joint Contracts Tribunal, London.

Johannessen, B., 2008. *Building Rural Roads.* International Labour Organization, Bangkok.

KSSP and HMP Liverpool, 2013. *Quantities of Facing & Common Bricks.* http://archive.excellencegateway.org.uk/.

LaBonde, S., 2010. *Capital Project Delivery.* American Water Works Association, Denver.

Law, J., 2016. *Incoterms. A Dictionary of Business and Management.* Oxford University Press, Oxford.

Legislative Council, HKSAR, 2010. LC Paper No. CB(1)304/10-11 - *Updated Background Brief on Measures to Rationalize Utilization of Build-Operate-Transfer Tunnels.*

Leighton Holdings, 2013. *North Lantau Hospital.* http://www.leighton.com.au/our-business/projects/north-lantau-hospital.

Lu, W., Yuan, H., Li, J., Hao, J.J.L., Mi, X., Ding, Z., 2011. Empirical Investigation of Construction and Demolition Waste Generation Rates in Shenzhen city, South China. *Waste Management* 31, 680–687.

Menheere, S.C.M., Pollalis, S.N., 1996. Case Studies on Build Operate Transfer. *Delft University of Technology*, Faculty of Architecture, Netherlands.

Miller, J.B., 2000. *Principles of Public and Private Infrastructure Delivery.* Springer, New York.

MPFA, HKSAR, 2020. *Making Contributions.* http://www.mpfa.org.hk/eng/main/employer/making_contributions.jsp.

Murdoch, J., Hughes, W., 2008. *Construction Contracts: Law and Management, 4th Edition.* Taylor & Francis, Oxon.

National Joint Consultative Committee for Building, 1985. *Code of Procedure for Selective Tendering for Design and Build.* RIBA, London.

National Joint Consultative Committee for Building (Great Britain), Joint Standing Committee of Architects, Surveyors and Building Contractors in Scotland, Joint Consultative Committee for Building, Northern Ireland, 1997. *Code of Procedure for Single Stage Selective Tendering.*

NBS, 2019. *What is Building Information Modelling (BIM)?* https://www.thenbs.com/knowledge/what-is-building-information-modelling-bim.

O'Brien, W.J., Formoso, C.T., Ruben, V., London, K., 2010. *Construction Supply Chain Management Handbook.* Taylor & Francis, London.

Packer, A.D., 2014. *Building Measurement.* Routledge, Abingdon.

Palmer, K., 2000. *Contract Issues and Financing in PPP/PFI (Do We Need the "F" in "DBFO" Projects?).* Cambridge Economic Policy Associates Ltd., Report prepared for the IPPR Commission on Public–Private Partnerships.

Poon, C.S., Yu, A.T.W., Jaillon, L., 2004. Reducing Building Waste at Construction Sites in Hong Kong. *Construction Management and Economics 22*, 461–470.

Powell-Smith, V., 2000. *Contract Documentation for Contractors, 3rd Edition.* Blackwell Science, Oxford.

Project Management Institute, 2013. *A Guide to the Project Management Body of Knowledge (PMBOK® Guide), 5th Edition.* Project Management Institute, Inc, Newton Square.

Ramsey, V., 2007. *Construction Law Handbook.* Thomas Telford, London.

RICS, 2013. RICS Draft Guidance Note - *Developing a Building Procurement Strategy and Selecting an Appropriate Procurement Route.* https://consultations.rics.org/consult.ti/procurement/viewCompoundDoc?docid=2704532&sessionid=&voteid=&partId=2704948.

RICS, 2012. NRM 1 - *Order of Cost Estimating and Cost Planning for Capital Building Works, 2nd Edition.* RICS, Conventry.

Sanders, S., Thomas, H., 1993. Masonry Productivity Forecasting Model. *Journal of Construction Engineering and Management 119*, 163–179. https://doi.org/10.1061/(ASCE)0733-9364(1993)119:1(163).

Scott, J.S., 1993. *Dictionary of Civil Engineering, 4th Edition.* Chapman & Hall, New York.

Seeley, I.H., 1988. *Building Quantities Explained, 4th Edition.* Macmillan Education Ltd., London.

Seeley, I.H., Winfield, R., 1999. *Building Quantities Explained, 5th Edition.* Red Globe Press, London.

Sinclair, D., 2020. *What's Behind the Updates to the 2020 Plan of Work. The RIBA Journal.* https://www.ribaj.com/intelligence/updates-to-the-riba-plan-of-work-2019-dale-sinclair-gary-clark.

Sundberg, B., Silversides, C.R., 1988. *Operational Efficiency in Forestry, Vol. 1: Analysis.* Springer, Dordrecht.

Surman, J., Kebergang, K., 2003. *Financing of Environmental Infrastructure in Hong Kong.*

Tabassi, A.A., Ramli, M., Bakar, A.H.A., 2012. Effects of Training and Motivation Practices on Teamwork Improvement and Task Efficiency: The Case of Construction Firms. *International Journal of Project Management 30,* 213–224. https://doi.org/10.1016/j.ijproman.2011.05.009.

The National Council for Public-Private Partnerships, 2013. *Types of Public-Private Partnerships.* http://www.ncppp.org/howpart/ppptypes.shtml.

Tolson, S., Glover, J., Sinclair, S. (Eds.), 2013. *Dictionary of Construction Terms.* Routledge, Oxon.

Trenter, N.A., 2001. *Earthworks: A Guide.* Thomas Telford, London.

U.S. Department of the Army, 1999. *Concrete, Masonry and Brickwork: A Practical Handbook for the Homeowner and Small Builder.* Courier Dover Publications, New York.

West Kowloon Cultural District Authority, HKSAR, 2018. LC Paper No. CB(1)115/18-19(03) - *Update on the Progress of the West Kowloon Cultural District Development.*

Bibliography

Ahuja, H.N., Campbell, W.J., 1988. *Estimating: From Concept to Completion*. Prentice-Hall, Englewood Cliffs, N.J.

Aqua Group, 1999. *Tenders and Contracts for Building*. Blackwell Science, Malden.

Architectural Services Department Hong Kong, 2019. *Schedule of Rates for Term Contracts for Building Works, 2019 Edition*.

Architectural Services Department, HKSAR. (2014). *Standard Structural Drawings*. http://archsd.gov.hk/media/11407/e203.pdf.

Ashworth, A., Hogg, K., 2002. *Willis's Practice and Procedure for the Quantity Surveyor, 11th Edition*. Blackwell Science, Oxford; Malden, MA.

Assaf, S. A., Bubshait, A. A., Atiyah, S., Al-Shahri, M., 1999. Project Overhead Costs in Saudi Arabia. *Cost Engineering, 41*(4), 33–38.

British Standards Institution, 1989. *BS4466:1989, Specification for Scheduling, Dimensioning, Bending and Cutting of Steel Reinforcement for Concrete*.

British Standards Institution, 2005. *BS8666:2005, Scheduling, Dimensioning, Bending and Cutting of Steel Reinforceme for Concrete Specification*.

British Standards Institution, 2011. *BS 8201:2011, Code of Practice for Installation of Flooring of Wood and Wood-Based Panels*.

Cartlidge, D., 2011. *New Aspects of Quantity Surveying Practice, 3rd Edition*. Spon Press, London; New York.

Census and Statistics Department, Hong Kong, 2019. *Average Wholesale Prices of Selected Building Materials*.

Chan, C.T.W., Pasquire, C., 2002, September 2). Estimation of Project Overheads: A Contractor's Perspective. *Proceedings 18th Annual ARCOM Conference*. 18th ARCOM Conference, Newscatle, U.K. http://www.arcom.ac.uk/-docs/proceedings/ar2002-053-062_Chan_and_Pasquire.pdf.

Chan, D.W.M., Chan, A.P.C., Lam, P.T.I., Lam, E.W.M., Wong, J.M.W., 2007. *An Investigation of Guaranteed Maximum Price (GMP) and Target Cost Contracting (TCC) Procurement Strategies in Hong Kong Construction Industry*. The Hong Kong Polytechnic University, Hong Kong.

Civil Engineering and Development Department, HKSAR, 2006. *General Specification for Civil Engineering Works, Vol. 1 and Vol. 2, 2006 Edition*. https://www.cedd.gov.hk/eng/publications/standards-spec-handbooks-cost/index.html.

Civil Engineering and Development Department, HKSAR, 2012. *Project Administration Handbook for Civil Engineering Works, 2012 Edition*. http://www.cedd.gov.hk/eng/publications/standards_handbooks_cost/stan_pah.htm.

Department of Justice, HKSAR, 2020. *Hong Kong e-Legislation*. https://www.elegislation.gov.hk/.

Development Bureau, HKSAR, 2017. *Practice Notes for New Engineering Contract (NEC) – Engineering and Construction Contract (ECC) for Public Works Projects in Hong Kong*.

Development Bureau, HKSAR, 2020. *Standard Contract Documents*. https://www.devb.gov.hk/en/publications_and_press_releases/publications/standard_contract_documents/index.html.

HKSAR, 2018, July 23). *Press Releases: Levy Thresholds Under Three Construction Industry-related Ordinances to be Raised Next Monday.* https://www.info.gov.hk/gia/general/201807/23/P2018072000654.htm.

Hong Kong Institute of Architects, 2005. *Agreement & Schedule of Conditions of Building Contract for Use in the Hong Kong Special Administrative Region, Private Edition: With Quantities.* Hong Kong Institute of Architects; Hong Kong Institute of Construction Managers; Hong Kong Institute of Surveyors, Hong Kong.

Hong Kong Institute of Surveyors, 2005. *Hong Kong Standard Method of Measurement of Building Works, 4th Edition.* Pace Publishing Limited, Hong Kong.

Hong Kong Institute of Surveyors, 2013. *Corrigenda to HK Standard Method of Measurement of Building Works (SMM4).* http://www.hkis.org.hk/ufiles/smm4-c201409.pdf.

Housing Department, HKSAR, 2013. *Manual for Design and Detailing of Reinforced Concrete to the Code of Practice for Structural Use of Concrete 2013.*

Institution of Structural Engineers, 1989. *Standard Method of Detailing Structural Concrete.* Institution of Structural Engineers, London.

Khosrowshahi, F., 1999. Neural Network Model for Contractors' Prequalification for Local Authority Projects. *Engineering, Construction and Architectural Management 6*(3), 315–328.

Langdon & Seah Hong Kong Limited, an Arcadis company. *Training Materials.*

Lock, D., 2007. *Essentials of Project Management, 9th Edition.* Gower Publishing Limited, Aldershot; Burlington.

MPA The Concrete Centre, 2016. *Structural Design of Concrete and Masonry: A compendium of technical papers.* http://www.concretecentre.com.

NEC, 2020. *NEC in Hong Kong.* https://www.neccontract.com/About-NEC/NEC-around-the-World/Hong-Kong.

Picken, D.H., Drew, D.S., 1991. *Building Measurement in Hong Kong: Worked Examples.* Hong Kong Polytechnic, Hong Kong.

Pratt, D., 2012. *Estimating for Residential Construction, 2nd Edition.* Delmar, New York.

RICS, 2012. *NRM 1—Order of Cost Estimating and Cost Planning for Capital Building Works* (2nd Edition). RICS. http://www.rics.org/uk/shop/NRM-1-Order-of-Cost-Estimating-and-Cost-Planning-for-Capital-Building-19143.aspx.

RICS, Building Employers Confederation (Great Britain), 1988. *Standard Method of Measurement of Building Works: Authorised by Agreement Between the Royal Institution of Chartered Surveyors and the Building Employers Confederation, 7th Edition* RICS, BEC, London.

Royal Institution of Chartered Surveyors, 1996. *The Procurement Guide: A Guide to the Development of an Appropriate Building Procurement Strategy.* RICS, London.

Royal Institution of Chartered Surveyors (Hong Kong Branch), 1979. *Hong Kong Standard Method of Measurement for Building Works, 3rd Edition.* Pace Publishing Limited, Hong Kong.

Seeley, I.H., 1988. *Building Quantities Explained, 4th Edition.* Macmillan Education Ltd, London.

Smith, A., 1995. *Estimating, Tendering and Bidding for Construction Work.* Macmillan Press Ltd., London.

Solomon, G., 1993. Cost Analysis, Preliminaries, the Marketplace Effect. *Chartered Quantity Surveyors October,* 9–11.

Turner, D.F., 1995. *Design and Build Contract Practice, 2nd Edition.* Longman, Harlow.

Wheeler, R.J., Clark, A.V., 1992. *Building Quantities: Worked Examples.* Newnes, Oxford.

Willis, A., Trench, W., 1998. *Willis's Elements of Quantity Surveying, 9th Edition.* Blackwell Science, Oxford.

Appendix

Abbreviations used in taking-off

A	
adjustment	adj.
aggregate	agg.
all round	a/r
alternate	alt.
aluminium	alum.
architrave	archve
around	ard
as before	a.b.
as before described	a.b.d.
asphalt	asph
average	av.
B	
backing	backg
bedding	bddg
bitumen	bit.
block	blk
both ends	b.e.
both sides	b.s.
brick	bk
brickwork	bwk
British Standard	BS
building	bldg
C	
cement mortar	c.m.
centre to centre	c/c
circular	circ.
coat	ct
column or colour	col.
common	comm.
concrete	conc.
curved	curv.
cut and fill	c. & f.
D	
damp-proof course	dpc
damp-proof membrane	dpm
deduct	ddt
deep	dp
diameter	dia. or ø
difference	diff.
distance	dist.
door	dr
drawing	dwg
E	
each	ea.
emulsion	emuls.
engineering	eng
English	Eng.
excavate	exc.
excavation	excavn
exceeding	ex.
external	extl
extra over	e.o.
F	
fair face	f.f.
filling	fillg
finish	fin.
formwork	fwk
foundation	fdn
frame	fr.
G	
galvanised	galv.
general	gen.

girth	gth
glazing	glzg
ground	grd
ground level	g.l.
H	
hardcore	h.c.
hardwood	hwd
height	ht
horizontal	horiz.
I	
including	incl.
internal	intl
irregular	irreg.
J	
jamb	jb
joint	jt
joist	jst
K	
kilogram	kg
kilometre	km
knot, prime and stop	k.p.s.
L	
length	len.
level	lev.
linear	lin.
lining	ling
M	
make good	m/gd
manhole	m.h.
material	matl
maximum	max.
measured separately	m/s
mechanical	mech.
metre	m
mild steel	m.s.
minimum	min.
N	
necessary	nec.
not exceeding	n.e.
number	nr
O	
one side	o.s.
opening	opg
ordinary	ord.
P	
paving	pavg
perforation	perf.
perimeter	peri.
plaster	plas.
prepare	prep.
projection / projecting	proj.
provisional	prov.
Q	
quality	qual.
quarry tile	q.t.
R	
radius	rad.
rainwater	r.w.
rainwater pipe	r.w.p.
rebated	reb.
rectangular	rect.
reduced	red.
reference	ref.
reinforced concrete	R.C.
required	reqd
rolled steel	r.s.
S	
section	sec.
selected	sel.
skirting	sktg
smooth	smth
soffit	soff.
softwood	swd
square	sq.
structure	struct.
surface	surf.
surplus	surp.
T	
tanking	tankg
temporary	temp.
thick	thk
tongued and grooved	t. & g.

trench	tr.
triangular	triang.
U	
undercoat	u/c
underside	u/side
upstand	upstd
V	
vertical	vert.
W	
with	wi.
working space	w/s
wrought iron	w.i.

Mensuration formulae

Section	Area
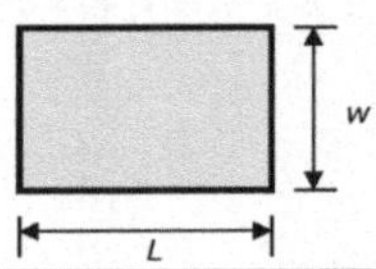	$A = Lw$
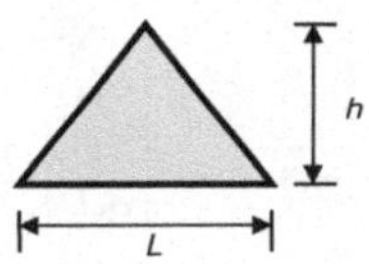	$A = \frac{Lh}{2}$
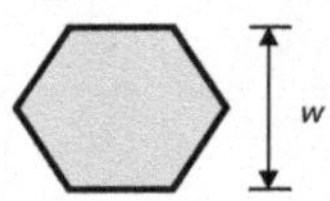	$A = 0.866w^2$
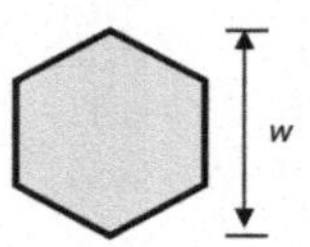	$A = 0.65\ w^2$
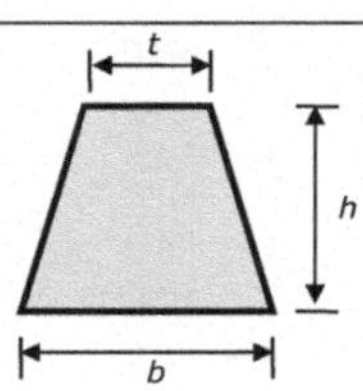	$A = \frac{h(t+b)}{2}$

Section	Area
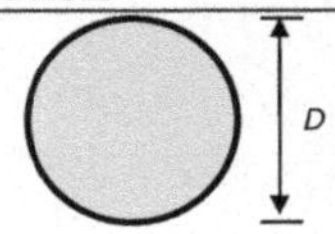	$A = \frac{D^2}{4}\pi$
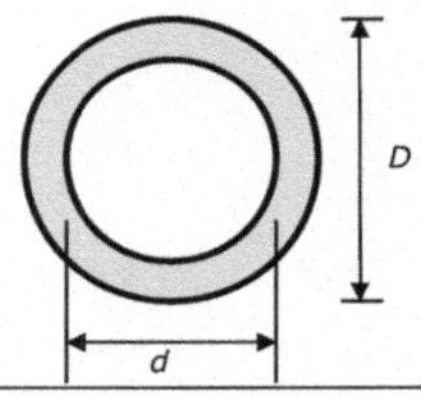	$A = \frac{D^2 - d^2}{4}\pi$
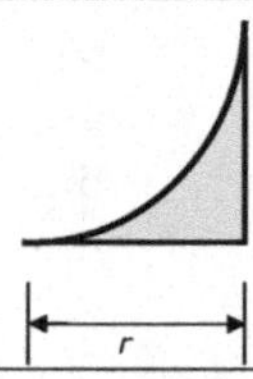	$A = r^2 - \frac{r^2\pi}{4}$
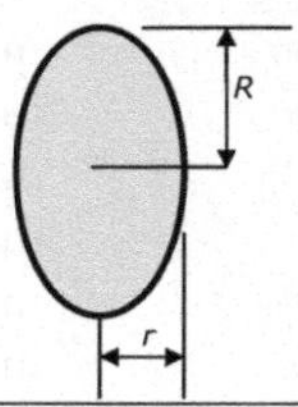	$A = Rr\pi$

Solid	Volume
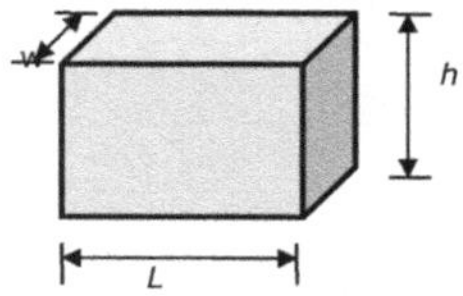	$V = Lwh$
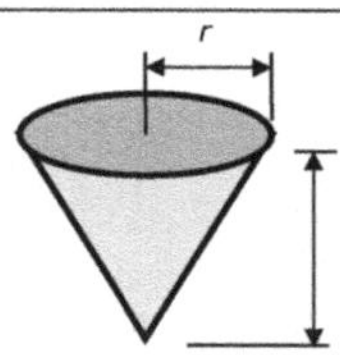	$V = \frac{r^2 h\pi}{3}$
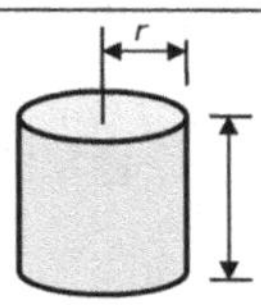	$V = r^2 h\pi$

Solid	Volume
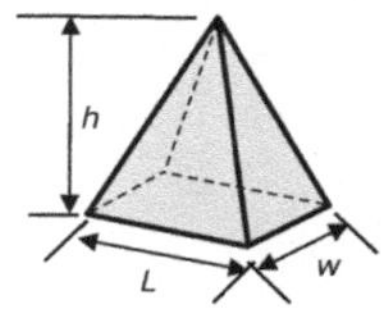	$V = \frac{Lwh}{3}$
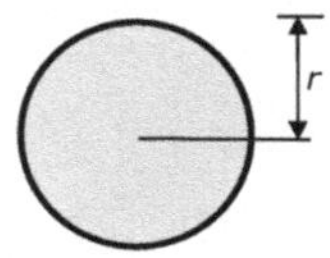 sphere	$V = \frac{4\pi r^3}{3}$

Index